Tourism, Hospitality & Event Management

This book series covers all topics relevant in the tourism, hospitality and event industries. It includes destination management and related aspects of the travel and mobility industries as well as effects from developments in the information and communication technologies. "Tourism, Hospitality & Event Management" embraces books both for professionals and scholars, and explicitly includes undergraduate and advanced texts for students. In this setting the book series reflects the close connection between research, teaching and practice in tourism research and tourism management and the related fields.

More information about this series at http://www.springer.com/series/15444

Roland Conrady • David Ruetz • Marc Aeberhard
Editors

Luxury Tourism

Market Trends, Changing Paradigms, and Best
Practices

 Springer

Editors
Roland Conrady
University of Applied Sciences Worms
Worms, Germany

David Ruetz
Head of ITB Berlin
Messe Berlin GmbH
Berlin, Germany

Marc Aeberhard
Luxury Hotel & Spa Management Ltd.
Zürich, Switzerland

ISSN 2510-4993   ISSN 2510-5000   (electronic)
Tourism, Hospitality & Event Management
ISBN 978-3-030-59892-1   ISBN 978-3-030-59893-8   (eBook)
https://doi.org/10.1007/978-3-030-59893-8

This Springer imprint is published by the registered company Springer Nature Switzerland AG.
The registered company address is: Gewerbestrasse 11, 6330 Cham, Switzerland

Preface

Tell me how you travel, and I'll tell you who you are ...

Like hardly any other activity, travelling is suited to depict the zeitgeist of an epoch. This was true in the nineteenth century as well as today: a mixture of curiosity, the desire for adventure, necessity, prestige and status is motivating an audience of billions to move on and around the globe in various ways. But the last two decades in particular have confused many parameters of the previously very stable travel patterns: rapid digitalization, widespread availability of low-cost travel, overtourism in sensitive destinations, the cry for ecological sustainability, political turmoil, etc. All this and much more have heralded a paradigm shift, especially in the most sensitive but also the most opinion-leading segment: luxury tourism.

Reason enough, then, to take up this phenomenon. The topic of luxury travel has been occupying a large part of tourism stakeholders for some time, but so far the clarification of concepts such as "premium luxury", "ultra luxury", "luxese" and "high-end luxury" and their classification in science and practice has still been the issue and confusion is spreading instead.

This book now attempts to close this gap. It provides a comprehensive overview of the phenomenon of luxury tourism, which has hardly been explored to date, and sheds light on the increasingly changing perceptions in society, economy, politics and culture. Market structures as well as demand and supply sides of international luxury tourism are explained from different perspectives. Particular attention is paid to those trends that will shape the luxury market in the future. Concrete recommendations for measures for luxury suppliers are also given. Selected examples from the most important segments of the tourism market round off the explanations.

On the one hand, this book addresses executives and employees of travel providers (hotels, destinations, tour operators and intermediaries, carriers, etc.) in the international arena and, on the other hand, members of universities with a tourism focus. Since developments in the luxury segment will continue to dominate the mainstream in the future, this book is ultimately relevant to all target groups concerned with tourism developments and strategy concepts in this sector.

The ITB Berlin—The World's Leading Travel Trade Show®—is one of the world's largest and most comprehensive platforms for tourism and at the same time an ideal breeding ground for current discussions about trends and tendencies. Together with the Worms University of Applied Sciences and Luxury Hotel & Spa Management, Zurich—an internationally renowned luxury hotel operator—this book was created. The editors were able to recruit renowned executives from luxury providers and renowned scientists as authors. Authorship is largely made up of the editors' networks and the ITB Convention. The editors and authors hope that this book will appeal both to practitioners interested in concrete design recommendations and to readers involved in scientific research.

Worms, Germany   Roland Conrady
Berlin, Germany   David Ruetz
Zürich, Switzerland   Marc Aeberhard
January 2020

Contents

Editors and Contributors

About the Editors

Roland Conrady Since 2002, Prof. Dr. Roland Conrady has been a professor at the tourism/transport department of the Worms University of Applied Sciences. His research and teaching focuses on aviation, tourism and digitalization. Since 2004, Roland Conrady has also been the Scientific Director of the world's largest tourism convention, the ITB Berlin Convention. He was president of the German Society for Tourism Science (DGT) e.V. and is a book author (among others Conrady, R./Fichert, F./Sterzenbach, R., Luftverkehr, Munich 2019). Roland Conrady is a member of various advisory boards of companies and politics. Previously, he was head of the study programme "Electronic Business" and Professor of General Business Administration at the Heilbronn University of Applied Sciences. After graduating as Dr. rer. pol. from the University of Cologne in 1990, he held various management positions at Deutsche Lufthansa AG until 1998.

David Ruetz (Head of ITB) has headed the ITB Berlin, the World's Leading Travel Trade Show®. Under his direction, ITB Berlin has seen many achievements, including the Innovation Award of the Federal Association of the German Tourism Industry (BTW). As part of the management of ITB, he was significantly responsible for the establishment of the ITB Asia in Singapore (since 2008) and also helped to introduce the ITB China in Shanghai (premiere in May 2017). After his studies in

Zurich and Berlin, David Ruetz entered tourism and the event industry and joined Messe Berlin in 2001. He is editor, author and co-author of numerous studies and books on trends and developments in the exhibition, travel and tourism industry, among others the annual World Travel Trends Reports. His interests include digitalization in the trade fair and event sector. As a guest lecturer, he regularly works at universities in Germany and Austria and is also a member of the advisory board of R.I.F.E.L. e. V. (Research Institute for Exhibition and Live Communication) and the IUBH—International University (until 2018) and is a member of the Executive Board of Research Association for Holidays and Travel (FUR).

Marc Aeberhard founded Luxury Hotel & Spa Management Ltd. in Zurich in 2004 and has been acting as Managing Director ever since. The company has access to a global network of travel trade partners, lifestyle and travel media and (U)HNWI and works closely with public relations and sales and marketing agencies in Frankfurt, Munich, Paris, Dubai, Milan, New York, Hong Kong and London. Furthermore, the native Swiss is a member of the consulting networks of the Gerson Lehmann Group, USA, and Hotellerie Suisse, Bern. He also takes on an active role in the "luxury" task force of the management of ITB Berlin, Germany. As author and co-author of various specialist publications, his name can be found regularly. He also holds guest lectures in Berlin, Istanbul, Lausanne, Lucerne, Munich, Singapore, Stuttgart, Thun, Vienna, Worms, Zurich, etc. The graduate hotelier graduated with distinction from the Ecole Hôtelière de Lausanne (EHL) and previously completed his studies in business administration as lic. rer.pol. (MBA) at the University of Bern. The luxury hotelier has more than 20 years of experience in the fields of hotel opening, management and renovation/refurbishment in Abu Dhabi, Germany, France, Maldives, Morocco, Seychelles, Sri Lanka, Switzerland, Thailand, Ukraine and Cyprus of small hotels in high and top end. Many of the hotels have been awarded international prizes. All projects are based on the definition of new luxury and work according to the principles of the triple bottom line.

Contributors

Marc Aeberhard   Luxury Hotel & Spa Management Ltd., Zürich, Switzerland

Magda Antonioli Corigliano   ACME Management School, Bocconi University, Milan, Italy

David Bosshart   Gottlieb Duttweiler Institute, Rüschlikon, Switzerland

Sara Bricchi   SDA Bocconi School of Management, Bocconi University, Milan, Italy

Andreas Caminada   Genusswerkstatt GmbH Schloss Schauenstein, Fürstenau, Switzerland

Sergio Comino   Jesolo International Club Camping, Lido di Jesolo, Italy

Roland Conrady   University of Applied Sciences Worms, Worms, Germany

Stefan Gössling   School of Business and Economics Linnaeus University, Kalmar, Sweden

Dirk Gowin   Select Luxury Travel GmbH, Berlin, Germany

Stephan Grandy   Premium Products & Lufthansa Private Jet, Frankfurt, Germany

Hannes Gurzki   European School of Management and Technology, Berlin, Germany

Stephan Hagenow   Pfarrschaft der Region Bern, Bern, Switzerland

Dorothea Hohn   Global Communication Experts GmbH, Frankfurt, Germany

Thomas P. Illes   Zürich, Switzerland

Juliet Kinsman   BOUTECO, London, UK

Keiko Kirihara   Worms University of Applied Sciences, Worms, Germany

Marcus Krall   Ocean Independence AG, Düsseldorf, Germany

Mario Krause   Deutsches Zentrum für Individualisierte Prävention und Leistungsverbesserung, Hannover, Germany

Brett McDonald   Flame of Africa, Kasane, Botswana

Antonella Mei-Pochtler   Vienna, Austria

Jörg Meurer   KEYLENS Management Consultants, München, Germany

Adam Parken   Blacklane GmbH, Berlin, Germany

Norbert Pokorny   Art of Travel GmbH, Munich, Germany

Thomas Reimann   thr media, Hamburg, Germany

David Ruetz Head of ITB Berlin, Messe Berlin GmbH, Berlin, Germany

Philipp Schmid avintas:schmid, Leuk-Stadt, Switzerland

Daniel Schönbächler Kloster Disentis, Disentis, Switzerland

Hasso Spode TU Berlin, Berlin, Germany

Ralf Vogler Hochschule Heilbronn, Heilbronn, Germany

Marco Walter ECOCAMPING Service GmbH, Konstanz, Germany

Maria Wenske Düsseldorf, Germany

Jens Wohltorf Blacklane GmbH, Berlin, Germany

David M. Woisetschläger TU Braunschweig, Braunschweig, Germany

Verena Zaugg-Faszl St. Niklausen, Switzerland

Introduction

Roland Conrady, David Ruetz, and Marc Aeberhard

What exactly is "luxury"? The term luxury is currently used in an inflationary manner without being sufficiently precise. All too often, it is just an empty phrase that everyone molds as they see fit.

The term luxury comes from Latin and means "excessive effort," "pomp," "excess." As early as 1913, Werner Sombart defined luxury as follows: "Luxury is any effort that goes beyond what is necessary" (Sombart 1992, p. 71).

Over the course of the last decades, a multitude of definition and explanation approaches to luxury have emerged.[1] The common denominator of these definitions are the following conceptual elements:

1. **Abundance or waste:** From an economic perspective, this conceptual element refers to the supply side. Luxury means that the characteristics of an offer have a degree of expression that goes beyond what is necessary. In most cases, this involves high production costs, perfectionist demands, and a corresponding price level. For the majority of consumers, such offers are hardly affordable or unattainable. The "excessive" or even "wasteful" nature of these supply characteristics means that the needs of a large proportion of customers are over exceeded.

---

[1] See for example the overview in Kolaschnik (2012).

R. Conrady (✉)
University of Applied Sciences Worms, Worms, Germany
e-mail: conrady@hs-worms.de

D. Ruetz
Head of ITB Berlin, Messe Berlin GmbH, Berlin, Germany
e-mail: ruetz@messe-berlin.de

M. Aeberhard
Luxury Hotel & Spa Management Ltd., Zürich, Switzerland

© Springer Nature Switzerland AG 2020
R. Conrady et al. (eds.), *Luxury Tourism*, Tourism, Hospitality & Event Management, https://doi.org/10.1007/978-3-030-59893-8_1

2. **Satisfaction of needs or desire:** From an economic perspective, this conceptual element refers to the demand side. Needs are feelings of deficiency that people strive to eliminate. Goods are often a means of eliminating feelings of deficiency. This applies to everyday goods (such as food to eliminate hunger and thirst) as well as luxury goods (such as luxury yachts to eliminate the lack of social recognition or self-fulfillment). The *sensations of lack* differ from person to person and are also (sub-)culturally different; therefore, it is often pointed out that luxury is highly individual ("luxury is in the eye of the beholder"). In addition, the *suitability* of goods for the satisfaction of needs is also perceived individually and also (sub-)culturally differently. However, the ability to satisfy needs does not constitute a luxury good. Only the suitability to clearly *exceed higher value* needs makes a luxury product. Only here does desire arise.

Therefore, luxury is unmistakably linked to the economy. It is no exaggeration to claim that luxury has always had an unmistakable relevance on product markets[2] (Cf. Veblen 2015). For many years, luxury was closely associated with goods. Only recently, for the first time in history, has the concept of luxury been transformed by material abundance.[3] Now immaterial characteristics such as silence, unspoiled nature, or free time are also associated with luxury.

Nevertheless, luxury continues to play a major role in product markets. Turnover with yachts, super sports cars, villas, haute couture, and many other luxury products has been showing remarkable growth rates for years. There are numerous studies on the size and structure of luxury markets, but these focus almost exclusively on material goods.[4] Most definitions and explanatory approaches also relate to tangible assets.[5]

There is a lack of market studies and, in particular, of scientific explanatory approaches for intangible goods such as services and, in particular, travel and tourism services.[6] Much of the knowledge gained about tangible goods can hardly be transferred to services. How, for example, can knowledge about the symbolic content of material goods be transferred to the remoteness of an island resort? How can one transfer insights into the importance of outstanding craftsmanship to the authentic experience of untouched nature and deep cultural experiences? The findings on brand management are also difficult to transfer to goods such as trips by small tour operators, overnight stays in boutique hotels, or tours with yacht charter companies.

---

[2]See for example the work of Veblen (2015).

[3]See Kühne and Bosshart (2014) on the change in the understanding of luxury.

[4]See for example the studies by Bain and Company (2017), The Boston Consulting Group (2017), Deloitte (2017).

[5]Nevertheless, there are significant gaps in research. See Meurer (2012, p. 330 ff) for the status of business research on luxury brand management.

[6]Among the few studies and publications available are ITB (2017), Amadeus (2016), Euromonitor (2017), Eye for Travel (2017), Pangaea Network (2018), Sabre (2017), Keylens and Inlux (2018).

Little is known about tourism's luxury needs and consumer demands. There are hardly any behavioral science explanatory approaches in German-speaking countries. One would like to see more research being done in the behavioral sciences (especially in the social sciences—here again especially in sociology and social psychology—and psychology) on luxury consumption behavior in general and on tourism markets in particular.[7] The (sub-)cultural imprinting of tourism's luxury needs and consumer demands also has serious shortcomings in terms of knowledge.

The lack of in-depth knowledge about luxury tourism is particularly regrettable, as a number of studies have shown that, especially in recent times, tourism offers are considered particularly suitable for satisfying luxury needs. Meurer states: "At the top [of personal luxury experiences—author's note] with 29%, named by almost every third person, stands 'travel and tourism' ..." (Meurer 2012, p. 326).

It is therefore not surprising that consumers are increasingly receptive to luxury offers in tourism and that considerable efforts are being made by a large number of suppliers in tourism markets to position offers in the upper market segment. In addition, luxury markets are often "pioneer markets" which, after a trickle-down process, also shape other market segments. Product design in mass markets is often based on the characteristics of luxury products. In general, methods of corporate management used by luxury providers also serve as models for other market segments. The lack of research and knowledge on luxury tourism is therefore all the more serious.

This book is the only comprehensive work on luxury tourism to date. It aims to provide a comprehensive overview of all components of luxury tourism. It is also a claim to describe trends that will change or shape the markets of luxury tourism in the future. The aim is to provide the reader with orientation aids for entrepreneurial decisions in the field of brand management of (luxury) tourism products.

We would be pleased if this work would bring an increase in knowledge to the many readers and if we could initiate further research on central questions.

Literature

Amadeus (2016) Shaping the future of luxury travel. Future traveller tribes 2030. https://amadeus. com/documents/en/travel-industry/report/shaping-the-future-of-luxury-travel-future-traveller-tribes-2030.pdf. Zugegriffen: 29 Oktober 2018

Bain & Company (2017) Luxury goods worldwide market study, Fall-Winter 2017. http://www2. bain.com/Images/BAIN_REPORT_Global_Luxury_Report_2017.pdf. Accessed 29 Oct 2018

Deloitte (2017) Global power of luxury goods. The new luxury consumer. https://www2.deloitte. com/global/en/pages/consumer-business/articles/gx-cb-global-powers-of-luxury-goods.html. Zugegriffen: 29 Oktober 2018

Euromonitor (2017) Global luxury travel trends report. https://www.euromonitor.com/global-lux ury-travel-trends-report/report . Accessed 29 Oct 2018

---

[7]The question arises as to the reasons for this research gap despite its unmistakable practical relevance and everyday evidence. Is it frowned upon to occupy oneself with luxury consumption?

Eye for Travel (2017) The global luxury travel consumer. https://www.eyefortravel.com/distribu
    tion-strategies/global-luxury-travel-consumer. Zugegriffen: 29 Oktober 2018
ITB (2017) New travel luxury. https://www.itb-berlin.de/media/itb/itb_dl_en/itb_itb_berlin/
    laender_und_segmente/ITB2017_Luxury_Manifest_Flyer_quer_e.pdf. Accessed 29 Oct 2018
Keylens and Inlux (2018) Consumer generations 2018. Premium and luxury study. With industry
    report tourism. http://www.keylens.com/wp-content/uploads/2018/05/Konsumgenerationen-
    2018_Branchenreport-Touristik.pdf. Accessed 29 Oct 2018
Kolaschnik A (2012) The figure of luxury. In: Burmann C, König V, Meurer J (eds) Identity-based
    luxury brand management. Basics—strategies—controlling. Gabler, Wiesbaden, pp 183–200
Kühne M, Bosshart D (2014) The next luxury. Gottlieb Duttweiler Institute, Rüschlikon
Meurer J (2012) Ebony or ivory—how brilliant is the future of luxury in Germany? Critical
    reflections on luxury brand management. In: Burmann C, König V, Meurer J (eds) identity-
    based luxury brand management. Basics—strategies—controlling. Gabler, Wiesbaden, pp
    321–336
Pangaea Network (2018) Luxury travel trends, 5th edn. https://pangaeanetwork.com/international-
    luxury-travel-trends-v-edition-full-report/ . Accessed 29 Oct 2018
Sabre (2017) The future of luxury travel. http://www2.sabrehospitality.com/future-of-luxury.
    Accessed 29 Oct 2018
Sombart W (1992) Love, luxury and capitalism. Wagenbach, Berlin
The Boston Consulting Group (2017) The true-luxury global consumer insight (4 Aufl). https://
    altagamma.it/media/source/BCG%20Altagamma%20True-Luxury%20Global%20Cons%
    20Insight%202017%20-%20presentata.pdf. Zugegriffen: 29 Oktober 2018
Veblen (2015) Theory of fine people. An economic study of institutions, 3rd edn. Fischer, Frankfurt
    a. M.

Roland Conrady Since 2002, Prof. Dr. Roland Conrady has been a professor at the tourism/
transport department of the Worms University of Applied Sciences. His research and teaching
focuses on aviation, tourism and digitalization. Since 2004, Roland Conrady has also been the
Scientific Director of the world's largest tourism convention, the ITB Berlin Convention. He was
president of the German Society for Tourism Science (DGT) e.V. and is a book author (among
others Conrady, R./Fichert, F./Sterzenbach, R., Luftverkehr, Munich 2019). Roland Conrady is a
member of various advisory boards of companies and politics. Previously, he was head of the study
programme "Electronic Business" and Professor of General Business Administration at the Heilbronn University of Applied Sciences. After graduating as Dr. rer. pol. from the University of
Cologne in 1990, he held various management positions at Deutsche Lufthansa AG until 1998.

David Ruetz (Head of ITB) has headed the ITB Berlin, the World's Leading Travel Trade Show®.
Under his direction, ITB Berlin has seen many achievements, including the Innovation Award of
the Federal Association of the German Tourism Industry (BTW). As part of the management of
ITB, he was significantly responsible for the establishment of the ITB Asia in Singapore (since
2008) and also helped to introduce the ITB China in Shanghai (premiere in May 2017). After his
studies in Zurich and Berlin, David Ruetz entered tourism and the event industry and joined Messe
Berlin in 2001. He is editor, author and co-author of numerous studies and books on trends and
developments in the exhibition, travel and tourism industry, among others the annual World Travel
Trends Reports. His interests include digitalization in the trade fair and event sector. As a guest
lecturer, he regularly works at universities in Germany and Austria and is also a member of the
advisory board of R.I.F.E.L. e. V. (Research Institute for Exhibition and Live Communication) and
the IUBH—International University (until 2018) and is a member of the Executive Board of
Research Association for Holidays and Travel (FUR).

Marc Aeberhard founded Luxury Hotel & Spa Management Ltd. in Zurich in 2004 and has been acting as Managing Director ever since. The company has access to a global network of travel trade partners, lifestyle and travel media and (U)HNWI and works closely with public relations and sales and marketing agencies in Frankfurt, Munich, Paris, Dubai, Milan, New York, Hong Kong and London. Furthermore, the native Swiss is a member of the consulting networks of the Gerson Lehmann Group, USA, and Hotellerie Suisse, Bern. He also takes on an active role in the "luxury" task force of the management of ITB Berlin, Germany. As author and co-author of various specialist publications, his name can be found regularly. He also holds guest lectures in Berlin, Istanbul, Lausanne, Lucerne, Munich, Singapore, Stuttgart, Thun, Vienna, Worms, Zurich, etc. The graduate hotelier graduated with distinction from the Ecole Hôtelière de Lausanne (EHL) and previously completed his studies in business administration as lic. rer.pol. (MBA) at the University of Bern. The luxury hotelier has more than 20 years of experience in the fields of hotel opening, management and renovation/refurbishment in Abu Dhabi, Germany, France, Maldives, Morocco, Seychelles, Sri Lanka, Switzerland, Thailand, Ukraine and Cyprus of small hotels in high and top end. Many of the hotels have been awarded international prizes. All projects are based on the definition of new luxury and work according to the principles of the triple bottom line.

Development of the Macro-environment of the (Luxury) Tourism Market

Roland Conrady

1 Preliminary Remarks

The basis of entrepreneurial decisions in the areas of target planning, strategy development, and marketing measures are the observation and analysis of the corporate environment. An analysis of the environmental factors is essential because they are critical to success for companies, they shape the opportunities and risks of entrepreneurial activities. Environmental factors change over time, so they need to be analyzed on a continual basis. In addition, forecasts are to be prepared regarding the future development of environmental factors, as entrepreneurial decisions pertain to the future of the company and ensure that the company fits the future environmental factors.

It has recently been highlighted that the corporate environment is characterized by significantly increased volatilities, uncertainties, complexities, and ambiguities (so-called "VUCA-World": Volatility–Uncertainty–Complexity–Ambiguity), which makes forecasting developments in environmental factors considerably more difficult.

The analysis and prognosis framework is usually referred to in the economic literature as a "model of the corporate environment" (See Meffert et al. 2015, pp. 43 ff., Kerth et al. 2015, pp. 118 ff .; Freyer 2011, pp. 120 ff., 148 ff.). The corporate environment is initially divided into the macro-environment (economy as a whole) and the micro-environment (particular section of the economy), see Fig. 1. The macro-environment has indirect effects on companies in many industries; factors of the macro-environment cannot be influenced by individual companies or only to a very limited extent. The micro-environment, on the other hand, has a direct impact on individual companies; micro-environmental factors can be more or less

R. Conrady (✉)
University of Applied Sciences Worms, Worms, Germany
e-mail: conrady@hs-worms.de

© Springer Nature Switzerland AG 2020
R. Conrady et al. (eds.), *Luxury Tourism*, Tourism, Hospitality & Event
Management, https://doi.org/10.1007/978-3-030-59893-8_2

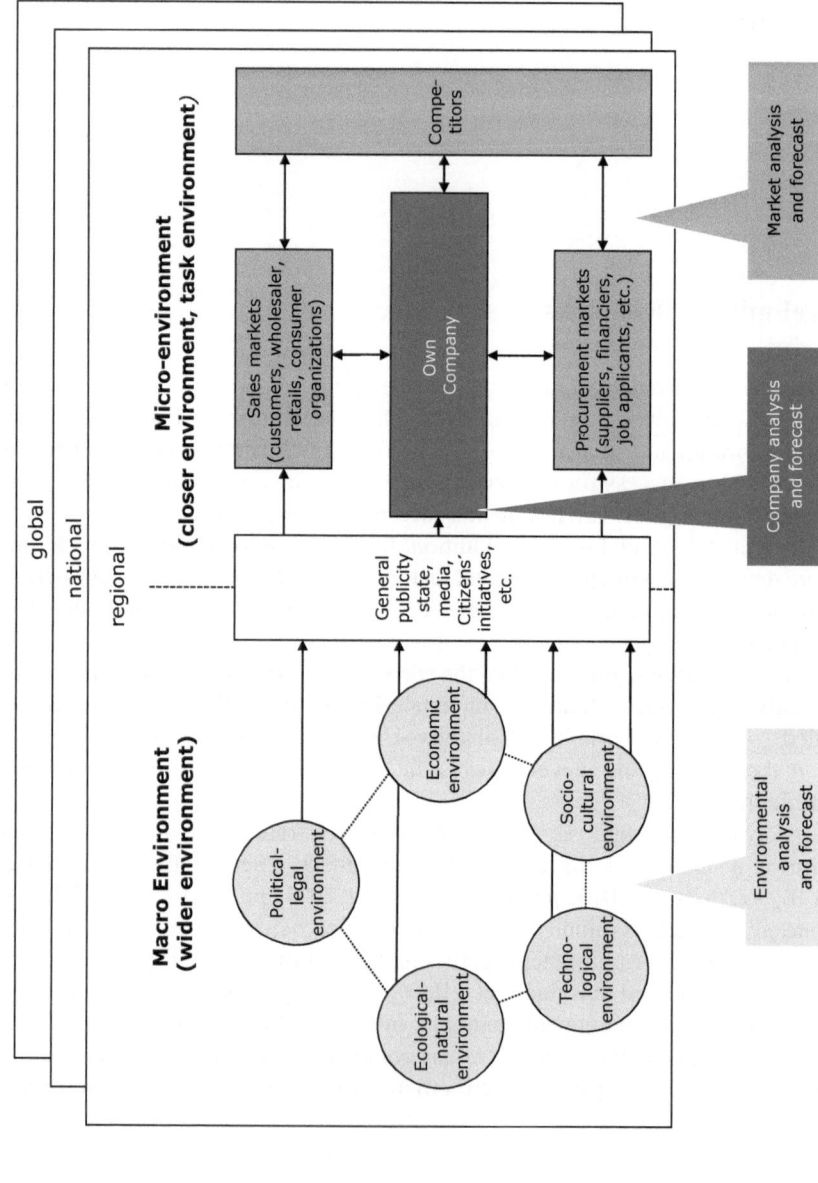

Fig. 1 Model of the corporate environment

Fig. 2 System of the
50 environmental factors

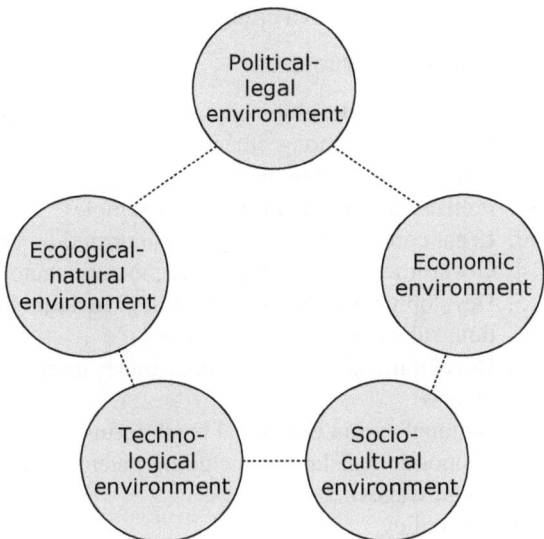

well influenced by the individual company. However, there are also factors clearly derived from the macro-environment but have direct effects on tourism companies (see, for example, the travel warnings issued by the Foreign Office of the Federal Republic of Germany). Other factors between the macro- and micro-environments are difficult to classify because, depending on the circumstances, they can be attributable to both macro- and micro-environments.

There is an almost unmanageable number of environmental factors.[1]

The following overview systematically presents the 50 most important environmental factors. Not all environmental factors are equally relevant in terms of tourism or luxury tourism, so targeted environmental analysis and forecasts are required.

The macro-environment is divided into different "environments," which, however, are interdependent in many ways. Often, five environments are listed: the political–legal, the economic, the sociocultural, the technological, and the ecological–natural environment (see also Fig. 2). The analysis of these environments is often referred to as PESTE analysis (Political–Economical–Societal/Cultural–Technological–Ecological Environment).[2] The micro-environment is divided into sales markets, procurement markets, and competitors.[3] A special component of the micro-environment is also the companies themselves.[4] In most cases, the environmental factors vary depending on the region.

---

[1] See, for example, the list in Nagel and Wimmer (2014, p. 129 ff.) And the graphic facts in www. ourworldindata.org.

[2] The analysis of the macro-environment is usually referred to as "environmental analysis."

[3] The analysis of the micro-environment is usually called "market analysis."

[4] The analysis of one's own company is usually called "enterprise analysis."

System of the 50 Environmental Factors

Political–legal environment

1. Geopolitical developments: regional, national, and international conflicts (terrorist attacks among others)
2. State of development and stability of the social and political system (the protection of human rights for example)
3. Legal certainty (property rights, corruption, etc.)
4. Government of the country (democratic, autocratic, dictatorial)
5. Developments in economic policy (including plan or market economy orientation, subsidies, industrial policy)
6. Developments in international trade: integration, protectionism (duties among others)
7. National and international legal norms
8. Economic legislation (including patent law, producer liability, labor law, minimum wages)
9. Tax policy
10. Regulation/deregulation (including freedom of travel/visa requirements)

Economic environment

1. Overall performance of the economy: gross domestic product (GDP), disposable income
2. Asset balance of the economy
3. Economic development
4. Monetary performance: consumer prices, wholesale prices, commodity, and producer prices
5. Foreign Trade Development: Terms of trade
6. Public finances: government ratio, debt, subsidies
7. Investment activity of the private and public sector
8. International currency and interest rate developments (including exchange rates)
9. International debt
10. Seasonal fluctuations

Sociocultural environment

1. Population development
2. Population structure: life expectancy, age structure, educational level, etc.
3. Marital status and household structure
4. Ethnic mix
5. Regional distribution of the population (including urbanization)
6. Social values and norms (among others regarding justice/equality, sustainability, hedonism, desire to possess, etc.)
7. Religion, morality, and ethics
8. Social motive structure and attitudes (amid other things self-realization, demonstration/privacy, work/leisure, safety/risk, etc.)

9. Consumption habits
10. Mobility and leisure behavior (including vacation entitlements)

Technological environment

1. General technology development
2. Traffic engineering and infrastructure
3. Building and civil engineering technologies
4. Material technologies
5. Energy and raw materials technologies (in particular oil, gas, electricity, coal, renewable energy)
6. Technologies of ecological resources: soil, water, air, light
7. Mechanical engineering and production technologies: automation, process technologies (for example, 3D printing)
8. Information and communication technologies (especially digitalization)
9. Government and private development investments
10. Substitution or rationalization technologies (in particular substitution of the job)

Ecological–natural environment

1. Natural space and topography
2. Climate change and weather extremes
3. Air pollution (in particular CO_2 emissions and nitrogen oxides)
4. Natural disasters
5. Water shortage
6. Water pollution
7. Ozone
8. Biodiversity
9. Soil sealing and pollution
10. Food supply

With regard to the present subject of luxury tourism, the environmental factors of the macro-environment particularly relevant or effective for tourism companies in general and luxury tourism companies in particular are considered below.[5]

Many plausible descriptions and explanations of trends exist. These trend developments are described below, followed by the derivation of implications for tourism or luxury tourism companies. The aim of the following remarks is to gain insights into trends in luxury tourism markets from forecasts in the macro-environment.

---

[5] An up-to-date analysis of the country-specific nature of a variety of relevant environmental factors for the tourism industry can be found in the Travel and Tourism Competitiveness Report of the World Economic Forum, see https://www.weforum.org.

2 Political–Legal Environment

Tourism companies, in general, should consider the following factors of the political and legal environment:

1. **Geopolitical developments**: Of particular importance are regional, national, and international conflicts, which are carried out through hostile methods. Such conflicts fuel fear in people for their own safety and often lead to the avoidance of travel to the relevant regions. Particularly sharp declines in travel are caused by terrorist attacks. In recent years, attacks in traditional tourist destinations such as Tunisia, Egypt, and Turkey have led to a decline in tourist arrivals of up to 60% compared to the previous year. The travel warnings issued by the Foreign Office of the Federal Republic of Germany as well as comparable warnings of the foreign ministries of other countries are playing a significant role in this as they affect the business activities of the tour operators. As a result of geopolitical problems, migration processes occur. These also cause emigrants to travel back to their home countries almost certainly with delay.

2. **State of development and stability of the social and political system**: A factor of the development status is the respect or preservation of human rights. Human rights violations by governments are becoming less and less accepted by tourists from highly developed countries, leaving the tourist with a damaged image of the respective country. In addition, as fear of their own safety increases, human rights violations make tourism development more difficult.

3. **Legal certainty**: The supply side of the tourist market is hampered by inadequate property rights as it scares investors and hinders entrepreneurship. Part of legal security includes the fight against corruption.[6]

4. **Government of the country**: For decades, tourists have shown little interest in the government of the countries visited. Many acted in a very opportunistic way: if the tourism offer was adequate, the countries were visited. Over the past few years, a change of mind has been emerging. People are increasingly judging holiday destinations according to how sympathetic the political leadership is to them or how much they agree with the government. Dictatorial regimes or autocratic states are increasingly avoided by tourists.

5. **Developments in economic policy**: Economic policy may be more or less conducive to tourism development. For example, states that want to promote tourism in a targeted manner, and thus want to pursue an active sectoral industrial policy, will specifically develop tourist regions, grant subsidies, and initiate other support measures.

6. **Developments in international trade**: International trade exerts influence on business travel activities. On one hand, international trade presupposes business travel activity and, on the other hand, it causes business travel activities.

---

[6]See the level of corruption in different countries according to Transparency International's Corruption Perception Index at www.transparency.org.

7. **National and international legal norms**: These have little relevance to tourism.
8. **Economic legislation**: Economic laws can affect the tourism industry directly or indirectly. Minimum wages indirectly affect labor-intensive, low-margin industry segments, such as the hotel and restaurant industry, by potentially increasing prices and reducing competitiveness. The European Package Travel Directive directly affects tourism companies such as tour operators and intermediaries.
9. **Tax Policy**: Tax policy may be more or less harmful to the overall economic development of countries. Industry-specific taxes such as the aviation tax or the municipal bed tax in Germany can affect tourism. Conversely, tax benefits, such as the reduction of VAT on hotel accommodation that took place in Germany a few years ago, can make the German hotel industry more competitive with neighboring countries.
10. **Regulation/Deregulation**: Particularly relevant here are the visa requirements of countries, as they directly influence the freedom of travel.

With regard to luxury tourism, only a few factors in the political and legal environment can be described as relevant:

1. A country's tourism policy strategy may make it difficult or beneficial to position itself as a luxury destination. A mass tourism orientation makes positioning as a luxury destination difficult, even if individual regions and providers may be able to decouple from the country's image. Conversely, a restrictive tourism policy, which could include a limitation of tourist numbers, may favor exclusive positioning (see, for example, the Seychelles).
2. Recently, tendencies toward autocratic forms of government, often accompanied by the restriction of liberties, have been observed worldwide. Throughout the world, more people live in an environment in which they are less able to behave freely.[7] The ability to behave freely is therefore increasingly perceived as a luxury. Countries that grant these freedoms are perceived as luxury destinations by tourists of autocratic states.

3 Economic Environment

For tourism companies in general, the following factors of the economic environment should be considered:

1. **The overall performance of the economy**: Tourism demand is heavily dependent on economic performance. A high level of economic performance usually implies a high level of prosperity among the population, which can allow travel

---

[7]The impairment of the LGBT segment (Lesbian-Gay-Bisexual-Transgender), for example.

because of higher incomes.[8] Also, high overall performance of the economy means that many business trips are being made. For aviation, it has been proven that there is a significant declining relationship between the GDP level of states and the number of air travel (measured per capita) (see Airbus 2018).

2. **Financial assets**: High stocks of liquid assets are conducive to tourism demand as they allow travel. Here, however, the distribution of wealth is to be considered. The more assets are distributed among the population, the more people can afford to travel.

   A strong asset concentration, on the other hand, allows only a smaller portion of the population to travel.[9] The development of share prices and dividends influences tourism demand, especially in the upscale market segment.

3. **Economic development**: A positive economic development has a directly proportional relationship with tourism demand. Wages, salaries, and capital incomes rise during periods of economic upswing while unemployment is lower and future prospects are perceived as more optimistic. Expenditure possibilities and readiness are therefore higher in large parts of the population.

4. **Monetary value development**: Consumer prices play an important role here. Consumer prices determine the purchasing power of tourists in the respective countries. High consumer prices, such as those prevailing in Switzerland or the Scandinavian countries, have a dampening effect on tourism demand from foreign source markets. Furthermore, the price of oil influences tourism demand. High oil prices directly impact airfares as kerosene is one of the biggest cost blocks for airlines. Monetary stability or inflation should also have an impact on tourism demand. High inflation drives capital flight, reduces the propensity to save, and fuels the consumption of experience and investments in tangible assets.

5. **Foreign trade development**: The extent of international trade shapes business travel demand. Procurement, production, and sales activities abroad are still strongly linked to travel, despite powerful information and communication technologies (e.g., email and messenger communication, video conferencing).

6. **Public finances**: See the remarks in Sect. 2 under point 5 on subsidies.

7. **Private and public sector investment activity**: Investment activity has an impact on the supply side of the tourism market. An active investment activity increases the quantity and quality of the tourism offer. In the tourism market, supply and demand are in an interdependent relationship. Thus, an extensive and attractive tourism offer leads to an increase in tourism demand.

8. **International currency and interest rate developments**: Exchange rates play an important role here. They determine the purchasing power of tourists in the

---

[8]Empirical surveys on the so-called "income elasticity of tourism demand" clearly show positive correlations

[9]In 2017, 70.1% of the world's population had assets under $10,000, 21.3% from $10,000 to $100,000, 7.9% from $100,000 to $1 million, and had 0.7% Assets of more than 1 million US dollars (See Credit Suisse 2018, p. 24).

respective countries. Unfavorable exchange rates reduce demand and therefore lead to a shift in travel flows. For example, if the Sterling falls in relation to the Euro, traveling to the UK will be cheaper for tourists from the Euro area, while traveling to the Euro area for British people will be more expensive. Strong currencies drive outgoing tourism, weak currencies drive incoming tourism. So far, little research has been done on the influence of interest rates on tourism demand. Low interest rates are likely to reduce the propensity to save[10] and increase willingness to spend in the tourism segment as well.

9. **International indebtedness**: There are still no findings on the correlation between the debt–equity ratio of an economy and tourism activity.
10. **Seasonal fluctuations**: They are closely related to the tourism industry. In addition to the construction industry, tourism is one of the industries that causes seasonal economic fluctuations.

With regard to luxury tourism, the following factors of the economic environment are relevant:

1. The income and wealth situation has a strong impact on luxury demand for obvious reasons. "The richer people are, the more important luxury gets" (Papon 2018). The wealth of people is growing globally to a considerable extent, which can be seen in the increasing number of millionaires and billionaires, as well as in the increase in global financial and property assets.[11] It can be assumed the income and wealth increase in the classic luxury goods industry (clothing, watches, jewelry, cars, yachts, real estate, etc.) also benefits the tourism industry. Presumably, the tourism industry even benefits disproportionately, because the understanding of luxury is changing worldwide. Instead of material goods, wealthy people are increasingly seeking intangible goods which include (travel) experiences.
2. However, the impressive extent of material wealth in the world leads to one other assumption. Since the financial possibilities are plentiful, other luxury forms will exist in the future. Intangible and "money can't buy" luxury traits will become more important.

---

[10]Since the financial crisis in 2008, ECB policy has led to historically low interest rates in the euro area. In Germany it can be observed that in parts of the population the propensity to save is decreasing.

[11]In Germany, between 2209 and 2017 the number of billionaires increased from 99 to 187. In terms of wealth development in the world, it is predicted that the number of ultra-rich people (people with assets of more than $50 million) will increase from 148,000 (in 2017) to over 193,000 (2022), about 30%. (see Credit Suisse 2018) Global wealth for 2017 is estimated at approximately $280 trillion. By 2022, it will rise to $341 trillion (see Credit Suisse 2018). Lastly, it should be noted that data on the wealth of people are based on partly questionable estimates, see article in the Frankfurter Allgemeine Sonntagszeitung of December 24, 2017, p. 31.

4 Sociocultural Environment

Tourism companies, in general, should consider the following factors of the socio-cultural environment:

1. **Population development**: In quantitative terms, the world population is expected to grow from about 7.55 billion (as of 2017) to an estimated 11.18 billion by the year 2100 (see Statista 2018a). A larger number of people in the world will also result in a higher number of trips taken.
2. **Population structure**: The structure of the population is changing in most countries. There is a widespread increase in life expectancy as a result of improved medical care, while at the same time, fertility rates are decreasing in most countries. Hence, a growing proportion of the population now falls into older age groups. The aging of the population is often referred to as "demographic change." Demographic change is a megatrend that will shape economic and social structures profoundly and sustainably over many years. Demographic change is altering the travel behavior of the population. Older people have different needs than younger people and therefore have different travel needs. The older population prefers other travel destinations, places more emphasis on comfort, care, safety, and medical care, and they (still) have a higher spending propensity and ability. They are also more flexible in their choice of travel time, making seasonal effects less likely in the future. For obvious reasons, older people emphasize the importance of health causing all forms of wellness tourism to become more valuable. The educational level of the population of many countries is also increasing.
3. **Marital status and household structure**: Family and household structures are changing. Households are becoming smaller as fewer generations live under one roof, parents have fewer children and singles are increasing ("isolation of society").
4. **Ethnic mix:** It is likely due to migration processes that ethnic groups will increasingly mix in many countries. Immigrants, however, often reflect on their origins and travel to respective homelands.[12]
5. **Regional distribution of population**: In macro-geographical terms, above-average population growth is expected in African and Middle Eastern countries. By contrast, the population in Europe and North America is underperforming or even shrinking. From a micro-geographical point of view, an above-average increase in the urban population is expected. Today, 50% of the world's population is living in cities. By 2030 about 60% of the world's population will live in cities (so-called "urbanization"). Urbanization is also changing tourism demand. Urban residents usually have a greater need for recreation due to the increased pressures of the urban environment—more noise, poor air, tightness ("density stress"), etc. In addition, the travel options are superior, since

---

[12]In aviation, this segment is referred to as "ethnic traffic."

the transportation infrastructure is more developed and oftentimes the urban population has a higher level of prosperity.

6. **Social values and norms**: In numerous countries, we have been experiencing a social "change of values" for many years.[13] Traditional values are increasingly losing their behavioral control. Instead of work discipline, leisure activities and the so-called "work–life balance" are foremost and equality is gaining in importance. Instead of striving for material ownership, the strive for adventure and hedonism are becoming paramount (so-called "adventure society"). As an alternative to ownership, sharing is increasingly considered (what benefits the "sharing economy"[14]). In addition, sustainability[15] (Rein and Strasdas 2017, p. 13) is gaining significance as a value. Values and norms have a remarkable impact on tourism markets. First of all, it should be noted that the social change in values is resulting in growth for tourism demand. As material ownership becomes less significant and experience and leisure orientation become over-riding, the demand for holidays and travel increases. In addition, the demands on travel are changing.

   Values and norms are shaped by the essential customs of cultures and sub-cultures. At the end of the 1960s, Geert Hofstede developed the "model of cultural dimensions." The model now includes six dimensions of culture: power distance, individualism/collectivism, masculinity/femininity, uncertainty avoidance, long- or short-term orientation, compliance/domination (see Hofstede 2017). Other well-known models of value classification and measurement are available from Schwartz (see Schwartz and Bilsky 1987, 1990; Schwartz and Sagiv 1995; Inglehart 1995).

7. **Religion, morality, and ethics**: Travelers' expectations are strongly influenced by religion, morality, and ethics. For example, travelers of the Muslim faith expect prayer opportunities, specific food offers and soft drinks (so-called halal tourism) (see Mastercard and HalalTrip 2017). Religion, morality, and ethics are also highly influenced by and an essential part of culture.

8. **Societal motivational structure and attitudes**: Motivational research deals with what motivates human behavior and why.[16] The most prominent example of need classification is Abraham Maslow's hierarchy of needs or Theory of Human Motivation (1954).

---

[13]Values are regarded as desirable or morally well-considered qualities that are attributed to objects, ideas, practical or moral ideals, circumstances, patterns of behavior, character traits, or in short "states of desirability." Norms are behavioral expectations. The degree of binding distinguishes between mandatory, target, and optional standards.

[14]To put it bluntly: from useless possessions to usefulness without possessions.

[15]In 1998, the Enquete Commission "Protection of Man and the Environment" of the German Bundestag defined sustainability as follows: "Sustainability is the concept of a sustainable development of the economic, ecological and social dimension of human existence" (German Bundestag 1998).

[16]For the terms motive, motivation, need, attitude, see Kroeber-Riel and Gröppel-Klein (2013, p. 178 ff.).

Attitudes are relatively permanent judgments of objects (people, subjects, or products) (see also Kroeber-Riel and Gröppel-Klein 2013, p. 232 ff.). As motives and attitudes are strongly influenced by culture, differences among them influence behaviors. Regarding travel behavior, there are multiple classifications of needs.[17]

9. **Consumer habits**: There are habits in societies hardly ever questioned. For example, in Germany, holiday trips are habitual.[18]

10. **Mobility and leisure behavior**: In modern industrialized societies, a high number of annual vacation days is the norm (for example, in Germany an average of 27 days), which is a prerequisite for one or more vacation trips per year. A higher number of bank holidays and a reduced workweek also favor the mobility and leisure behavior of the population.

With regard to luxury tourism, the following factors of the sociocultural environment can be described as relevant:

1. Demographic change should almost automatically lead to above-average growth in luxury tourism. Older people have been able to accumulate higher assets compared to younger people because of their longer lives and are therefore able to spend more. Often, they are willing to spend money, as various studies on the saving behavior of different age groups show (see o. V. 2018a). It can be assumed the readiness for luxury consumption among older people is also higher. The increasing number of older people is driving luxury demand.

2. On one hand, the change in values is influencing the quantitative side of tourism demand. As material assets become less important and intangibles become more, demand for goods suffers, while demand for services benefits. This should also boost the demand for luxury travel. On the other hand, the change in values influences the qualitative side of tourism demand. In recent years, luxury is understood differently in tourism. Material components of the trip, such as the opulent furnishing of hotel rooms, have become less important. Intangible components of a trip, such as relaxation, the experience of nature, space, have gained in importance.

3. The definition of luxury developed in this book also refers to people feeling as if they are missing something. Maslow's needs pyramid provides initial clues to identifying deficiencies. It is plausible, for example, that luxury tourists perceive a poor satisfaction of higher-value needs.

Social demands such as getting together and communicating with life partners are often neglected in stressful everyday working life (third level), independence and freedom in everyday life remain on the line (fourth level) and self-realization and

---

[17]See, for example, Walker and Walker (2011), Bosshart and Frick (2006) or the systematics for the empirical research of the Forschungsgemeinschaft Urlaub und Reisen FUR, presented by Lohmann (2016) and Flash Eurobarometer (2015).

[18]For the travel behavior of Europeans, see, for example, Eurostat's tourism statistics at http://ec. europa.eu.

individuality become more difficult (fifth level). Personal attention, personal care, and high touch are thus elements of luxury offers. In the middle of the last century, Maslow could not have imagined how fundamentally different the modern world of social media and complete surveillance of individuals would be. He did not have the need for privacy and anonymity in view. Today, this need is one of the strongest drivers of luxury tourism (so-called "Hidden Luxury").

5 Technological Environment

Tourism companies, in general, should consider the following factors of the technological environment:

1. **General technology development**: Only has indirect links to the tourism industry.
2. **Traffic engineering and infrastructure**: Since movement is an integral part of tourism, developments in traffic engineering are highly relevant for tourism in all modes of transport (air traffic, shipping traffic, rail traffic, road traffic). Means of transportation, traffic routes, and traffic stations are ultimately technical elements. Highly developed traffic engineering and a developed transport infrastructure are conducive to tourism development. Developments in traffic engineering and transport infrastructure since the Second World War have made the worldwide boom in tourism possible. Work is currently being carried out on radically new traffic engineering that cannot be classified in either of the four existing modes of transport. Entrepreneur Elon Musk's Hyperloop is expected to enable rapid, eco-friendly travel in capsules shot through vacuum tubes in the near future. Flying cars combine the advantages of motorized private transport with those of aviation (see o. V. 2018b).

   (a) In the aviation sector, the introduction of jet engines and wide-bodied aircraft in the late 1960s has led to an increase in range, an increase in capacity, and a reduction in unit costs, enabling mass tourism over long distances. Work is currently underway on the second generation of supersonic aircraft. In contrast to the Concorde, the new generation will be more environmentally friendly and cheaper. Extremely demanding from a technology standpoint are solar drives for aircraft. Here, the Solar Impulse project is making an impressive pioneering effort.[19] Cost-effective individual transport in the air is also being pursued vigorously in many places. Passenger drones, airline taxis, etc. are in numerous conception and prototype phases.[20] Airspace monitoring has also become more efficient thanks to

---

[19]See also https://solarimpulse.com.
[20]See for example the multitude of projects on www.evtol.news/aircraft.

modern information and communication technologies, enabling a drastic increase in air traffic volumes to be managed.

(b) In shipping, the construction of cruise ships is considered to be technically demanding but permits ships to be constructed with enormous passenger capacity. The world's largest cruise ships, the Oasis ships (Oasis of the Seas, Allure of the Seas owned by Royal Caribbean), hold approximately 6300 passengers. The considerable size and the rapidly growing number of cruise ships have enabled the cruise segment to outperform the market development.

(c) In rail transport, high-speed trains present technical challenges. The German ICE, the French TGV, or the Japanese Shinkansen promises extremely fast connections between metropolitan regions. Thus, they represent a competition to aviation and favor business travel and city tourism.

(d) In road traffic, the road network has expanded exponentially. Since the Second World War, an enormous increase in the number of modern, easily manageable passenger cars has been observed, first in industrialized countries and later in emerging markets. Thus, the mobilization of broad sections of the population was achieved. Today, cars are the preferred means of transportation for leisure travel in many countries. The most important topic of road traffic of the future is autonomous, electrified driving.

3. **Civil engineering**: An efficient transport infrastructure is based on modern civil engineering (for example, bridge and tunnel construction, road and track construction, seaports, and airports). The hotel industry is benefiting from the associated building technologies (especially high-rise construction).

4. **Material technologies**: Modern materials such as composites, carbons, etc. are used in aircraft and vehicle construction, among other benefits are weight savings, fuel consumption optimization and, where appropriate, cost reduction.[21]

5. **Energy and raw materials technologies**: Means of transport such as airplanes, passenger cars, locomotives, and cruise liners are still largely fueled by fossil fuels. Adequate availability of fossil fuels and affordable raw material prices are conducive to tourism development.[22] In the future, the use of renewable energies (biofuels, solar energy, etc.) will become predominant and electric motors will be a crucial element of drive technology. Should vehicle price, loading time, and range problems be solved, it can be assumed that the development toward a more environmentally friendly transport system will not affect tourism.

6. **Technologies of ecological resources**: For tourism, intact ecosystems are indispensable. Clean, undisturbed natural spaces are strong attractions for tourists. Soil, water, and air pollution control are often prerequisites for tourism

---

[21]The Boeing 787 ("Dreamliner") is largely made from composite materials.

[22]See, for example, fracking technology, which leads to an increase in supply on the oil market and thus has a dampening effect on the price of oil.

development. An attractive flora and fauna (forests, wildlife, etc.) is dependent on and part of undamaged landscapes.

7. **Mechanical engineering and production technologies**: Noteworthy here is 3D printing which allows decentralized and individualized production. It is conceivable that in the future, hiking shoes or diving equipment, for example, will be printed in the hotel. The field of sensor technology has high potential, as sensors are becoming smaller and cheaper and are now installed in devices such as smartphones, watches, etc.

   They measure, for example, air temperature, pressure, humidity, speed, acceleration, fingerprints, geoposition, brightness, geomagnetic field, noise, pulse rate, etc. In addition, infrared technologies, face and voice recognition, or even chemical spectrometers are used (Cf. Kreutzer and Land 2015, p. 57 ff.).

8. **Information and communication technologies**: For decades, information and communication technologies have had a tremendous impact on the tourism industry. The foundation of information theory was formed by Claude Shannon in 1937 and eventually the first computers were built.[23] In the early 1960s computer technology, in the form of Global Distribution Systems (GDS), made its way into the tourism industry. GDS optimized and streamlined the marketing communication and sales process, first in the aviation industry and later in other segments of global tourism. In the mid-1990s, the industry experienced the next technology push manifested as the Internet, which was commercially "developed" in 1995. Consumers gained access to an enormous wealth of information and travel offers through the Internet, which significantly changed the information and booking behavior of travelers. About 10 years later, another development spurt—Web 2.0—occurred in which social media such as Wikipedia and Facebook were created. Back then, the Spiegel (German business magazine) called this very vividly the "participatory web." The technological innovation of the smartphone by Apple in 2007 is another notable milestone.

   At the moment, the new "megatrend" is called "digitalization," which is shaping the discourse in society, politics, economy, and science like almost no other.

   In the meantime, all "ingredients"[24] are available for an explosive development of information technology possibilities. It is very clear today we are in a

---

[23]The thesis of the mathematician and engineer Claude Shannon is still regarded as the most influential student thesis of all time. (see Lenzen 2018, p. 37)

[24]The six prerequisites of digitalization in society and the economy are: (1) high global distribution of devices; (2) high-performance technical elements of the devices (for example, cameras, sensors, GPS positioning); (3) comprehensive networking of the devices via the Internet; (4) high computer performance, which develops according to "Moore's Law"; (5) adequate network bandwidths; and (6) low costs of devices, data processing, transmission, and storage.

phase of revolutionary change in the economy and society around the world.[25] We are experiencing a hitherto unknown impact of digitalization, which is often referred to as "disruption." Technology giants[26] are investing heavily in modern information technologies, transforming our way of living, working, and conducting business. Communication and leisure behavior, business processes, and models are profoundly changing. In the tourism industry as well, we are experiencing the dismantling and reorganization of existing value chains. Marketing and distribution processes, too, are being completely overhauled.[27]

9. **Government and private development investment**: Not very relevant to the tourism industry.
10. **Substitution or rationalization technologies**: Artificial intelligence and robot technology, in particular, have the potential to substitute human labor. The groundbreaking studies by Frey and Osborne have further initiated a large number of studies for individual countries.

It is reasonable to assume that technological developments will replace a considerable number of higher-quality jobs. However, new jobs will be created, which is customary when using new technologies. Nevertheless raising the question—unanswered until now—will the technological change take off so rapidly in the future that the necessary adaptation processes of society will not be able to cope?

Short Excursion: Technology Trends
Currently, the following technology trends dominate discourse, strategy, and actions of politics, society, media, science, and business:

1. **Mobile device usage**: Powerful devices connected to the Internet, such as smartphones, are almost ubiquitous worldwide. In 2017, 4.3 billion smartphone contracts existed worldwide, and by the year 2023, 7.2 billion are expected (see Ericsson 2018). At the same time, the media usage behavior of the population is changing massively toward the use of smartphones. For example, the daily use of mobile devices in the USA increased from 0.3 hours daily in 2008 to 3.1 hours daily in 2016. Similar developments took place in other industrialized and emerging economies.
2. **Global networking**: Currently 4.2 billion people, and thus 54% of the world's population are using the Internet (as of December 31, 2017) (see Miniwatts Marketing Group 2018). In addition, exponentially more things are connected to the Internet. Cisco estimates that in 2020, 50 billion objects will be connected to the Internet. In addition to smartphones, tablets, notebooks, and desktop computers, these are items such as cars or other means of transportation, household appliances, machines, clothing, etc. The Internet is thus becoming an "Internet of Things" (IoT), in reference to a connected industry the term Industry 4.0 is used. This global network enables entirely new products, business processes, and business models (see, for example, Oliver Wyman 2015).

---

[25]Often, these processes are referred to as "industrial revolution" whose effects are on a par with the first industrial revolution of the second half of the eighteenth century. See for example Brynjolfsson and McAfee (2015), Kreutzer and Land (2015).

[26]Especially the so-called GAFA (Google, Amazon, Facebook, Apple).

[27]For more details see, for example, Conrady et al. (2013, p. 472 ff.), Kreutzer (2018) and publications on e-tourism, online travel, etc.

3. **Cloud computing**: Cloud computing refers to the provision of IT infrastructure such as storage space, computing power, or application software as a service on the Internet. Important providers of cloud computing are Amazon or Microsoft. Cloud computing offers many benefits such as lower costs, higher computing power, and more powerful applications. Data security, however, poses large challenges.

4. **Social media**: Social media refers to digital technology applications that enable people to network through Internet platforms[28] and publish information easily and cost-effectively, when necessary and with a large reach. Information can be text messages, pictures, videos, songs, or the like.[29]

   Important social media are Facebook (with approximately 2.23 billion monthly active users worldwide (2nd quarter 2018) and approximately 30 million active monthly users in Germany in May 2017 (see Statista 2018b), Pinterest, YouTube, Twitter, and Instagram. Social media are relevant to the tourism industry in many ways: they serve at the beginning of the so-called customer journey and provide inspiration to consumers. Subsequently, customer review sites (such as TripAdvisor or HolidayCheck) help with the selection of offers, traveling pictures and updates are being shared on social media and finally, the vacation is rated on customer review sites. Destinations and attractions are increasingly judged by how well they are suitable as a photo motif (so-called "instagramability").

5. **Security and privacy**: More and more areas of public life and business are infused with digital technologies. The use of digital technologies makes it possible to influence people and leaves their mark (so-called "digital shadows"). In addition, production and other economic processes are already almost completely controlled by digital systems. Businesses and especially so-called "critical infrastructures," such as energy and water supply or traffic systems, are thus more susceptible to hacker attacks. Data protection and cybersecurity are therefore of fundamental importance. Biometric techniques (such as iris or fingerprint scanning and facial recognition) could provide a higher level of security in the future and maybe accompanied by questionable surveillance of the total population.

6. **Virtual and augmented reality**: Virtual Reality (VR) refers to the representation and simultaneous perception of reality and its physical properties in a real-time computer-generated, interactive virtual environment. Augmented Reality (AR) refers to the computer-aided extension of the perception of reality. VR has three features: (1) Latency (control of the scene by motion sensors), (2) Presence (the brain accepts the VR world as real, the real world disappears), and (3) Immersion (VR teleports the observer, lets him "immerse" in virtual and interactive worlds). VR makes places, products, and processes tangible regardless of time and place. VR leads to more intense experiences, stronger emotional reactions, and lasting memories. VR applications can be found in almost all segments of the tourism industry: tourist destinations and sights, cruise ships and aircraft equipment, hotel rooms, and amusement parks. It is also speculated VR can substitute travel completely. AR applications can be found when visiting sights (for example, information about history, technical data, etc. may be displayed when a tourist visits the Brandenburg Gate in Berlin).

---

[28] At the moment, the development of a so-called "platform economy" is being critically discussed (see Brynjolfsson and McAfee 2018).

[29] The German Association for the Digital Economy (BVDW) describes social media as follows: "Social media are a variety of digital media and technologies that allow users to interact and create media content individually or collectively. The interaction involves the mutual exchange of information, opinions, impressions and experiences as well as contributing to the creation of content. Users actively refer to the content through comments, ratings, and recommendations, building a social relationship with each other. The line between producer and consumer is blurring. These factors distinguish social media from traditional mass media. As a means of communication, social media, individually or in combination, use text, images, audio and/or video and can be platform-independent" (BVDW 2017, p. 80).

7. **Blockchain**: For several years, blockchain technology has been credited with revolutionary potential (Cf o.V. 2016a, b, 2017b, c). Although cryptocurrencies such as bitcoins have their basis in it, blockchain applications go far beyond the financial markets. The blockchain technology is already being used in the tourism industry (see Rose 2016).
8. **Artificial Intelligence**: At present, probably the most significant component of digitalization is Artificial Intelligence (AI), which often includes algorithms, machine learning, and big data.[30] Big technology giants (like Google, Amazon, Facebook, Apple, Microsoft, and also SAP) are currently investing heavily in Artificial Intelligence. The term Artificial Intelligence was "invented" by John McCarthy in 1955 at the Dartmouth Conference. The Dartmouth Conference is regarded as the starting signal for AI research. At first, AI received massive funding from the research department DARPA of the US military. In the last few years, computers have proven they can outperform humans (for example, chess, jeopardy, go, poker, etc.), see also the results of the Turing test in Lenzen 2018, pp. 25 ff.).

To date, there is no uniform definition of AI due to the very different fields of application. It is still a wide, interdisciplinary, unclear area. According to the AI Glossary by National Geographic, AI is a "branch of computer science that explores the mechanisms of intelligent human behavior and simulates them using computer systems" (o.V. 2017a, p. 69). This raises the question of what intelligence actually is. Intelligence measurement techniques focus on solving computational problems, logic puzzles, turning complex geometric bodies using imagination, and memory performance. Emotional, social, artistic intelligence and body intelligence have also been included recently.

AI has a strong interdisciplinary character. Disciplines involved are Computer Science, Psychology, Logic, Philosophy, Mathematics, Statistics, Economics (Decision and Game Theory), Neuroscience, Linguistics, Pedagogy, Ethnology, Biology, and Engineering. AI is regarded very ambivalently. In a BBC interview in 2014, the famous physicist Stephen Hawking said that AI could be the best or the worst thing to happen to humanity.

With AI today, three objectives are being pursued: (1) The solution of concrete problems (application-oriented approach aiming at specific, individual solutions to problems); (2) Development of a general artificial intelligence (occasionally also referred to as "strong AI", since it aims at very different problem solutions[31]); and (3) Develop a better understanding of human cognition. Today, autonomous cars are the most advanced application of AI. As future perspectives of AI, the so-called "singularity" (convergence of man and machine) by Vernon Vinge, the merger of robotics and AI, and the invisibility of AI are being discussed and explored.

One of the best-known terms of AI is the algorithm. An algorithm is a provision that describes how to achieve a goal step by step (cooking recipes, directions, and Lego instructions can also be called algorithms). Algorithms automate problem-solving processes.

In AI, algorithms are rules for solving (mathematical) problems. A program is also an algorithm written in a programming language.

Another important subfield of AI is Machine Learning (ML). The core of ML is (as in human learning) to learn something from past experience that can be used to orient oneself in the present. Central to this is pattern recognition. In ML, a system uses examples to identify patterns or regularities it can use to classify new data (closely related to data mining). There are three main types of learning methods: (1) supervised learning, (2) unsupervised learning, (3) reinforced learning. For example, in supervised learning, the system should learn to

---

[30]See Lenzen's monograph (2018) for the various facets of AI.

[31]Today, AI applications are able to solve individual problems better than humans. For example, an AI system may play better chess than a human being, but it cannot simultaneously be better at driving or understanding speech. A general AI system could do this and be much more similar to humans.

distinguish dogs from cats. The system first gets to see many pictures of dogs and cats, which were previously marked as such.

In unsupervised learning, the system autonomously tries to recognize patterns and regularities.

Reinforced learning is used when the system attempts to learn actions. The system keeps trying and receives positive or negative feedback and adjusts its behavior accordingly (for example, a robot doing a somersault makes its own corrections after it has fallen down).

Today, "deep learning" is based on "artificial neural networks" (ANN) which are modeled after the brain. ANN can handle both structured and unstructured data. ANN are used today in very different fields of application: for face recognition, sorting of photos, search engines, autonomous driving, speech recognition, translation, personalization of advertising, buying recommendations, etc.

The personalization or individualization of marketing communication (including advertising), product offerings and prices is one of the most important operational fields of application. Customized advertising messages are designed to increase advertising effectiveness and prevent wasting resources. Tailor-made products are aimed at increasing customer satisfaction and brand loyalty, while individual prices are designed to exploit the consumer surplus. Significant advances in psychometrics will make tailor-made travel suggestions possible in the near future.[32]

On one hand, the large data sets existing today are a prerequisite for machine learning, since otherwise, supervised learning does not work. On the other hand, they require AI for their analysis. Data mining, which is the discovery of new connections, only succeeds with artificial intelligence and powerful algorithms.

Big data is the collective term for large, complex amounts of structured and unstructured data (for example, in the zettabyte or terabyte range) that come from different sources (for example, internal data such as CRM data and external data such as social media data). Big data software makes it possible to aggregate, analyze, and deliver this vast amount of data from multiple sources at a very high speed. The focus of the analysis is put on the determination of relationships and patterns (see BVDW 2017, p. 76.).

9. **Robotics**: Industrial robots have been used widely for years. In 2016, more than 1.8 million industrial robots, mainly in the automotive industry, were in use. For the year 2020, the use of nearly 3.1 million industrial robots is forecasted (see ifr 2017). There are hardly any robots in the service sector. However, first users can be found in the tourism industry. Hotel chains like Marriott and Holiday Inn are experimenting with robots to greet guests or provide room service.[33] Cruise companies such as Costa Cruise Lines and Royal Caribbean International as

---

[32]Online actions (visits to websites, movement and communication patterns, likes on Facebook, etc.) allow startlingly accurate conclusions about their authors. The influential psychometrist Michal Kosinski is constantly refining his models. By analyzing only 68 Facebook likes of a person he could determine their skin color with 95% accuracy, the sexual orientation (88%), whether they were a Democrat or Republican (85%). Intelligence, religious affiliation, alcohol, cigarette and drug consumption can also be calculated. Kosinski's program only needs ten Facebook likes to rate a person better than an average work colleague. Seventy likes are enough to know more about a friend, 150 to overshadow parents, and with 300 likes, the machine can predict behavior better than the person's partner. And with even more likes, you can surpass what people think they know about themselves. The exploitation of psychological behavior analysis, big data, and ad-targeting was done by Cambridge Analytica. In the election campaign of Donald Trump, up to 175,000 different nuances of an advertising message were sent to US Americans (tailor-made advertising messages, "targeting") (see Krogerus and Grasegger in: Magazine No. 48, cited after o. V. 2016c).

[33]The Japanese Henn-na Hotel is spectacular, it is operated entirely by mostly human-looking robots.

well as airlines such as KLM at Amsterdam Airport are using multilingual service robots. Empirical studies show that people are open to the use of service robots in the tourism industry. This is particularly true for respondents from Asia and for situations where one may assume better service from robots (for example, for foreign-language guests or information that requires access to large databases).[34] Robots are anticipated to look much more human in the future ("humanoid robots") and have considerable motor skills.[35] As already mentioned, a combination of artificial intelligence and robot technology is to be expected.

With regard to luxury tourism, only a few factors of the technological environment can be considered relevant:

1. The described innovations in the field of traffic engineering (Hyperloop) and air traffic (supersonic jets, passenger drones, air taxi) will initially be available only to a small circle of customers and be perceived as luxury offers.
2. The role of technology in the luxury segment will be, at best, that of an enabler. Technology will enable luxury products, services, and business processes. However, it will be virtually "invisible" in the background, or in other words: "High Tech" will at best optimize "High Touch," but in no way replace it. Of paramount importance in the luxury segment will always be the personal contact and the personal interaction, as many contributions of this book prove.
3. Digital technologies are ubiquitous in today's private and business life exerting influences that are increasingly understood as compulsion. Total digital abstinence ("digital detox") can be perceived as a luxury while time and self-determination are regained.

6 Ecological–Natural Environment

Tourism companies, in general, should consider the following factors of the ecological environment:

1. **Natural space and topography**: Nature and topography are of paramount importance for tourism, as they are considered outstanding attractions. Seas and beaches, rivers and lakes, high and low mountain ranges, forests and meadows—in short, "the landscape" has always exerted a strong attraction on tourists (so-called "pull factor" of tourism). Notingly, natural deficiencies in the everyday environment of tourists exert strong "push effects." They drive tourists into regions of high natural attractiveness.
2. **Climate change and extreme weather**: The biggest problem of the ecological environment is climate change. If nothing changes, the temperature of the earth's atmosphere will increase anywhere from 0.9 to 5.4 °C (on average

[34]See the extensive empirical study by ITB and Travelzoo (2016) at www.itb-berlin.com.

[35]See, for example, the humanoid robot ChihiraKanae by Toshiba or Atlas by Boston Dynamics.

about 4 °C) until the year 2100 compared to the pre-industrial time.[36] An increase of 1.5–2.0 °C is considered barely controllable for life on our planet.

The cause of climate change is the massive increase in CO_2 emissions since the beginning of industrialization due to the burning of fossil fuels such as oil, natural gas, and coal. An increased CO_2 concentration in the earth's atmosphere causes the greenhouse effect. This increase is due to human activities known as "anthropogenic climate change."

The consequences of climate change are diverse and dramatic. The increased temperature of Earth's atmosphere creates several adverse effects and critical weather systems: sea levels rise due to ice melt (particularly in the Antarctic), decreases in snowfall, increases in droughts, water scarcity, and extreme winds such as hurricanes, heavy rains, and floods. In short, the overall ecological system is out of balance, and life on our planet is severely affected as the foundations are being destroyed. The international community has therefore been trying, more or less seriously, for decades to implement measures limiting the rise in Earth's temperature to 1.5–2.0 °C.

The weather extremes caused by climate change could turn into crisis situations for tourism companies. Tourism-related infrastructure may be impaired, holiday-makers want or need to change their travel plans, and repeatedly affected regions are being continually and unsustainably damaged.

Global tourism is both the cause and the victim of climate change. On one hand, traffic flows, which are particularly associated with tourism, make a significant contribution to the increase in CO_2 emissions.[37] On the other hand, tourism is dependent on an intact nature. Too much heat in southern destinations and temperature increases in previously cool regions, destruction of beaches and coral reefs, water scarcity, decrease in snowfall in alpine regions, etc. lead to a shift in tourist flows and changes in previous forms of holiday. Tourism will also be heavily impacted if measures are taken to curb traffic growth. For example, the inclusion of aviation in emissions trading schemes (see the European Emission Trading Scheme EU-ETS) would lead to an increase in the costs of airlines and thus to price increases, which in turn would bring demand-dampening effects (see Conrady and Bakan 2008).

The tourism and transport sector is taking the "adaptation and mitigation" approach, i.e., adaptation to climate change and mitigation of climate change.

3. **Air pollution (CO_2 emissions and nitrogen oxides)**: In addition to the aforementioned impacts of CO_2 emissions on climate change, air pollution also affects the tourism industry in other ways. Increased air pollution, for example in urban areas, reduces the attractiveness as a tourist destination and increases the incentive for its inhabitants to spend their vacations in destinations with clean air.

---

[36]See the research results of the Intergovernmental Panel on Climate Change IPCC at www.ipcc.org and the tourism references in Rein and Strasdas (2017, p. 45 ff.).

[37]The transport sector accounts for almost one-fourth of global CO_2 emissions.

4. **Natural disasters**: Natural disasters such as volcanic eruptions, tsunamis, earthquakes, or even forest fires destroy regions and their infrastructure, thereby affecting tourism. Natural disasters are a specific type of crisis. The tourism industry is preparing by constructing action plans and implementing appropriate crisis management strategies and measures.

5. **Water scarcity**: Access to sufficient and clean water is one of the most important foundations of life. Due to various causes (climate change, population growth, accelerated water consumption, etc.), we will increasingly be experiencing water scarcity. Sufficient clean water is essential for tourism regions. Consequently, extensive competition with the local population for the scarce water recourses in many tourist regions is expected.

6. **Water pollution**: Water pollution affects the flora and fauna of the water, the food supply, and the health of humans and animals.[38] The tourism industry relies on clean water. Much of global tourism takes place on the beaches of lakes and oceans, as "sun-and-beach vacations" are the most prominent form of vacation.

7. **Ozone**: A few decades ago, the expansion of the ozone hole over the Antarctic caused great concern. Due to the ban on CFCs, the ozone hole seems to be closing and the thinning of the ozone layer is currently no longer a hot topic.

8. **Biodiversity**: Biodiversity refers to "the variability among living organisms of all origins, including but not limited to: terrestrial, marine and other aquatic ecosystems and the ecological complexes to which they belong; this includes diversity within species and between species and diversity of ecosystems" (Dickhut 2017, P. 100). Tourism and biodiversity are closely interrelated.

   Biodiversity is often an attraction of tourism (see, for example, Africa's wildlife). Tourism, however, can negatively affect biodiversity, in particular through the overuse of natural space. Nevertheless, tourism can also set in motion measures to conserve biodiversity, as the economic benefits of biodiversity conservation are clearly visible and desirable.

9. **Soil sealing and pollution**: Soil sealing can lead to increased flooding, a contaminated food supply, and damage to the health of humans and animals.

10. **Food Supply**: An adequate food supply is essential for tourism regions. Considerable efforts to achieve sustainable economic activity communicates that the supply of food for tourists from local production is becoming increasingly important.

With regards to luxury tourism, the following factors of the ecological–natural environment can be described as relevant:

1. Due to global warming as a result of climate change, the desirability of cooler destinations will increase. Greenland, Iceland, and Lapland, for example, are not yet considered to be luxury destinations, but they have the potential for future

---

[38]Water pollution from plastic waste has recently been increasingly discussed.

luxury tourism, especially from source markets where it will be unbearably hot in the future.

2. The growth of the world's population and steady urbanization mean that fewer and fewer people live in undisturbed natural areas. Unimpaired nature, tranquility, space, and a small number of people are gradually becoming rare and are therefore increasingly seen as a shortcoming. In the future, luxury tourism will progressively include the features just mentioned.

3. The growth of the world's population and urbanization are also making unimpeded passenger transport more difficult. In many cities around the world, the traffic collapse is threatening to happen in the near future or it has already occurred. An imperative feature of luxury tourism will be the unimpeded transportation of passengers in the tourist destination. For example, excursions by helicopter, hotel transfers with drone taxis, flights with private planes but also tours on foot or with e-bikes are gradually becoming part of luxury tourism.

4. The luxury tourism segment can hardly escape the increasing burdens a deterioration of the ecological environment entails. It can be assumed that the awareness for sustainability is also strengthened in this segment and features of sustainable tourism are becoming much more significant components of luxury tourism.

7 Conclusion

In many cases, future environmental factors can be reasonably predicted or scientifically proven. The development of a number of environmental factors favoring luxury tourism is shown. Therefore, it can be assumed that the luxury tourism market segment will continue to project above-average growth rates in the future.

Literature

Airbus (2018) Global market forecast 2018–2037. https://www.airbus.com/aircraft/market/global-market-forecast.html. Zugegriffen: 14 Sept 2018

BMU (2017) Kernbotschaften des Fünften Sachstandsberichts des IPCC. https://www.bmu.de/fileadmin/Daten_BMU/Download_PDF/Klimaschutz/ipcc_sachstandsbericht_5_teil_1_bf.pdf. Zugegriffen: 16 Okt 2018

Bosshart D, Frick K (2006) Die Zukunft des Ferienreisens – Trendstudie. Gottlieb Duttweiler Institute, Rüschlikon

Brynjolfsson E, McAfee A (2015) The second machine age, 2. Aufl. Börsenmedien AG, Kulmbach

Brynjolfsson E, McAfee A (2018) Machine, platform, crowd: Wie wir das Beste aus unserer digitalen Zukunft machen. Börsenmedien AG, Kulmbach

BVDW (2017) Social media Kompass 2017/2018. Düsseldorf. https://www.bvdw.org/themen/publikationen/detail/artikel/social-media-kompass-20172018/ . Zugegriffen: 14 Sept 2018

Conrady R, Bakan S (2008) Climate change and its impact on the tourism industry. In: Conrady R, Buck M (eds) Trends and issues in global tourism 2008. Springer, Berlin, pp 27–40

Conrady R, Fichert F, Sterzenbach R (2019) Luftverkehr. Betriebswirtschaftliches Lehr- und Handbuch, 6 Aufl. Oldenbourg, München

Credit Suisse (2018) Global wealth report 2017. http://publications.credit-suisse.com/tasks/render/
    file/index.cfm?fileid=12DFFD63-07D1-EC63-A3D5F67356880EF3 . Zugegriffen: 13 Sept
    2018
Deutscher Bundestag (1998) Abschlußbericht der Enquete-Kommission "Schutz des Menschen und
    der Umwelt – Ziele und Rahmenbedingungen einer nachhaltig zukunftsverträglichen
    Entwicklung". Drucksache 13/11200 vom 26.06.1998. http://dip21.bundestag.de/dip21/btd/
    13/112/1311200.pdf. Zugegriffen: 13 Sept 2018
Dickhut H (2017) Tourismus und Biodiversität. In: Rein H, Strasdas W (eds) Nachhaltiger ·
    Tourismus. Einführung, 2 Aufl. UVK, Konstanz, pp 99–136
Ericsson (2018) Ericsson mobility report. Stockholm. https://www.ericsson.com/en/mobilityreport.
    Zugegriffen: 14 Sept 2018
Flash Eurobarometer (2015) Preferences of Europeans towards tourism. Report of the European
    Commission, Brüssel
Frey CB, Osborne MA (2013) The future of employment: How susceptible are jobs to
    computerisation? https://www.oxfordmartin.ox.ac.uk/downloads/academic/The_Future_of_
    Employment.pdf. Zugegriffen: 16 Okt 2018
Freyer W (2011) Tourismus-marketing, 7 Aufl. Oldenbourg, München
Hofstede G (2017) Lokales Denken, globales Handeln. Interkulturelle Zusammenarbeit und
    globales Management, 6 Aufl. dtv, München
ifr (2017) World robotics report 2017. https://ifr.org/downloads/press/Executive_Summary_WR_
    2017_Industrial_Robots.pdf. Zugegriffen: 14. Sept. 2018.
Inglehart R (1995) Changing values, economic development and political change. Int Soc Sci J 47
    (3):379–403
ITB & Travelzoo (2016) Robots and artificial intelligence in the hotel industry. https://www.itb-
    kongress.de/media/itbk/itbk_dl_de/itbk_dl_de_itbkongress/itbk_archiv_2016/itb_hospitality_
    day_3/Robots_And_Artificial_Intelligence_In_The_Hotel_Industry_Singer.pdf.   Zugegriffen:
    8 Okt 2018
Kerth K, Asum H, Stich V (2015) Die besten Strategietools in der Praxis, 6 Aufl. Hanser, München
Kreutzer RT (2018) Praxisorientiertes Online-Marketing, 3 Aufl. Springer, Wiesbaden
Kreutzer RT, Land K-H (2015) Dematerialisierung – Die Neuverteilung der Welt in Zeiten des
    digitalen Darwinismus. Future Vision Press, Ort
Kroeber-Riel W, Gröppel-Klein A (2013) Konsumentenverhalten, 10 Aufl. Vahlen, München
Lenzen M (2018) Künstliche Intelligenz. Was sie kann und was uns erwartet. Beck, München
Lohmann M (2016) Präsentation der Ergebnisse der Forschungsgemeinschaft Urlaub und Reisen
    FUR auf dem ITB Berlin Kongress am 11.03.2016
Maslow AH (1954) Motivation and personality. Harper & Row, New York
Mastercard & HalalTrip (2017) Muslim millennial travel report. Singapore. https://www.halaltrip.
    com/halal-travel/muslim-millennial-travel-report/. Zugegriffen: 13 Sept 2018
Meffert H (1998) Marketing. Gundlagen marktorientierter Unternehmensführung, 8 Aufl. Gabler,
    Wiesbaden
Meffert H, Burmann C, Kirchgeorg M (2015) Marketing: Grundlagen marktorientierter
    Unternehmensführung Konzepte – Instrumente – Praxisbeispiele, 12 Aufl. Springer, Wiesbaden
Miniwatts Marketing Group (2018) Internet world stats. https://www.internetworldstats.com/.
    Zugegriffen: 14 Sept 2018
Nagel R, Wimmer R (2014) Systemische Strategieentwicklung: Modelle und Instrumente für
    Berater und Entscheider, 6 Aufl. Schäffer-Poeschel, Stuttgart
Oliver Wyman (2015) The internet of things. Disrupting traditional business models. https://www.
    oliverwyman.de/content/dam/oliver-wyman/global/en/2015/jun/2015_OliverWyman_
    Internetof-Things.pdf. Zugegriffen: 14 Sept 2018
o. V. (2016a) FAZ vom 23.5.2016
o. V. (2016b) FAZ vom 13.6.2016
o. V. (2016c) FAZ vom 11.12.2016
o. V. (2017a) National Geographic. Juli 2017, p 69

o. V. (2017b) FAZ vom 29.8.2017

o. V. (2017c) Handelsblatt vom 19.7.2017

o. V. (2018a) FAZ vom 23.02.2018

o. V. (2018b) FAZ vom 20.07.2018, S. 27

Papon K (2018) Die Lust am Luxus lässt die Aktionäre strahlen. FAZ Online. http://www.faz.net/aktuell/finanzen/finanzmarkt/die-lust-am-luxus-laesst-die-aktionaere-strahlen-15684356.html. Zugegriffen: 13 Sept 2018

Rein H, Strasdas W (eds) (2017) Nachhaltiger Tourismus. Einführung, 2 Aufl. UVK, Konstanz

Rose N (2016) Will blockchain disrupt travel distribution and settlement? Phocuswright. https://www.phocuswright.com/Travel-Research/Technology-Innovation/Will-Blockchain-Disrupt-Travel-Distribution-and-Settlement. Zugegriffen: 14 Sept 2018

Schwartz JLK, Bilsky W (1987) Toward a universal psychological structure of human values. J Pers Soc Psychol 53(3):550–562

Schwartz JLK, Bilsky W (1990) Toward a theory of the universal content and structure of values: Extensions and cross-cultural replications. J Pers Soc Psychol 58(5):878–891

Schwartz JLK, Sagiv L (1995) Identifying culture-specifics in the content and structure of values. J Cross Cult Psychol 26(1):92–116

Statista (2018a) Prognose zur Entwicklung der Weltbevölkerung von 2010 bis 2100 (in Milliarden). https://de.statista.com/statistik/daten/studie/1717/umfrage/prognose-zur-entwicklung-der-weltbevoelkerung/. Zugegriffen: 13 Sept 2018

Statista (2018b) Anzahl der monatlich aktiven Facebook Nutzer weltweit vom 3. Quartal 2008 bis zum 2. Quartal 2018 (in Millionen). https://de.statista.com/statistik/daten/studie/37545/umfrage/anzahl-der-aktiven-nutzer-von-facebook/. Zugegriffen: 14 Sept 2018

Walker JR, Walker JT (2011) Tourism—concepts and practices. Pearson, Upper Saddle River, NJ

Prof. Dr. Roland Conrady has been a professor since 2002 at the tourism/transport department of the Worms University of Applied Sciences. His research and teaching focus on aviation, tourism, and digitalization. Since 2004, Roland Conrady has also been the Scientific Director of the world's largest tourism convention, the ITB Berlin Convention. He was president of the German Society for Tourism Science (DGT) e.V. and is a book author (among others Conrady, R., Fichert, F., Sterzenbach, R., Luftverkehr, Munich 2019). Roland Conrady is a member of various advisory boards of companies and politics. Previously, he was head of the study program "Electronic Business" and Professor of General Business Administration at the University of Applied Sciences Heilbronn. After graduating as Dr. rer. pol. from the University of Cologne in 1990, he held various management positions at Deutsche Lufthansa AG until 1998.

Analysis of the Luxury Phenomenon

David Bosshart, Hannes Gurzki, Dorothea Hohn, and Antonella Mei-Pochtler

1 Definition and Concepts of Luxury

David Bosshart

Luxury, it seems, is no longer elitist. A bottle of champagne, a Burberry scarf, or a night at a five-star hotel, a broad European middle class is able to afford this now and then. But when luxury is democratizing, the central question is what does this mean for its further development? Luxury is by definition the rare and therefore desirable. Consequently, when prestige becomes a commodity, a "masstige": Does not luxury then disappear as a category?

At least nobody can escape the fascination of luxury. "Luxury is the inherent tendency of every human being to improve living conditions and, under certain circumstances, to do so in a concentration of consumption in a specific field," says cultural sociologist Reinhard Knoll (2009). This raises the question of where this consumption will be concentrated in the future. Dana Thomas tries to become more concrete in her book "Deluxe: How Luxury Lost its Luster." "Luxury is not how much you can buy. (...)"; Luxury is buying the *"right thing,"* she quotes a

D. Bosshart (✉)
Gottlieb Duttweiler Institute, Rüschlikon, Switzerland
e-mail: info@gdi.ch

H. Gurzki (✉)
European School of Management and Technology, Berlin, Germany

D. Hohn (✉)
Global Communication Experts GmbH, Frankfurt, Germany
e-mail: hohn@gce-agency.com

A. Mei-Pochtler
Vienna, Austria
e-mail: mei-pochtler.antonella@advisor.bcg.com

© Springer Nature Switzerland AG 2020
R. Conrady et al. (eds.), *Luxury Tourism*, Tourism, Hospitality & Event Management, https://doi.org/10.1007/978-3-030-59893-8_3

The phases	INFANTILE PHASE	ADOLESCENCE PHASE	MATURITY PHASE	SENIORITY PHASE
The principle	More is more	More is a must	More is less	Less is more
The motivation	Advancement and pent-up demand	Recognition and affiliation	Differentiation and Distinction	Devotion and self-transcendence
The prototypical clientele	young and hungry	performance- and status-oriented	mature and saturated	aging and sense seeking
The appearance	Kitsch and children's dreams	Status Objects and Positional Goods	Adventures and experiences	Indulgence, leisure and memories

Fig. 1 The phases of luxury (source: Bosshart and Kühne 2014, p. 14)

knowledgeable luxury consumer (Thomas 2008, p. 363 f.). But what is the "right thing" today? And what will it be tomorrow?

The challenge is to bring together the various manifestations of luxury—in its purely material, ostentatious form as swank and splendor as well as in its (often) immaterial form as time and leisure. A step-by-step model seems obvious, as a first glance into the world of luxury has already shown that the understanding of luxury changes with increasing luxury exposure and experience.

The Phases of Luxury

In order to be able to orientate ourselves at all in the diverse world of luxury, to bring together the different manifestations of luxury, and to be able to better classify the changes in luxury consumption, we resort to an ideal–typical model (Cf. Bosshart and Kühne 2014). It is oriented toward the different phases of life—infantile phase, adolescence phase, maturity phase (see also Fig. 1)—and transfers them metaphorically to the world of luxury. Looking to the future, the next step is the seniority phase. It is important for the understanding of the model that the described maturing process describes on the one hand the change of the concept of luxury in individual biographies, but on the other hand also the change of entire societies.

– **The infantile phase**

The first phase of luxury development is characterized by a hunger for consumption that is satisfied with what is offered. Figuratively speaking, the child—or the small consumer—takes everything that is "fed" to him and what makes his eyes glow. The predominant principle: "More is more." This consumption, which is marked by children's dreams, can be observed in young, up-and-coming luxury markets. A backlog demand and the desire to rise is eminent. At the same time, there is a lack of knowledge about how and for what lifestyle the newly acquired wealth should be used.

– **The adolescence phase**

The next phase requires solvency but is dominated by increased competitive pressure (peer pressure). The dream of (further) social advancement increasingly gives way to a fear of social decline. Now the "more" becomes the "must." Goods with a signaling effect are gaining in importance: questions about how and where one lives, how big the car and the second car is or which school the children attend. The concern to be able to keep up, especially in the sense of comparison with neighbors or social peers ("Keeping up with the Joneses"), is driving a broad middle class, especially in the USA.

- **The maturity phase**

In this phase, luxury fatigue sets in. It is characterized by the decreasing marginal utility of the material products. Therefore, the realization that happiness when purchasing a product decreases the more often and without hindrance it is possible. Or in short: "More is (always) less." Consequently, luxury consumption shifts from the product to the experience level. Because experiences can be infinitely increased—from a simple restaurant visit to a luxurious wellness weekend to the ultimate adventure trip. Today, the majority of saturated affluent societies find themselves in this phase of the luxury life. And a new phase is already in the offing. Because: Those who have everything long for less. All the more pressing the question: What is next?

- **The seniority phase**

If we continue the ideal–typical model, we are now on the threshold of the seniority phase. The term fits in two senses: what has long since become apparent are the demographic facts. Our society is inevitably getting older, increasingly so in the coming decades. In the figurative sense of the model, however, seniors also include those who, regardless of their age, have long belonged to the group of luxury consumers. This also includes younger generations of consumers who are characterized by material abundance, the possibilities of which they have quickly exhausted. Those who have carried out the experience of the preceding stages in their individual biographies, so to speak, in rapid succession and who are now wondering which target values should now be shaping their life. A stronger awareness of sustainability and dwindling tolerance of waste are two new, formative attitudes with which younger generations are growing up today. With digitalization, new questions will arise: How will increasing dematerialization affect our relationship to luxury products? What is clear is that rapid technological progress combined with economic and social change will significantly change our idea of luxury, our awareness and appreciation of handicrafts.

2 Understanding of Luxury in the Course of Time

David Bosshart

A glance at history shows that luxury has always been a versatile concept. The fact that the idea of what luxury is changes over time is not new in itself. What is considered luxury has always been dependent on the zeitgeist—depending on how rare a good was. In ancient times and the Middle Ages, the transport of pepper from the West Indies was not only long but also dangerous; consequently, pepper was scarce and therefore precious and expensive. Today, pepper, like many other oriental spices, is a standard product in Western cuisines; distinction is possible through different varieties and quality levels alone. The development of salmon is similar: smoked salmon has arrived in the cheap range with large breeding farms and has lost its status as a noble luxury fish. Another parallel development from the exceptional phenomenon to the common good can also be found in technical devices such as refrigerators, cars, mobile phones, etc.

If one looks at the conceptual history, then the Latin term Luxuria, the "removal of old Roman austerity and the life of pleasure," had the connotation of abundance and waste. In the literal sense, luxuriare also means "to be luxuriant" and "to grow luxuriant." The French term luxure contains the moralizing component of desire and debauchery even more strongly. Common to all terms is the wastefulness that both the French philosopher Georges Bataille and the German sociologist Werner Sombart later emphasize. "Luxury is any effort that goes beyond what is necessary," says Sombart. And continues: "Luxury in the quantitative sense is synonymous with the waste of goods" (Sombart 1992, p. 85).

The English word to luxuriate includes strolling, the French invention of idly walking, promenading. Thorstein Veblen (2007) also defines idleness in his "Theory of the Fine People" as central to the demonstrative expression of wealth. Veblen asserts the need for recognition and prestige and considers wealth not from the perspective of the rich but from that of others. Conspicuous leisure and conspicuous consumption are the two pillars of his distinction theory. The wealthy show their wealth through (ostentatiously) idle time or opulent consumption—and thus demonstrate to the outside what others cannot afford.

The idle pastime does not serve the experience, but according to Veblen (2007) is a strategy to make wealth visible—in the form of nonproductive time not dedicated to the acquisition of money. Other luxury theories, however, assume that in luxury the aspects of showing and experience belong together. Showing—that is: for others—has always belonged together with the experience—that is: for itself. However, with increasing maturity, the focus shifts.

What remains despite all the changes in the concept of luxury is people's striving for recognition and belonging, but also for differentiation and distinction, for self-reward and self-realization. The classic motives of luxury consumption remain: the desire to indulge oneself, to impress others, to belong to the illustrious circle, to distance oneself from the bottom, and to underpin one's social status.

The controversial discussion also remains about when the boundaries of morality and ethics are crossed, when luxury becomes wasteful, decadent, or even reprehensible. This question has a long tradition: the excessive lifestyle was supposed to be curbed again and again throughout history with luxury bans. In former times in Europe, the ban on luxury applied to nobility and clergy, in today's China to politics and the military. This illustrates how morally charged the term has always been.

In Europe, the relationship to luxury did not ease until the eighteenth century. With the French Revolution, the dissolution of social class boundaries, and the emergence of a new class, the bourgeoisie, luxury suddenly appears in a new light.

With growing trade relations and the establishment of large manufactories, even "ordinary mortals" (i.e., non-nobles) can become wealthy, and so the attitude toward valuable and exclusive goods changed among many. The new science of political economy no longer puts the phenomenon of luxury under such a strict moral microscope. The British economist James Stewart, for example, already stated in his "Principles of Political Economy" of 1767 that luxury was a subtlety in taste and lifestyle (cf. Steuart 1848). Rather, the advantages of luxury as a driving force for demand, technological progress, increased employment, and the prosperity of society are emphasized.

3   Economic Relevance of Luxury Markets[1]

Antonella Mei-Pochtler and Hannes Gurzki

The luxury industry is a defining industry and an important driver of the European economy. The cultural and creative sector alone is a strong contributor. According to the European Cultural and Creative Industries Alliance, European high-end cultural and creative brands account for over 70% of the world's market and create employment opportunities for over 1.7 million people in Europe (ECCIA n.d.). Moreover, they are a strong driver of promoting creativity, arts, crafts, cultural heritage, and setting standards across industry sectors with regard to quality and customer experience. Luxury brands contribute to setting high performance and quality standards, thereby they foster economic value added. They are also important cultural actors as they promote creativity and traditions. Without luxury brands, many artisans and their techniques would struggle to survive in a competitive market place. Moreover, the close relationship between luxury brands and art makes them an important cultural patron and driver of innovation.

The global personal and experiential has an expected size of nearly 1 trillion € in 2017 serving over 400 million consumers. Driving this luxury market are 375 million consumers who regularly or occasionally purchase luxury goods (BCG 2018a). We will briefly illustrate five different perspectives to view the luxury market, highlight its economic relevance, and discuss current and emerging trends.

Luxury Categories: What Is Being Purchased?
In 2017, the core luxury market consisting of personal and experiential luxury was expected to be worth 913 billion €, with personal luxury making up 327 billion € and experiential luxury the remaining 585 billion €. Personal luxury consists of watches and jewelry (126 billion €), accessories (83 billion €), apparel (68 billion €), and perfumes and cosmetics (50 billion €). Experiential luxury includes hotels and exclusive vacations (455 billion €), food and wine (76 billion €), and furniture (55 billion €) (BCG 2018b).

---

[1]All figures if not quoted otherwise come from BCG Global Consumer Insight Study (BCG 2018a).

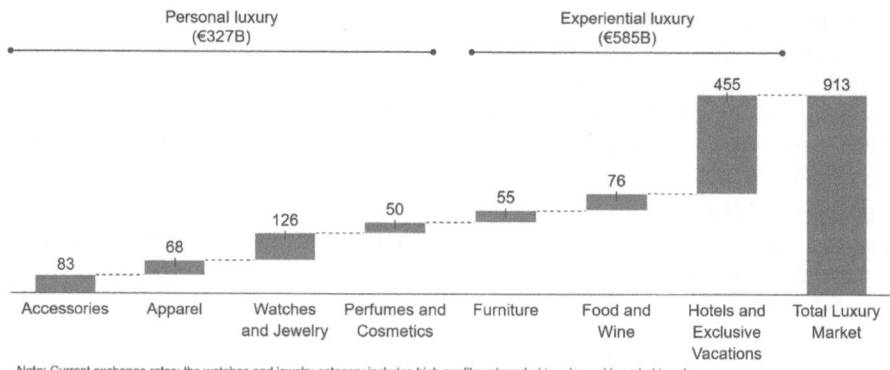

Note: Current exchange rates; the watches and jewelry category includes high quality unbranded jewelry and branded jewelry

Fig. 2 The global luxury market (source: BCG 2018b)

Experiential luxury is growing fast and will account for two-thirds of the core luxury segments by 2024, significantly increasing its share from the current 50% (BCG 2018a). Experiential luxury categories such as high-end food and wine, luxury hotels, and exclusive vacations will be the main drivers. This increase is strongly driven by a maturation of the luxury market and a shift in consumer values from owning to being. Yet we also expect personal luxury categories such as accessories or apparel to continue their growth trajectory as they will remain relevant lifestyle categories, particularly for the younger generations. In addition, new luxury categories, particularly luxury tech will emerge and command their share of the luxury market (Fig. 2).

The economic relevance of the luxury segment is easily seen when we look at the price point that we use to qualify as luxury. For example, for handbags it is 1000 € per bag, for hotel stays it is 450 € per night, for restaurant meals 200 € per person, for cars 100,000 €, and for yachts 750,000 € based on European price levels. And the threshold is likely to increase over the next years.

Luxury Consumers by Spend Cluster: How Much Are Luxury Consumers Spending?

Today, we have approximately 400 million luxury consumers globally, and the number is growing. The top spenders, the 18 million "true luxury" consumers spending more than 5000 € per year, make up 30% of the global luxury market spend. In addition, we also have a layer of ~20 million top aspirational consumers with around 7% of global luxury spend. The largest group in terms of consumers and spend are the 375 million other aspirational luxury consumers, making up more than 60% of luxury spend.

Looking at the expanding consumer base, the luxury market is poised for growth. We expect all spend clusters to increase over the next years to nearly 500 million consumers with a global spend of 1100 billion € by 2024 on personal and experiential luxury alone. The latest BCG research forecasts the number of true luxury consumers to be 23 million with a spend of nearly 400 billion € on personal and

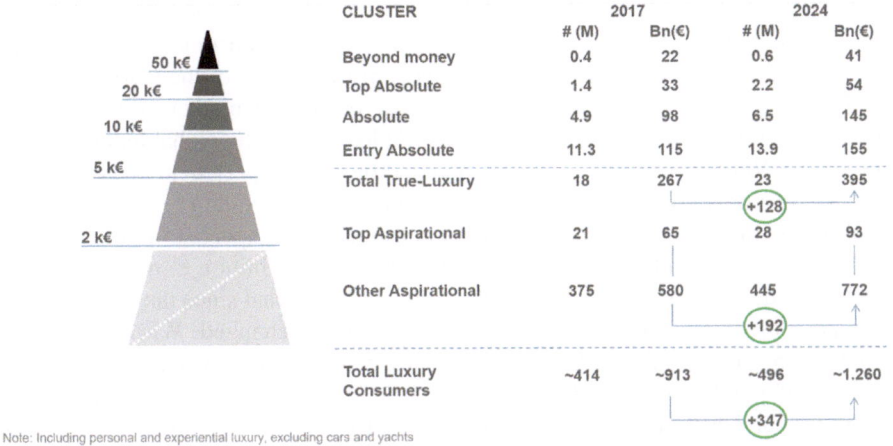

CLUSTER	2017		2024	
	# (M)	Bn(€)	# (M)	Bn(€)
Beyond money	0.4	22	0.6	41
Top Absolute	1.4	33	2.2	54
Absolute	4.9	98	6.5	145
Entry Absolute	11.3	115	13.9	155
Total True-Luxury	18	267	23 (+128)	395
Top Aspirational	21	65	28	93
Other Aspirational	375	580	445 (+192)	772
Total Luxury Consumers	~414	~913	~496 (+347)	~1.260

Note: Including personal and experiential luxury, excluding cars and yachts

Fig. 3 Global luxury consumers by spend cluster (source: BCG 2018b)

experiential luxury in 2024. Similarly, the aspirational consumers are expected to grow significantly, making up a spend of nearly 900 billion € by 2024, driven by 445 million other aspirational and 28 million top aspirational consumers (Fig. 3) (BCG 2018b).

Luxury Consumers by Age Group: How Old Are Luxury Consumers?
Looking at the core luxury consumers in the key markets, spend is driven by Generation X, aged 36–50 years old. They account for nearly 40% of spend today, followed by Millennials with ~30% (aged 21–35 years), Baby Boomers with ~25% (aged 51–70). Older consumers and Gen Z consumers (aged 5–20) make up less than 3% of personal luxury spend each.

Yet this picture is going to change substantially over the next 5 years. In 2024, Millennials will account for 50% of the global personal luxury market and will drive 130% of market growth. This means that this generation will more than double their personal luxury spend from ~100 to ~200 million € within the next 7 years. This will be accompanied by slight increases in spend by Generation X and Generation Z consumers to compensate a lower spend by the older generations. In 2024, we expect baby boomers and older generations to account for 10, Generation X consumers for 35% and Generation Z for ~7% of personal luxury spend.

Luxury Consumers by Nationality: Where Do Luxury Consumers Come from?
It is important to distinguish the place of purchase from the consumer nationality for luxury goods. Many purchases take place abroad and consumers engage in luxury shopping trips globally. As the consumer is in the focus of our studies, we take a view of the consumer nationality rather than the place of purchase. Moreover, while shopping destinations often change due to fashion trends, personal travel preferences, or government policies, the consumer nationality is more stable over time and thus better suited to identify longer-term trends.

Spend today is driven by Chinese consumers (32%), Americans (22%), and Europeans (18%). Japanese consumers (10%), other consumers from APAC (11%), and other global consumers (7%) account for the rest of the market.

Chinese consumers are on the rise. Seventy percent of market growth will be driven by Chinese consumers who will account for 40% of the global luxury market by 2024. While consumers from all nationalities are expected to increase their absolute personal luxury spend, they all lose share to the spending power of Chinese consumers. In absolute numbers, Chinese consumers are expected to increase their personal luxury spend from ~105 billion € today to ~160 billion € by 2024.

The focus on consumer nationalities is especially relevant since the traveling and shopping behavior of Chinese consumers has recently changed. Whereas in 2013 Chinese consumers only made 30% of their purchases locally, the number increased to 65% of purchases in 2016. Before 2015, a strong yuan and a high price gap between purchases in China and abroad has led to a large number of foreign purchases. Phenomena such as Daigou, that is overseas personal shopping, have been a challenge for luxury brands in China. After 2015, with the more consistent implementation of global pricing, government initiatives such as duty rates and stricter border controls to fight illegal and non-taxed imports, as well as a weak yuan have led to a shift toward purchases in Mainland China. Moreover, with the introduction of anti-corruption laws in 2012, the luxury market has changed substantially. While particularly luxury watches have lost share directly after the introduction, experiential luxury has gained share. Today, digital is the new playing field in China, where China is leading the global digital revolution. Sixteen percent of the local purchases of Chinese consumers happen online and cross-border e-commerce is a new trend that emerged. Players such as Tencent and Alibaba are strongly driving the market and redefining the luxury consumer's path to purchase, for example, with social commerce.

Luxury Consumer Segments: Which Luxury Segments Are the Most Economically Relevant?

BCG (2018a) categorizes the global luxury market into 12 luxury segments based on multiple behavioral characteristics such as their personality and lifestyle, their attitude toward luxury, or their purchase behavior.[2] The four largest segments, the Absolute Luxurer, Megacitier, Experiencer, and Social Wearer account for over 60% of luxury market spend and represent ~35% of all true luxury consumers. And they are growing: Between 2013 and 2016, nearly 100% of the personal and experiential luxury market growth came from these segments, thereby compensating the decline in spend from other segments. And we expect their influence to continue as these segments most strongly represent emerging market nationals and younger generations. The Rich Upstarter, Megacitier, #LITTLEPRINCE, and Absolute Luxurer segments represent 70% of Millennials' spend. Thus, luxury brands need to understand these segments in depth, in order to win with their brand within these target

---

[2]For an illustration of the consumer segments see https://www.youtube.com/watch?v=Jp1ARkFWBrE.

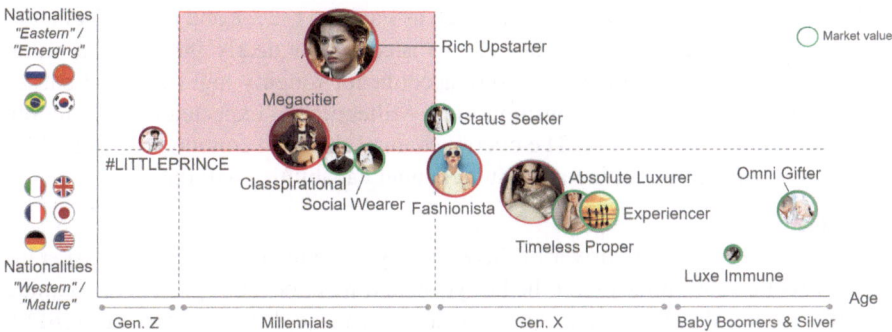

Fig. 4 Luxury consumer segments (source: BCG 2018b)

groups. Moreover, of the large and growing segments, most are strongly present in emerging markets. Rich Upstarters, Megacitiers, and Fashionistas make up for 55% of Chinese personal and experiential luxury spend. Only the absolute luxurer has a strong footprint in mature markets and among older customer groups. Nevertheless, also niche strategies targeting the smaller segments can be successful. For example, targeting the Fashionista segment in China with a focused approach using the local digital ecosystem, influencers, and social media can give brands an edge in a growing market niche. This positioning can have a positive halo effect on related segments, such as #LITTLEPRINCE or Classpirational consumers in the case of Chinese consumers (Fig. 4).

Luxury Channels: Where Do Consumers Purchase?

More than 60% of luxury purchases are digitally influenced today, a number that we expect to increase even further over the next years. While the store remains central in the overall path to purchase, only 40% of purchases are made only in-store without any digital touchpoints along the path to purchase. More than 80% of luxury consumers expect a seamless experience across channels, particularly the younger generations. And this is particularly challenging as they are digital natives and their expectations for luxury brands are the highest. Social media has become the key influence lever for true luxury consumers, followed by magazines and brand websites. For social media, particularly Instagram is gaining importance for Western luxury consumers, whereas Facebook and YouTube are losing share. In China, Wechat, Weibo, and QQ are the key social media platforms used to interact with luxury brands. Nevertheless, there is ample room for improvement in the digital and omnichannel approach of many luxury brands

Digital will drive future luxury growth, yet the store and personal experience will still remain a central element of the omnichannel experience of luxury brands. We expect 40% of the luxury market growth until 2020 to come from E-Commerce, thereby clearly outperforming growth in traditional channels. Moreover, we expect 60% of E-Commerce sales to be additional purchases that would not have taken place offline. In China the share is even higher with 70%, reaching to new customer groups with limited store presence or infrequent store visits. Monobrand websites

(33%), full-price multi-brand websites such as Farfetch (32%), and generalist market places such as Amazon or T-Mall (23%) account for nearly 90% or the online purchases. While mobile devices today account for roughly half of the purchases, the growing share of Gen. Z consumers and Millennials is expected to drive mobile adoption to 80–90% by 2024. The overall channel shift will continue and only those companies with strong online and Multichannel capabilities will thrive.

Outlook

Luxury has been a strong driver of the economy since its inception. The market has been growing substantially over the last years and we expect this growth to continue. Yet the growth drivers for the future of luxury will be different from those of the past. Moments and memories matter more than material possessions across all age groups. Experiences will be driving future market growth with a growth rate of 5% until 2014, compared to 3% for personal luxury. In 2024, Millennials will account for 50% of the luxury market (29% today). Chinese consumers will account for 40% of the market (32% today). Megacitiers and Absolute Luxurers will be key segments. Digital will further increase its reach and become a central part of the path-to-purchase and the consumer journey. Yet, the store will still be central to the physical experience of the brand. Technology will play a major role, for example through the ability to personalize experience through the use of artificial intelligence and data-driven personalization engines. In terms of the overall contribution of this sector to the identity and economic growth of Europe, we think that it will further be central and a major force into the future.

Particularly for luxury tourism, the outlook is positive. A growing and increasingly mobile group of luxury consumers is ready to embark on their quest for meaningful experiences. Sustainability and authenticity are values on the rise and particularly important for luxury tourism. And we are only at the beginning of a technological revolution that will continue to change the way we travel. While human touch will still be key for all experiences, complementing it by technology greatly increases the possibilities and opportunities to personalize the experience throughout the entire customers and foster a personal relationship with consumers in the moments that matter.

4   Future Scenarios of the Luxury Phenomenon

David Bosshart

Let us return to our phase model of luxury. We remember that the majority of people in saturated affluent societies are in the maturity phase, but many are already on the threshold of seniority. To put it in a nutshell, you know where you want to go in the maturity phase, and you also know how to get there in the seniority phase. "Less is more" is the leitmotif here describing the change in luxury in its various aspects.

At first, this is a shift from waste to more simplicity. If we take "less" in the narrower sense, what counts here is not only a more informed view, which is able to perceive more nuanced in terms of quality, value, origin, etc. but at the same time also the introspection, which in the new phase is becoming more important than the display of luxury and the status image to the outside world. Or more precisely: it is no longer about being seen, but about the "art of seeing" (Führer 2008, p. 237).[3] The art of seeing also means the ability to see the other as something unique—and not in the boredom of abundance simply as more of the same, as a variation of what one already knows. Connoisseurship, knowledge, and skills will therefore become even more important. And finally, the art of seeing also encompasses the new question of meaning and the desire to see behind things—or in other words: the desire for enlightenment.

So what the seniority phase is about is the expansion of consciousness. It is marked by the desire to change. One does not want to find oneself in things, but to grow beyond oneself through means of things. The luxury experience should be a challenge: "However, a luxury product is not only expected to self-affirm, it should also expand the self by being emotionally and intellectually activating and inspiring. A luxury product also wants to be a muse"(Führer 2008, p. 237). In return, comparing oneself to others is no longer of relevance; the question of how others value what one does and consumes loses importance.

The "less" of the seniority phase refers to the essential in the sense of what has become rare: time, space, leisure. At the same time, it refers to the ability to derive maximum benefit from what is necessary and simple. More precisely, the ability to experience, decode, and enjoy the reduced and essential is what distinguishes this new phase. The (demonstrative) renunciation of their own car by many younger people suggests it: "The new status symbol, isn't what you own—it's what you're smart enough not to own"(Jurich o. J.). Sharing is considered smart especially among younger people and therefore is cool.

The strategy of the new luxury is demonstrative renunciation—as a testimony to consumer criticism, modesty, or asceticism. Here lies the distinction between the experience in the maturity phase and the experience as it becomes more important in the seniority phase. Experience means that the expected comes true. It entertains, is pleasant, exciting, captures the senses. What it does not contain is the moment of absolute uncertainty. But this is precisely what makes the luxury product a muse, a source of inspiration. The muse does not do to me what I already know. It takes me into new territory.

But more precisely: What characterizes this new experience? The shift toward deceleration and renunciation of consumption, as can be observed in an accelerated present, can simply be described as a necessity: "The clever hedonist cannot help but live renunciation of consumption at many points in his life. We are on the verge of exhaustion and on the verge of the meaningful". In the maturity phase, the strategy

---

[3]Führer (2008) diagnoses this nuance: "In the new luxury, the art of seeing is more important than being seen."

was to know how to deal with things, to keep up with the technical refinement of things. If there is no time for that, things stay foreign. At the end of the maturity phase, the foreignness of things becomes a basic experience. A new strategy comes into play. If you ask the people in Switzerland and Germany whether they prefer a break or an expensive purchase (car, home, yacht, etc.), the answer is clear. A majority (66%) prefers the break. In Switzerland, 73% (compared to 59% in Germany) would prefer free time (Cf. Bosshart and Kühne 2014).

So, the greatest luxury is time. But what expertise does it take to live it? The new luxury still clearly has a distinctive function, even if the signs have become much more subtle. In the seniority phase, the habitus comes into play: it is a matter of dealing naturally with the social rules of the game, codes, and cultural techniques, not of striving to follow them. The casual handling of status signs is important. For example, knowledge about eating and drinking has become not only a means of identification but also a distinguishing feature. The connoisseur, who knows about wine or tea, meat, or vegetables and can tell a story about every ingredient, is widely acknowledged today (Cf. Hauser et al. 2013). This also applies to data in the age of digitalization: We must first acquire the wisdom to deal with them. "Having the brains"—the "savviness"—will in the future become a testament of one's own future viability. And thus the central prerequisite for advancement and recognition. But anyone who wants to acquire cultivated taste, education, or even sophisticated hobbies must invest: not necessarily money but in any case time. In this way, through cultural capital, time becomes a sign of distinction in the seniority phase.

What distinguishes this lifestyle from the previous lifestyle is the relationship it has to authenticity. This new foundation is particularly evident in the manufacturing of a product. The new major topic in the luxury industry is the design process— which today means transparency throughout the entire production chain. This longing for the genuine, unadulterated is not only evident in the manufacturing of luxury products, but also in retail, gastronomy, tourism—and last but not least in a survey conducted by our institute: "Would you rather spend a night in a trendy eight-star hotel or in untouched nature under the stars?", was the question. The majority opted for the pure, ascetic version: the starry sky (Cf. Hauser et al. 2013).

What do people want to experience before they die? Or rather: Which luxury do they want to treat themselves to? Buy a Hermès bag, write a bestseller, or father another child? We also wanted to know this in the survey mentioned above and compiled a luxury "bucket list" from the answers—a list with the great wishes of life. It turns out that no matter how many years one still has ahead of oneself, one topic clearly dominates the wish list—travel (see Fig. 5).

Fig. 5 "What luxury would you like to treat yourself to in your life?" GDI survey on luxury (10 top categories, $n = 1003$) (source: Bosshart and Kühne 2014, p. 4)

The „bucket list" of luxury

1.	Holidays and travel	291
2.	Housing and home furnishings	122
3.	(Free) time	58
4.	Mobility (car, bicycle, camper van, etc.)	54
5.	Health and Fitness	30
6.	A family, partner, children	15
7.	Being financially independent	14
8.	Work less / no longer	12
9.	Education and training	7
10.	Others	32

5 Study Results on the Travel and Booking Behavior of Luxury Travel Customers

Dorothea Hohn

The worldwide agency network *The Pangageanetwork,* which is represented in Germany by its founding agency Global Communication Experts GmbH, investigated the travel and booking behavior of luxury travel customers for the fifth time in winter 2017/2018. The survey of around 500 tourism experts once again showed the high relevance of this travel segment: luxury travel combines exclusive and unique experiences with personalized services. The buying impulse is first set by the destination, followed by the services offered on site and the accommodation—preferably an exclusive boutique hotel or international hotels on a high level. The number of high-budget trips is approximately three per year. These are mainly booked offline, with customers relying on the advice and recommendations of industry specialists.

The luxury travelers mainly consist of Generation X and are between 36 and 55 years old. They travel as couples, look for comfort and relaxation on their vacation, and attach great importance to special culinary experiences. Private excursions dedicated to nature and culture, as well as a reasonable price–performance ratio and a plausible added value have a higher priority than the price. They pay attention to ecological sustainability. Their decisions are based primarily on the advice of friends and relatives and online evaluations. The average budget is between 5000 and 10,000 euros per trip and person. In the coming years, they will be very demanding, informed, and sensitive to environmentally friendly travel solutions. In general, luxury travel is expected to increase by 6–10% over the next 2–3 years.

In 2017/2018, the market share of high-price luxury travel was 7%, with annual double-digit growth rates (Cf. Buck and Ruetz 2018). But: "My villa, my yacht, my sports car" is out. You do not show off your own wealth anymore. The luxury market is experiencing a blatant paradigm shift and is no longer defined by "bling bling" only, but rather by the demands of travelers for authenticity and exclusivity.

The Core Target Groups of Luxury Tourism and Their Travel Behavior

The generation of the so-called baby boomers aged 55 and over is regarded as one of the most attractive target groups in the entire tourism sector and above all in luxury tourism: they have travel experience, are wealthy and healthy. The baby boomers have their own houses or apartments and an average net income of over 3600 euros per month. This generation is considered to be extremely pleasure-oriented. Primarily the youngest among them spend 80% of their income on consumer goods. But especially in their online behavior, the baby boomers are oriented toward the up-and-coming generation Z. The 16–22-year-olds are considered a luxury target group in their own right, still at the beginning of their careers, but they have an enormous impact on older generations. Generation Z is dematerializing luxury holidays: the goal is now to enjoy peace and quiet, strengthen health and find time for friends and oneself—expensive spa treatments and expensive gourmet restaurants are just a nice addition to a luxury vacation. This poses new challenges for luxury travel providers. Influenced by the younger generation, the originally conservative core groups now have a new requirement profile. The largest target group of luxury tourism, however, is the well-educated and ambitious generation X aged between 36 and 55 (62%).

This generation benefited from the economic development in the 1960s and 1970s, which was accompanied by a change in mentality—the baby boomers' credo of "living to work" became "working to live." The goal of the generation: a material security and being able to afford things. Generation X is striving for a high quality of life and attaches importance to work–life balance. All target groups prefer to travel with their partner: 73% of customers mainly travel as a couple. Family travelers are in second place (22%), travel with friends (4%) in third place. Only one percent travels alone.

Duration and Number of Trips Are Changing

Germans like to travel a lot: in 2017, over 54 million German holidaymakers took around 69.6 million holiday trips (Cf. Statista 2018). Luxury tourists book an average of two to three trips per year, about one-third go on one long trip. Twenty-seven percent of the tourism professionals surveyed said that their customers go on one big trip per year and the remaining 26% said that luxury travelers travel four or more times a year. However, the number of bookings declined compared to 2012 and 2014. At that time, the majority of customers booked more than four trips a year.

About half of the respondents (49%) indicated an average length of stay of 7–10 days for luxury trips, while 34% reported eleven to 13 days. Ten percent of holiday trips last longer than 2 weeks, while 7% last less than a week. In the same study 5 years ago, 46% of all luxury travelers had an average travel time of 10 days. Twenty-six percent traveled for 1 week, 15% for 2 weeks. Compared to 2010, travel time has decreased in percentage terms, with travel under 1 week becoming more popular.

All luxury travelers have not only time but also the suitable budget: The costs for an individual luxury vacation vary from market to market, on average however the expenditures lie between 5.000 and 9.999 euros per journey. In particular travelers from Spain and Scandinavia book in the lowest budget category between 5000 and 6999 euros. Travelers from the USA, the Netherlands, and Italy invest between 7000 and 9999 euros, while Brazil, Germany, Great Britain, and France spend between 10,000 and 14,999 euros per trip. More important than the price for many customers is the price–performance ratio, while for others the focus is on the added value for themselves: Consumers are increasingly spending their income on experiences rather than material goods. This leads to a paradigm shift within luxury tourism and the desire for extraordinary travel and offers.

Although, as already described, generation Z has such a significant impact on older generations and is very digital-affine, most consumers reject online sales of luxury travel. Despite digitalization, customers swear by travel agencies, specialized consultants, and their individual offers—87% of those surveyed said so. This figure appears to be rising, with only 54% of respondents saying in 2012 that the travel agency was the most popular booking channel.

Over 60% of online users use a desktop PC for their booking, while 38% use mobile technology. When booking on the Internet, travelers rely mainly on the websites of specialized agencies, as indicated by 43% of the panel. Direct bookings from hotels, airlines, and local service providers account for 32%, while the websites of large online agencies reach 21% of customers. The recommendations of friends and relatives are essential for 49% of luxury travelers. In second and third places are online ratings (36%) and prestigious awards (43%), which support the booking decision.

New Luxury: Opulence Was Yesterday
The golden fittings in the marble bathroom have had their day. If you want to appeal to luxury travelers in 2018, you have to offer more than the opulence of the past days and instead concentrate on offering immaterial rarities (i.e., broadening the customers' horizons). Back in the 1980s, the top 10,000 defined themselves through expensive cars and brand clothing—luxury had an impact on the outside world. In contrast, the 1990s were dominated by the new "get stingy" trend. The withdrawn lifestyle of the billionaire Aldi brothers is the most extreme example. In recent years, a new way of dealing with luxury has developed. With the revaluation of high-quality products, unique items, and exclusive services, luxury today is also showing its post-material side: Consumers increasingly spend their disposable income on experiences instead of goods. This has a significant impact on customers' travel decisions. For 30% of the respondents, uniqueness and exclusivity are the key factors of a luxury trip, while 29% of the respondents state that services are being adapted. Thirdly, the quality of the hotels (21%) and fourthly the flight class (12%) are important. In general, there is an interest in quality products and exclusive, tailor-made services. The quality of accommodation (67%), privacy/exclusivity (62%), and the reputation of the hotel/destination (59%) were the three top criteria influencing luxury travelers in 2012. Spa facilities had gained in importance compared to 2010.

One element of the New Luxury that should not be underestimated is sustainability—fair tourism is booming. As the needs of target groups for exclusivity, authenticity, and immaterial luxury are now more than bling-bling, eco-tourism, and corporate social responsibility are increasingly coming to the forefront. Just 5 years ago, 42% of respondents said that environmentally conscious tourism was not important, and 40% believed that it was important but did not affect travelers' choices. As a result, responsible tourism did not become a key factor for the luxury industry.

The experience-oriented requirement profile—which corresponds to the post-material mentality of luxury—also goes hand in hand with a form of focusing on the individual. The individuality-centered luxury traveler wants to feel good about what he does. No matter if consumers choose a villa on one of the Fiji Islands or a boutique hotel in a German nature park—the luxurious accommodations must be sustainable: collecting rainwater, planting vegetables, and supporting the local population. A kind of ecologization takes place, to which sociopolitical factors are contributing. If, for example, the catch of the day is served after the fishing excursion at the private dinner on the beach, this is exactly the kind of lasting and authentic experience that guests are longing for.

Motivation of the Luxury Customer

The decision-making of luxury travelers is primarily based on the destination (74%), followed by the overall experience (20%). Accommodation is crucial for 5% of travelers and thus ranks third among the criteria. In addition to sustainability, luxury travelers are particularly looking for relaxation and comfort in absolute privacy during their holidays. Gourmet experiences, tours and cultural visits on a private basis, innovations in destinations, products and services as well as tailor-made excursions into nature are a great incentive for customers to choose a destination.

As the most popular luxury destination, the Maldives rank as the most trendy destination in the luxury travel segment. The resorts on the islands in the Indian Ocean also meet most of the criteria listed and the concept "one island - one resort" is still something unique. Asia, the USA, the Caribbean, and South Africa are also still popular. The focus is once again on classic and well-known holiday destinations, while 5 years ago, for example, Far East countries such as Myanmar, Bhutan, Vietnam, China, and Cambodia were in great demand. In 2010, the Indian Ocean or the Caribbean and French Polynesia were among the most popular luxury destinations.

The trend is also moving away from large luxury hotels to smaller, individually designed houses: Lifestyle, design, and the individual feel-good factor are the buzzwords here. Boutique hotels combine uniqueness with luxury standards—they offer travelers exclusivity due to the low number of rooms, can look after each individual guest personally, and yet travelers do not miss the luxury they know from the well-known resorts. In terms of the type of accommodation, the majority of respondents (65%) therefore consider exclusive boutique hotels to be the number one, followed by large international chains (21%). Other places include private accommodation (6%), castles and historic buildings (2%), and other accommodation (5%). In particular, luxury travelers look for relaxation and comfort in absolute privacy (51%) during their holidays. Gourmet experiences (50%), tours and cultural visits on a private basis (49%) as well as innovations in the areas of destinations, products and services (45%), and tailor-made excursions into nature (43%) are an incentive for customers. Nevertheless, large international chains with good reputations are still popular and have a significant influence on 18% of respondents. A similar influence is achieved by specialized tour operators, travel agencies, and consultants with 15%. Their good image precedes them and the accommodation is often internationally standardized. Furthermore, it is especially important for Generation Z to be able to produce beautiful images for social networks. Hotels and destinations must be "instagrammable": Guests want to show off where they are, what they are experiencing, and what they can afford—post-materialism meets virtual materialism. But it is also about the question: Who was there before me? And this can be looked up, in case of doubt, on Instagram. After all, contributions from influencers serve almost half of consumers as inspiration for their next journey.

Growth Market of the Future
The luxury travel segment will continue to grow: Most respondents to The Pangaeanetwork's (2017) study expect bookings to rise over the next 2–3 years, with 42% expecting bookings to rise by 6–10%. As time goes by, luxury travelers themselves are becoming more and more demanding: interest in new destinations, products, and services emerges. They identify with their destination and inform themselves extensively. In their search for unique experiences, they also react very sensitively to ecological sustainability, as well as advice from acquaintances and expert recommendations. On the other hand, luxury consumption is becoming more widespread. Due to the positive economic situation of the past 10 years, more and more people can afford luxury goods. The luxury market has already reacted to this. Cheaper fashion lines from established luxury brands or cafes offering hot drinks made from organic coffee beans at an affordable price open the market for exclusive consumer goods to the masses. Hotel groups are also establishing young and cheaper brands on the market. Thus, luxury is becoming more suitable for everyday use. By 2020, the management consultancy Bain & Company forecasts a growing market with around 400 million luxury customers worldwide (Cf. D'Arpizio et al. 2017).

Bottom Line
"Luxury" is a term that cannot be precisely defined. What we perceive as luxury depends on our cultural, socialization-specific influences and is bound to our subjective perception. In addition, one can understand luxury as an orientation toward material goods or as immaterial if emotional needs are meant. In regards to this fact it is hard to agree on a concrete definition for the term "luxury travel." It is clear that the understanding of luxury travel is changing over times. While in the 1980s people proudly showed off their possessions and went on holidays in the most renowned five-star hotels, today's travelers attach way more importance to exclusivity and individuality.

Private excursions dedicated to nature and culture, as well as a justifiable price–performance ratio and a plausible added value have a higher priority than the final price. Luxury travelers are increasingly paying attention to what is ecologically sustainable. What is to be expected in the coming years: Travelers with high expenditures will be very demanding, informed, and sensitive to environmentally friendly travel solutions. In general, luxury travel is expected to increase by 6–10% over the next 2–3 years. This means that the requirements profile of the customers and the required competence from industry professionals are also increasing and that the industry has to meet these challenges.

Literature

BCG (2018a) 2018 true-luxury global consumer insight. http://media-publications.bcg.com/france/True-Luxury-Global-Consumer-Insight-2018.pdf . Accessed 28 Feb 2019

BCG (2018b) Growth in the luxury market. https://www.bcg.com/industries/consumer-products/luxury.aspx. Accessed 16 Nov 2018

Bosshart D, Kühne M (2014) The next luxury. What will be dear and dear to us in the future. Gottlieb Duttweiler Institute, Rüschlikon

Buck M, Ruetz D (eds) (2018) Boom or bust? Where is tourism heading? https://www.itb-kongress.de/media/itb/itb_dl_all/itb_presse_all/ITB_WTTR_A4_2018_interaktiv.pdf. Accessed 20 Nov 2018

D'Arpizio C, Levato F, Kamel M, de Montgolfier J (2017) Luxury goods world-wide: market study, Fall-Winter 2017: The new luxury consumer: Why responding to the millennial mindset will be key. http://www.bain.de/Images/BAIN_REPORT_Global_Luxury_Report_2017.pdf. Zugegriffen 20 Nov 2018

ECCIA (n.d.) Homepage. http://www.eccia.eu/ . Accessed 16 Nov 2018

Federal Statistical Office (2014) Focus series baby boomers. https://www.destatis.de/DE/ZahlenFakten/ImFokus/SerieBabyboomer.html . Accessed 20 Nov 2018

Führer B (2008) The new understanding of luxury as a consequence of the elite change and the implications for the market communication of luxury companies. Dissertation. Süddeutscher Verlag für Hochschulschriften, Riga

Hauser M, Bosshart D, Muller C (2013) Consumer spring: Beginning of a new eating awareness. GDI Study No. 40. Gottlieb Duttweiler Institute, Rüschlikon

Jurich L (o.J.) Why this CEO doesn't own a car: the rise of dis-ownership. www.fastcoexist.com/1681112/why-this-ceo-doesnt-own-a-car-the-rise-of-dis-ownership. Zugegriffen 21 Mai 2014

Knoll R (2009) Luxury is when expectations are exceeded. Goldsmith Newspaper, 10

Radical I (2007) "We don't know what we've got anymore." Why clever hedonists practice renunciation—and why only deceleration sharpens the eye for the essential. A conversation with the sociologist Hartmut Rosa. Time. https://www.zeit.de/2007/52/Interview-Rosa . Accessed 21 May 2014

Sombart W (1992) Love, luxury and capitalism—On the emergence of the modern world from the spirit of waste. Wagenbach, Berlin

Statista (2018) Statistics on the travel behaviour of Germans. https://de.statista.com/themen/1342/reiseverhalten-der-deutschen/ . Accessed 20 Nov 2018

Steuart J (1848) An inquiry into the principles of political oeconomy. Millar, London

The Pangaeanetwork (2017) Luxury travel trends (2010, 2012, 2018). https://lab.pangaeanetwork.com/latest-luxury-travel-trends . Zugegriffen 20 Nov 2018

Thomas D (2008) Deluxe: how luxury lost its luster. Penguin, London

Dr. David Bosshart has been the CEO of the Gottlieb Duttweiler Institute for Economics and Society since 1999. The Institute is an independent European think tank for trade, economy, and society. The philosopher has a doctorate and is the author of numerous international publications and an international speaker. His work focuses on the future of consumption, social change, digitalization (man–machine), management and culture, globalization, and political philosophy.

Hannes Gurzki is a consultant at The Boston Consulting Group. He is an expert in luxury, marketing, branding, and consumer behavior. He studied business administration and cultural management at the University of Mannheim and holds an MBA from ESSEC Business School. His research appeared in leading academic journals such as the Journal of Business Research or Psychology & Marketing.

Dorothea Hohn founded the communication and representation agency Global Communication Experts GmbH in 2009 and within a very short time built up an international customer portfolio in the travel sector, including destinations, airlines and airports, hotels and hotel chains, car rentals, cruise lines, tour operators, travel agencies, and trade fairs. At the same time, as President, she heads the international agency network The Pangaeanetwork, which regularly conducts studies, for example, on the development of luxury tourism. Dorothea Hohn studied business administration with a focus on tourism and started working as a travel journalist. She then began a 2-year trainee program at Deutsche Lufthansa AG in Frankfurt, Cologne, Hamburg, and New York. There she worked for another 4 years in corporate communications before she decided to switch to the agency side. As managing director of a leading German tourism PR agency, she developed the company over 15 years and advised more than 250 clients from various tourism segments before founding her own agency.

Dr. Antonella Mei-Pochtler is a senior advisor of The Boston Consulting Group. She is an expert on luxury, branding, and media and has served in multiple global functions for BCG, including being a member of the Global Executive Committee. She studied business administration at Ludwig-Maximilians-Universität München, and acquired her doctoral degree at Università degli Studi di Roma and holds an MBA from INSEAD, Fontainebleau with the Dean's List Award.

Behavioral Explanations of Luxury Consumption

Hasso Spode, Hannes Gurzki, David M. Woisetschläger, Marc Aeberhard, and Stephan Hagenow

1 Luxury in a Changing Sociocultural Environment

Hasso Spode

What would life be like without luxury? Certainly very sad and probably not even possible. And what would luxury be without its socio-cultural environment, its sociotope, its context? An absurdity. Without language, without comparison, without attribution, without judgment no luxury. Thus, it fulfills central functions in this context at the same time. In the animal world, one may speak of a silent, unconscious luxury. If the peacock's wheel, the play of colors of the butterfly, the antlers of the moose are simply accredited to courtship behavior, this ultimately explains nothing at all, as Darwin had already suspected when he used the term "sexual selection": We do not know the underlying reason for such sometimes even dysfunctional development of splendor in the respective biotope—here, apparently, an aesthetic of its own has developed: a whim of nature. But in the world of humans, whose "second

H. Spode (✉)
TU Berlin, Berlin, Germany
e-mail: spode@hasso-spode.de

H. Gurzki
European School of Management and Technology, Berlin, Germany

D. M. Woisetschläger
TU Braunschweig, Braunschweig, Germany
e-mail: d.woisetschlaeger@tu-braunschweig.de

M. Aeberhard
Luxury Hotel & Spa Management Ltd., Zürich, Switzerland

S. Hagenow
Pfarrschaft der Region Bern, Bern, Switzerland
e-mail: stephan.hagenow@refbejuso.ch

© Springer Nature Switzerland AG 2020
R. Conrady et al. (eds.), *Luxury Tourism*, Tourism, Hospitality & Event
Management, https://doi.org/10.1007/978-3-030-59893-8_4

nature" is culture,[1] silent luxury does not play a major role. Here luxury is merely a social construction, a discourse.

The Concept of Luxury

It is well known that terms are often used quite differently and are also highly versatile. Luxury is different: there is a considerable and long-standing consensus on how luxury should be defined. Strangely enough, in the middle of the eighteenth century, there is still no entry in the Zedler, the first comprehensive "Universal-Lexicon," and in 1801 the Encyclopedia of Krünitz rebutted luxury as "over-refinement of the sensual taste."[2] But then a more neutral, *grosso modo* definition became accepted that is still valid today. In 1840, the Staats-Lexicon defined the following: "Any effort exceeding necessity or the true need," whereby of course not the "absolute" but the "relative" need was decisive.[3] Accordingly, one read in the Brockhaus of 1895: "Luxury (Latin), strictly speaking any effort that goes beyond the basic needs," whereby these "needs" turned out to be very different, so that "luxury had no fixed limits."[4] Similarly, the great theorist and luxury historian, Werner Sombart, laconically said: "Luxury is any effort that goes beyond what is necessary." And he continued: "The term is obviously a concept of relation, which only gets more tangible when one knows what 'the necessary' is."[5] Newer definitions agree. In the Brockhaus of 1970 it was stated: "any personal effort which strikingly exceeds a lifestyle perceived as normal by the social environment", whereby this "varies according to cultural group and social class."[6] The Dictionary of Sociology defines 2007 succinctly: "Consumption or effort that—according to culturally changeable and historically as well as regionally specific norms—exceeds what is socially necessary and usual."[7] And finally in 2018 not fundamentally different but less concise the talkative successor of the Brockhaus, Wikipedia: "Behaviors, effort or equipment which go beyond the usual measure (the usual standard of living) or beyond the measure considered necessary or meaningful in a society."[8]

We, therefore, have a somewhat paradoxical situation: everyone knows—or at least suspects—what luxury means in abstract terms, but what does luxury mean in

---

[1]See Recital Knowledge Ethics 22(2011)1.

[2]Krünitz (1773 ff), vol. 82, p. 40.

[3]von Rotteck and Welcker (1834 ff), vol. 10, p. 293 f.

[4]Brockhaus (1894 ff), vol. 11, p. 406.

[5]Sombart (1922, p. 71 f) [first 1913]; similar to Roscher (1861, p. 408), from whom Sombart tried to set himself apart somewhat convulsively: "The concept of luxury is quite relative."

[6]Brockhaus (1966 ff), vol. 11, p. 722.

[7]Hillmann (2007, p. 515).

[8]http://en.wikipedia.org/wiki/Luxus (25.8.2018).

concrete terms? As the Staats-Lexicon already indicated, there cannot be any timeless-objective criteria. All attempts to purge a shopping basket of everything luxury and reduce it to a "subsistence minimum"—for example with the Hartz IV rule—are always a mirror of the respective society: expression and means of distribution struggles and struggles for moral interpretative power. So, it remains a "concept of relation." Relational, however, by no means individual-subjective or even arbitrary, according to the foolish motto: "You can't argue about taste." Whether something is considered a luxury or not is decided collectively in relation to the respective sociocultural environments, i.e., setting, mentality, and structure. In other words, luxury is a text that is created through the context of location, societal class and time, materially and discursively.

Sombart, like Wilhelm Roscher, distinguishes "quantitative" luxury (more than what is necessary, for example, one hundred servants instead of one) from "qualitative" luxury (the refinement of what is necessary, for example, silk instead of linen). Both versions "are mostly united in reality," but the latter (which I will discuss later) tends to gain the upper hand in highly developed civilizations, especially in "Western" civilizations. What connects both types of luxury? The "scarce good" imposes itself as the abstract *tertium comparationis,* whereby "good" needs to be understood more general here, not only in the economic and not only in the material sense. Luxury would then simply be a synonym for scarce goods.

Scarce goods are always sought-after goods, otherwise they would not be scarce.[9] At least in functioning market-based societies (where neither the guilds nor the state regulates production), it generally goes like this: what is luxury today is either *out* tomorrow—or it becomes a necessity. For the demand for highly sought-after tangible and intangible consumer goods encourages increased, often also technically improved and thus cheaper production of these goods or increased imports—until finally the scarce good has turned into the "socially necessary and usual." There have been and still are attempts at artificial shortages (when in Brazil, for example, the locomotives were heated with the luxury good coffee), and some things, especially raw materials and time, cannot be multiplied at will. But looking at the big picture, the democratization of luxury—the great political promise since the interwar period everywhere—is a contradiction in itself: As soon as it succeeds, luxury evaporates.[10] The norm is never luxury. This devaluation mechanism also suggests that luxury should be defined as a scarce commodity. Both are "relation concepts," both are subject to change over time.

---

[9]Fortunately, on the other hand, not everything desired needs to be scarce; however, with increasing availability, the value ascribed to it tends to fall, for example in drinking water or sex.

[10]The dictatorships of the twentieth century had to learn this above all: When the Nazi regime and then the communist states baited the "people" with social benefits, especially with cheap travel, an unspectacular normality set in after the first enthusiasm—the propaganda effect largely evaporated (see below). On the "transience" of luxury, see also Berry (1994).

Luxury in Times of Shortage

The Big Feast

In 1184, 50–60,000 people flocked to a huge tent town just outside of Mainz—
Barbarossa had called a court day, probably the most expensive of the whole Middle
Ages. On Whit Monday the emperor's sons were solemnly knighted. "This honor,"
according to a chronicler, "caused them as well as all princes and other noblemen to
give expensive gifts to knights, prisoners [!] and crusaders, and jugglers: Horses,
precious clothes, gold and silver."[11] In 1921, the Kwakiutl clan chiefs also held a big
feast in the forests of western Canada. Not only 50 seals were eaten, but also, as an
official report meticulously noted, gifts were distributed generously: "200 silver
bracelets, 7000 brass bracelets, 33,000 blankets, 54 deer skins, 8 canoes and
6 prisoners."[12]

Not only are these two events separated by 737 years, they also took place in
completely different parts of the world. And yet they are united by the act of boastful
generosity, the "demonstrative consumption" for which the Indian word "potlach"
has become established in ethnology.[13] As a rule, there was something for the
"people," but sometimes the luxury was only on display. Countless examples
could be cited. Of the sinfully expensive (and transparent) silk fabrics in which
Roman patricians wrapped themselves at the banquet, which, according to Pliny,
was said to have caused 100 million sesterces to flow into China every year (the
Senate, on the other hand, issued ineffective edicts); about the gigantic "show
dinners" in front of the eyes of the mob, with which since the late Middle Ages
the rich people tried to outdo each other, displaying illuminated table fountains from
which exquisite wine bubbled, golden sows and pheasants, whole landscapes of
sweets and hosts of cooks, servants and minstrels; or the month-long wedding
celebration that August the Strong organized for his son in 1719, at which the
king presented himself to the public in a purple robe adorned with jewels, estimated
at 2 million talers (corresponding to a quarter of the Saxon state budget); to the *jet set
of* the post-war period, when the millionaire heir Gunter Sachs arranged festivities
that would have honored any baroque prince (in which, however, the public could
only participate through magazines), and who in 1966 delivered a thousand red roses
on the estate of Brigitte Bardots in St. Tropez to win over her heart (which worked
out perfectly so that the "King of the Playboys" could soon adorn himself with the
possession of the "most beautiful woman in the world").

Such potlach-like waste is associated with a strong moment of competition. Gifts,
if they go to equals, are to be reciprocated, and often then the "exchange of gifts"
becomes a "battle of gifts." With the ostentatious annihilation of their own

---

[11]Moraw (1988, p. 72).

[12]Antweiler (2008, p. 70); such festivals were forbidden since 1876 and were therefore
documented.

[13]Cf. classical Mauss (1978) [first 1925] and Veblen (1958) [first 1899].

possessions, a whole new level was reached: if, for example, the chieftains of the aforementioned Kwakiutl competed to destroy more copper bowls from their possessions, burn more blankets, sink more canoes and kill more prisoners (after the copper bowls the most valuable possession of a chieftain), the winner was whoever held out the game the longest.[14] Or when the Rolls Royce dealer in London did not pay the due tribute to Brijendra Singh, the Maharajah of Bharatpur, he bought all the Phantom II in stock, shipped them to India and there had the luxury cars "polished to the finest detail and set up in rank and file in front of the palace"—in order to have them scrapped afterward in front of the eyes of the traveling salesman.[15]

The wealth and exaltation of Indian princes were proverbial[16] (later the "oil sheiks" took over the role of such fairy tale princes), but this money annihilation action was undoubtedly an "effort" far beyond the "usual." But also the "socially necessary"? Here in Germany such showmanship is considered irrational and seems more embarrassing than great. But for the Maharajah another social logic applied: He had the duty to save his face by publicly punishing a humiliation. And what could be better suited for this than the demonstration of an inexhaustible wealth that makes the Englishman look like a poor sausage?

So, we see that splendor and bling-bling can occur in the most diverse cultures, whether tribal society, antiquity, feudalism, or capitalism, whether in China, India, Old America, Africa, Russia, the Ottoman Empire, in the "West" or in the transnational circles of the rich and beautiful. In this respect, demonstrative waste is a ubiquitous phenomenon, largely independent of the concrete sociocultural context. However, in complex societies based on the division of labor, beginning with antiquity, it not only arouses admiration but is also sometimes judged very critically[17]—complex societies never quite agree on the "right life." Whereby the luxury criticism never comes from the people, but always from "educated" circles. In pre-Christian times, for example, Cynics and Stoics taught the contempt of material goods and praised frugality and self-restraint; even the preacher Solomon, after having collected silver, gold and women "en masse," recognized: "everything was vanity and just as senseless as chasing the wind" (Kohelet 2, 11). Christian theologians then regarded the waste of God's gifts such as wine—the *Luxuria*—as a mortal sin, and numerous monks and eremites (originally Indian inventions) renounced the earthly vice to the point of self-mortification (others, however, led a scandalous life of well-being). In addition, laws were repeatedly enacted that sought to suppress certain forms of luxury consumption by banning them and later also by means of

---

[14]Antweiler (2008, p. 71).

[15]Spiegel 34/1987, p. 125: Here two conflicting value systems collided: the symbolic capital of honor and the sober interest in profit. Brijendra Singh did not realize that Rolls Royce could not care less about the fate of the cars (in light of the "planned obsolescence," he should have been grateful).

[16]Carl Barks had turned this into a brilliant potlach story: Scrooge Duck challenges the Maharajah of Zasterabad to a battle for gifts after he had rupees thrown among the people during the magnificent entry into Duckburg: They donate huge monuments to the city until the Maharajah is ruined (Mickey Mouse 10/1952).

[17]Cf. Berry (1994); Grugel-Pannier (1996); Jäckel (2008).

extreme taxation, whereby prohibitions were almost always directed against rising classes, while luxury taxes were intended to prevent the outflow of foreign currency for imported goods.[18]

The most prominent place of abundance or waste was the lavishly laid banquet tables.[19] *Luxuria*, the consumption of the largest possible drinks and the devouring of the largest possible quantities of the spiciest food (the greed for exotic spices—and soon also for coffee, tea, chocolate, tobacco, and cane sugar—then became the driving force behind the early colonial expansion). In principle, the joy of gluttony was common to all classes. The main point of criticism of moral preachers, such as Berthold von Regensburg in the thirteenth century, was the ancient tribal tradition of the "archaic feast": the sporadic excess on the day of the feast, where it was common to eat until you vomit and most importantly get drunk. At the end of the Middle Ages, attempts were even made to ban drinking contests. The majority of people, whose short lives were always threatened by hunger and deprivation, were by no means critical of such waste. They were living in the moment. They dreamed of the luxury of the land of milk and honey, paradise, where nobody has to work but everyone eats and drinks plenty.[20] And for a few days throughout the year, for example, during the bacchanal of the "Pfingstbier," this was reality.

A Question of Taste

The fact that luxury has become a controversial topic in societies that are sufficiently complex, has had little effect on people's behavior: The people dreamed of luxury, the rich squandered it. They had to do this in order to represent their status. Since the end of the Middle Ages, however, a different pattern has also become visible in parts of the upper classes: in addition to or even in place of quantity, quality, as already indicated, is increasingly taking the place of quantity. This tendency was already evident among the elites of antiquity, but ended with the mass migration, when the Germanic barbarians preferred the foul-smelling beer over the finest vintage wines and did not even master Latin correctly, let alone Greek. But in the Renaissance, it awakened to new life, strengthened in the early modern period, and then shaped luxury up to the present day.

---

[18]Several special taxes have survived as relics (tobacco, petrol, sparkling wine, etc.); conversely, reduced rates were intended to privilege the physically necessary and culturally valuable (including mules and accommodation services), so that the standard VAT is actually a luxury tax.

[19]See Spode (1993); Montanari (1993).

[20]A substitute for Christian paradise, which—in contrast to the torments of hell—remained pale and vague: "eternal life" did not entice with luxury. Quite differently the Koran, which pleasurably paints out the delights that the faithful expect: an abundance with silk cushions interspersed with gold and "bowls of flowing wine" (a promise of course only for men were the "virgins with big black eyes," e.g., 56, 23). But in contradiction to this, as noted in the Kohelet, the "splendor, the addiction to glory and the desire for more wealth" is scourged (57, 21).

Sombart classifies this as a genuinely European phenomenon and dates the long phase of turning from quantity to quality from the thirteenth to the late eighteenth century, distinguishing four phases:[21] Firstly, the "domestication" of luxury consumption (wealth is less and less celebrated in front of a gawking, greedy crowd, but rather moved behind walls, into domestic privacy, which is only now gradually separating itself from the "representative public"); secondly, the "objectification" (an ever greater part of luxury expenditure flows into permanent goods, into fabrics, furniture, books, etc. instead of into pleasures of the table); thirdly, "sensualization and refinement" (these goods are becoming more and more refined, i.e., more labor-intensive to produce and design); and fourthly, the "crowding together" (luxury goods—from fine threads to locks—are being produced and consumed ever more quickly). One does not have to consistently follow Sombart here, but the growing weight of quality in the valuation of luxury goods, namely "refinement," remains undisputed: a growth in depth rather than in breadth, an aestheticization of consumption that can extend to the complete spiritualization of pleasures.

The magic word of this refinement was the *bon goût*, the "good taste." From the sphere of culinary art, "taste" had been transferred to the sphere of the art of living. Especially in France, where in 1696 the Abbé de Bellegarde wrote a bestseller about "ridiculousness": exuberant plaster, stiff etiquette, and pompous appearance were not noble, but embarrassing, revealed "bad taste." The translator first had to explain the term to the German reader and aptly described it with a lack of judgment power ("Beurtheilungs-Krafft").[22] Such aesthetic "judgement" (as Kant later said) does not fall from heaven—it must be learned.

Thus, with taste, a qualitative, intellectual type of capital came to the fore: in addition to quantifiable economic capital and social capital in the form of noble ancestry and useful relationships, cultural capital increasingly gained importance. Despite it being a scarce good, wasteful expenditure was not out of the question—however, it was a result of social negotiation processes, which claims general validity, but whose concrete value can nevertheless be judged very differently depending on the milieu.[23] Thus, it is closest to the relational definitions of luxury mentioned at the beginning.

The primary carrier of this development was the nobility. The various privileges of status acquired at birth were already a kind of luxury. But it now tasted stale: Forced by the prince under the thumb of court life, the rebellious warrior caste was disempowered in absolutism, emasculated and degraded to a parasite. And so the courtiers suffered from their loss of function and feared nothing more than *ennui*, boredom. There was a lot of time to kill. They sought their right to exist and the

---

[21]The driving force was the spoiled "female" (Sombart 1922, p. 111 ff). For the following see Elias (1969); Habermas (1975); Lepenies (1981); Spode (1993) and Bourdieu (1992).

[22]Zit.n. Spode (1993, p. 303).

[23]A part of cultural capital, educational capital, can be objectified through academic titles—but the nobility despised such bourgeois "pedantry" (Friedrich Wilhelm I of Prussia even made poor Professor Gundling a court jester).

meaning of life in the art of consumerism. The result was an aestheticization of the environment: the dark, barren rooms of the knight's castle were replaced by the bright, well-designed chambers of the palace; instead of reaching for the table with one's hands, one ate "daintily" with silver cutlery; instead of laughing at the court jesters, one joked "gallantly" in cultivated conversation; instead of hiring vagabond minstrels, one kept a staff of perfectly educated musicians. With such refined luxury, one demonstrated not so much one's economic wealth—which may have been based on debt making—but rather one's wealth of cultural capital, one's exquisite taste, the true *délicatesse* being to perceive even the finest differences. In this regard, one could—and had to—perform in this lazy courtly society. In the social micro-context, it was aimed at their peers who—if they did not want to lose face—had to make an effort to commensurate with their rank and in doing so constantly tried to outdo themselves. However, in the social macro-context, there are the common people, to whom the legitimacy of the state order willed by God was demonstrated: The nobility stylized itself as the better people—as the people "of quality."

That sounds arrogant and it was. And yet it was true in a certain way: While before the lord and servants were united in their tendency towards potlach-like waste, now the aristocracy[24]—at least a growing part of it—is separating itself in its consumer behavior and its standards of value from the broad mass of subordinate existences, where the archaic ideal of excessive expenditure was maintained. A deep trench opened up between top and bottom, deeper than ever before and ever after.

The Birth of Capitalism from the Spirit of Asceticism

That is only half the story, though. Between the aristocracy and the mob, there was a third component: the citizens. The refinement and spiritualization of luxury was not solely the fault of the court—the urban bourgeoisie also played its part. It has always been made up of merchants and educated people, the latter naturally placing the greatest value on the accumulation of cultural capital. Here luxury ideally consisted of the scarce good knowledge (which does not mean that tangible luxury would have been spurned). Looking at how to acquire the necessary judgment a potentially revolutionary punch line was developed in humanism: Cultural capital must not be a prerogative of a class. As early as 1531 Erasmus of Rotterdam stated that nobody could choose his parents, but everyone could acquire spirit and good morals.[25]

At that time, Europe was shaken by the Reformation. With Protestantism, which no longer knows the absolution of sin and therefore demands constant self-control, a new type of virtuous spiritualization entered the bourgeoisie, especially among Calvin's followers, also known as Puritans: Any luxury, any debauchery was frowned upon, pious asceticism should take the place of gluttony and pomp—

---

[24]One can summarize here the second state (nobility) with the first (high clergy): Closely interwoven in terms of personnel, they hardly differed in thinking and behavior.

[25]See Spode (1993, p. 62).

bourgeois thrift and diligence against feudal waste and idleness. But the irony of history wanted this ascetic "Protestant ethic" to make its followers rich—it was aimed at a place in heaven and yet proved to be an economically highly functional value system on earth.[26] This first became apparent in Holland and England. Investment instead of hoarding or squandering possessions, systematic improvement of production and distribution methods, and aggressive world trade allowed early capitalism to flourish in the seventeenth/eighteenth century—long before the steam engine—which produced something like a "consumer society."[27] Citizens could now buy goods that had previously been reserved for noblemen: the first steps toward the democratization of luxury. At the same time, merchants accumulated fantastic riches and promptly converted them into expensive luxury—even if they ran around dressed in godly black.

The Birth of Capitalism from the Spirit of Waste

While courtiers might have attested a lack of *délicatesse* to bourgeois luxury, they gradually had to realize that they were facing dangerous competition from this third class. Namely in two aspects: as owners of cultural and economic capital. Another irony of history, the consumerist aristocracy had first created this competition to a large extent and then fattened it to the best of its ability. Someone finally had to design and produce the luxury goods that decorated the class of lazy slackers. And somebody had to lend them money for it. Although the payment morale of the nobility, especially of the sovereigns, was not the best, by the eighteenth century at the latest the legal certainty had developed to such an extent that there was only a small danger that powerful debtors—such as the Welser around 1600—would actually expropriate them. And so merchants and manufactory owners made nice profits—and as a side effect, they paid people wages, who could now buy a few things above what was "necessary," such as cheap "fashion goods," which in turn helped others to prosperity. "Progress" became the buzzword of the time: state, science, and economy set a new dynamic of civilization in motion. Away from the guilds, a "proto-industry" developed, which was still organized by craftsmen but was already producing for larger markets, partly even for the world market. This resulted in a silent alliance between the second and third classes, endowed by luxury: while nobility and citizens mutually disdained each other—the one despised the acquisition of money, the other undeserved wealth—together they laid the foundations of modern capitalism.[28]

---

[26]Cf. klassisch Weber (2009) [first 1905], whose Protestantism capitalism thesis for the understanding of Western "rationalization" remains topical in principle. "Bourgeois" virtues (diligence, self-discipline, planning, rational calculation, thriftiness) are sometimes found in other times and cultures as well, but only Protestantism has made them a comprehensive and lasting social maxim.

[27]See Reith and Meyer (2003); Kriedtke et al. (1977).

[28]Cf. classic Sombart (1922), whose luxury capitalism thesis directed against Marx and Weber also remains topical despite justified objections (see note 27).

This alliance already attracted the attention of clever contemporaries. Pioneering Bernard Mandeville, whose teaching poem "Fable of the Bees: or, Private Vices, Publick Benefits" from 1714, anticipated the trickle-down theory: "Luxury served to preserve millions of poor ... / The need to be part of it / In clothing, housing and other things / Always made fun of but admired / The true driving force of trade."[29] Theologians, who insisted on the sin of wasting scarce goods, and economists, who analogously postulated the finiteness of the quantity of goods, attacked Mandeville equally fiercely (especially when he pleaded for public brothels). A little later, luxury was rejected even more fundamentally: Albrecht von Haller, Rousseau, and other protagonists of a "Back to Nature" movement lamented "progress" as a historical aberration that puts people in "golden chains"—which Voltaire countered with a mocking poem: "Pity the good old days if you want to / ... I love luxury, lush and comfortable."[30] Voltaire knew: Romanticism itself is a luxury of thought that requires a high level of prosperity. In the end, nobody really wanted to do without the achievements of "progress" and Mandeville's view prevailed, although a distinction was now made between good and bad luxury: In mercantilist thinking, luxury was good when it increased prosperity and work, bad when it triggered expensive imports. At the same time and across the board, it goes without saying that a distinction was made in terms of taste between good and bad luxury.

Demonstrative Consumption in the Bourgeois Age[31]

Then, in the nineteenth century, when the barriers of status fell and the Industrial Revolution took its course, the bourgeoisie became the ruling and leading social class. Hunger had been defeated since the middle of the century, and toward the end of the century the now fully developed capitalism opened up enormous opportunities for advancement for "intellectual workers"—in contrast to the mass of "manual workers" who found their modest livelihood in the new factories, on the fields, and as servants. In the empire, the bourgeoisie, including the "small people," accounted for almost one-fifth of the population. It split into a myriad of milieus, professional and income groups that did not get along well. The supposedly postmodern "pluralization of lifestyles" is an old hat. Ultimately, it was a bourgeoisie that was more felt and characterized by fundamental ethical principles[32] than a sociologically comprehensible bourgeoisie that united assistants and priests, civil servants and senior officials, merchants and doctors, master craftsmen and professors, artists and bankers.

---

[29]Here to Sombart (1922, p. 186); cf. Berry (1994); Grugel-Pannier (1996); Jäckel (2008).

[30]Le Mondain quoted http://correspondance-voltaire.de/html/luxus.htm (25.8.2018).

[31]From here on, I will concentrate exemplarily on the German development.

[32]See note 26; see Kocka (1987). Added to this was the aristocracy, which continued to occupy positions in the military and administration, but hardly differed sociologically from the bourgeois upper class.

The most important link, in addition to the "white collar" and minimal domestic furnishings, was the privilege of being able to participate in a new luxury pleasure: the holiday trip.[33] "The whole world is travelling," noted Fontane, referring of course only to his own bourgeois circles. Originally the elitist passion of some romantics and adventurers, the consumption of space and experience became a central component in the lives of the better-off "intellectual workers" and their wives and children—thanks to railways, the provision of holidays, and growing prosperity—toward the end of the nineteenth century. A limited democratization of travel. Within this limited social framework, a surprisingly wide range of holiday forms developed, from family seaside holidays to exquisite cruises, according to financial possibilities and preferences in taste. With devotion one milieu separated itself from the other: the "backpacker" from the spa guest, the spa guest from the summer visitor, etc. But as a whole, the holiday trip retained its function as a status symbol: An upscale tourist class, about a tenth of the people, sunbathed themselves in the consciousness of being something "better."

Similarly heterogeneous as the bourgeoisie around 1900, although numerically much smaller, one can picture the nobility of the Ancient Régime. In both times, luxury consumption served the purpose of "social distinction."[34] Of course, the dynamism of the industrial society was much greater than that of the agrarian-protoindustrial hierarchical society. The population groups in which the struggle for social recognition was raging were now broader as well. This fight was fought to a large extent through consumer decisions. In his famous theory of the leisure class, Thorstein Veblen had analyzed the "demonstrative consumption" as a modern potlach.[35] The economy now provided a huge arsenal of luxury goods. So huge that in the respective sociotope it was the taste and only the taste that decided on its value and thus the rank of the owner[36]—quite in contrast to the classic potlach, the child of a deficient society, where, if at all, one can speak of show-off taste. Now the taste of quality prevailed: the long process of replacing quantity by quality had reached its preliminary final stage. Cultural capital had not defeated economic capital, but without knowledge of the "right" use of money it was worth little. Anyone who spent the summer in Davos, which had now been overrun by Russians, earned pitiful looks at home instead of admiration; anyone who put a grand piano in their salon, but obtained it from a less reputable manufacturer—or even could only play it amateurishly—made a fool of himself among the knowledgeable; anyone that hosted a big dinner (which in better circles had to happen at least once a year) and did not know how to greet the guests "correctly," how to hold

---

[33] See Spode (2009) [first 2003]. The first scientific treatise on "tourism" defined tourists as "luxury travellers" (Stradner 1917, p. 9).

[34] Elias had coined the term for the aristocratic world in 1939; it became popular through Bourdieu's contemporary analysis (1992).

[35] Veblen (1958) shows humor, but it is not a "satire," as the translators noted.

[36] By attributing aesthetic judgments "ultimately" to money, Veblen provides only one theory, a theory of *bad* taste.

the cutlery and how to have a proper table conversation, caused their guests to cringe. "The Parvenü," they said, "is standing in the midst of his guests, sweating in fear."[37] The newly rich—back then due to upward mobility a more frequent figure than today—was a target of ridicule, if not hatred (a central aspect also of the emerging hostility toward the Jews). Luxury consumption could be a big burden. And it was also seen quite critically, at least if it exceeded the "necessary" according to one's rank. In large parts of the bourgeoisie, by no means only among the "small people," the classic virtue of thrift continued to apply—in these families the taste of modesty reigned in everyday life, which did not rule out making a detour to Paris and sending the son to the best boarding school.

All this did not affect the manual labor force. There was nothing to save or spend here. The need for necessity reigned and individual chances of advancement were low. But the workers' movement strengthened and turned the social issues into *THE* domestic political issue (and this was to remain so until the post-war period). The SPD (the Social Democratic Party of Germany) and trade unions had to operate in a legal gray area, but the longing for the luxury of the "ruling classes" allowed a proletarian associations to flourish. Strictly rejecting the bourgeois competitiveness, they eagerly began to acquire bourgeois educational capital, practiced bourgeois virtues, and copied bourgeois leisure activities in sports or theatre clubs, even including a small tourism club "Die Naturfreunde" (friends of nature).

Democratization of Luxury

World War I marked the end of the long, glorious era of luxury. Its beginning is lost in the darkness of history, in ancient times luxury had celebrated fantastic triumphs, in courtly society it had been driven to bizarre glory and then lived its golden years in the bourgeois era. Now the descent began. This does not mean that everyone suddenly confined themselves to what was "necessary." Of course, there was still potlach-like demonstrative consumption. It merely means that the foundations had now been laid for a new phase in the history of consumption: the age of mass consumption or the affluent society.[38]

---

[37]Trautwein (1919, p. 66). Like the distinction and the potlach, a timeless phenomenon, from the Roman *Homo novus* to the most powerful man in the world today - a pattern of tastelessness that makes him all the more popular with his followers, who thus revolt against the owners of cultural capital.

[38]See Triebel (1991); Maase (1997); Haupt and Torp (2009).

Interwar Period

While at the beginning I had defined luxury as a scarce commodity, the history of consumption so far had been marked by the fact that few had much and many little. Roughly speaking, this is the basic constellation of agricultural societies, which are also deficient societies: The masses of producers were confronted with a tiny group that buys a large part of the products and, despite low social productivity, accumulates enormous wealth; in return, they had to display this wealth and organize public festivals to symbolically restore cosmic unity. But gradually an expansion of the circle of those who did not have to live from hand to mouth had begun, especially in the booming empire, when the agrarian society became the industrial society. Admittedly, here it still encountered narrow, artificial barriers—relics of the agrarian society of the social classes. The "fourth class," the workforce, was politically excluded and benefited little from the growth of the national product.

In 1918, this exclusion formally came to an end. The state should now be a "people's state" overcoming the "division by class." To this end, the state was not only to ensure, as it had done partly before, a basic supply of the "necessary," but also to ensure distributive justice, which at least partially implies a democratization of luxury—the SPD promised that in future even the workers would sail the seas. In fact, the trade unions imposed a modest leave entitlement for workers. But a worker was hardly in the position to travel. In the crisis-ridden republic, real wages remained miserable. At the same time, the classic bourgeoisie was hit hard economically; silverware came under the hammer, and many Grand Hotels had to file for bankruptcy because the clientele failed to appear. Rather, the growing "new middle classes," especially the employees, became pioneers of mass consumption. Sociologically they belonged to the middle classes, but instead of saving, they developed a hedonistic taste of experience: the often quite scarce money was spent on fancy clothes and accessories, excursions and vacations, dances, and variety shows. This was still an expense that exceeded the "socially customary," i.e., a luxury: a breakthrough in mass consumption (as was evident in America until the stock market crash of 1929) was still a long way off. Thus, the travel intensity increased only slightly; it remained a privileged tourist class.

Little changed in 1933 when Hitler announced that he wanted to finally realize the "Volksgemeinschaft" (people's community).[39] Instead, wages were kept low. But to compensate for this, "people products"[40] produced under state control were to increase prosperity: Prestigious luxury goods—including televisions—were to become mass products through serial production à la Henry Ford, with cars, radios, and vacations leading the way. Due to the premature start of the war, the "Volkswagen" did not roll off the assembly line, but the "Volksempfänger" (radio receiver) became a huge success, as did the cheap package tours of the organization for leisure pleasure "Kraft durch Freude" (strength through pleasure). Under the

---

[39] See Haupt and Torp (2009); Torp (2012); Spode (2009).

[40] Heute wieder—without copyright—a label of the Bild-Zeitung.

motto "You too can now travel," KdF sent millions of "Volksgenossen" (German citizens) on holiday—even on cruises, as the SPD had promised before. Among them numerous workmen. A tremendous propaganda success. At the same time, unrestricted tourism grew strongly thanks to a good economic climate, and travel intensity was likely to have reached a good fifth of the population. But workmen, who made up half of the working population, remained clearly underrepresented; the increase in travel was primarily due to those who had already traveled before 1933. The same was true for all other consumer goods. The NS welfare society ("socialism of action") remained stuck in its beginnings.

Post-war Period

Only the post-war period brought the breakthrough of mass consumption.[41] This breakthrough has gone down in history as an "economic miracle". In the beginning, there was still poverty, and everywhere a prude-puritan taste of modesty was preached—quasi in recourse to the imperial period. But then the economy picked up speed again, although now, unlike under the Nazi dictatorship, the producers were involved in the progress in productivity: Wages and salaries rose. In 1957 the Minister of Economic Affairs Erhard issued the motto "Prosperity for all." And indeed, it was no longer only the upper and middle classes that could afford goods that had previously been considered luxury. At first, the focus was on old acquaintances: Cars, radios—or television sets—and travel. The Volkswagen now lived up to its name, a radio, and then also a "flicker box" were now also found in workers' households and the boundaries of the privileged tourist class began to dissolve. In 1958, a study was able to explain tourism as a "demonstrative consumption of experiences" for the purpose of increasing "social prestige."[42] By the mid-1960s, however, travel intensity had doubled to two-fifths of pre-war levels. Travel *per se* offered less and less social prestige.

At that time, real incomes increased once again strongly. Nevertheless, much was bought on credit; in view of the looming saturation of the markets, the economy fired the hunt for fast consumption. What was luxury a moment ago was soon considered miserable and obsolete. At the beginning of the 1970s, after half a century, the process of transition from elite to mass consumption was largely completed. The former luxury was democratized and therefore no longer a luxury. The vast majority of households now had a television and a telephone, one in three Germans[43] had a car and travel intensity exceeded the 50% mark. The "68ers" spoke strictly of the "consumer terror," rightly suspecting that the need for the "necessary" typical of the time was in principle met; in his wildest dreams, Karl Marx could not have imagined

---

[41]See Schildt (1995); Maase (1997); Torp (2012).

[42]Knebel (1958) following Veblen.

[43]To the GDR, where under contrary conditions mass consumption also started late, here only Haupt and Torp (2009).

the standard of living of a mason or bus driver. The "socially customary," of course, was different. Here the borders were even more blurred and the "68ers" were in the middle of it all when they bought a Stones record instead of a Heino record and flew to Crete instead of Mallorca. They were the ones who had become addicted in their own way to the hedonistic taste of the experience that had accompanied the development of mass consumption.

Luxury in Times of Abundance

Since then, not much has changed in principle. We are still living in the age of mass consumption. Unfortunately, the harmonious "Volksgemeinschaft" (people's community) has not taken shape yet. Instead, the social distinction that once shaped the lives of the courtiers and then the bourgeoisie has become an (almost) all-societal game: Everyone against everyone.[44] We let taste and the use of time and money determine social ranks and use it to assure ourselves of our identity. In addition to countless variants of the taste for experiences, the taste for quality and modesty are of course still to be found, also in countless variants.

But where has luxury gone? When one considers its central role in the structure of societies characterized by chronic shortages, luxury has disappeared: Today's society is no longer produced over the classic potlach. "Mass production has simultaneously brought luxury its greatest triumph and its downfall."[45] Fortunately, we no longer live in times of absolute shortage and public sphere. However, massive, possibly anthropologically conditioned relics of the latter have survived. Veblen rightly found "archaic traits" in demonstrative consumption.[46]

This, too, has changed little since Veblen's analysis. We still have the parvenu, who thinks he only needs money to demonstrate "a special taste."[47] The so-called luxury brands live from the uncertainty of taste in affluent ascending milieus. They owe their scarcity and thus their luxurious character to their price and limited editions: from quantity to quality. [48] In the respective sociotope, possessions bring much prestige as a seemingly irrational waste; outside, however, the prestige value weakens considerably—to the point of contempt that a golden Rolex or a white Lamborghini can trigger.[49]

---

[44]Cf. classical Schulze (1993); Bourdieu (1992).

[45]Enzensberger (1996, p. 116).

[46]Veblen (1958, p. 206).

[47]"It has always been a little more expensive to have a special taste," advertised the cigarette brand *Attika* in the 1960s with some irony.

[48]With some cutbacks also in tourism; on the luxury travel market, to which in particular quite banal "high travel costs" are ascribed in comparison to the exclusive trips of the imperial era, however, luxury tourism today seems conservative (Spode 2009).

[49]They are "above all pimps, gangsters and drug barons ... who place the greatest value on adorning themselves with exclusive shit" (cf. Enzensberger 1996, p. 117). Those were the days when

Sombart's qualitative and spiritual "refinement" continues to dominate. (Still) there is the society as the comprehensive sociocultural environment of consumption. In controversial negotiation processes, the diversity of tastes is being distilled into a mainstream, a vague basic consensus. This is not the "legitimate taste" of the super-rich and great thinkers; [50] there is no such thing and perhaps never was. At the latest since the expansion of education in the 1970s, the academic middle classes have set the tone. Accordingly, consumption in general and luxury in particular are judged according to the cultural capital coagulated therein.

This mainstream is characterized by the rapidly increasing moralization of consumption.[51] This is due to a sharpened distinction in the form of a "closing of the middle":[52] The educated classes compensate their loss of status security in the course of neoliberal reforms and increasing competitive pressure with the assertion of a superior morality, a morality of self-restriction and self-control. The hedonistic is thus replaced by an ascetic taste of experience or a new taste of modesty.[53] It is not necessary to be economical for personal well-being, but for the good of the whole, especially the "environment."

The same applies here: special taste is always somewhat more expensive. From this perspective, eco, fair trade, vegan and regional products are a luxury—even if they come across as critical of luxury: Such a "return to nature" à la Rousseau has the democratic and at the same time hegemonic claim to one day become the "socially customary" (which would then inevitably result in the trendsetters switching to other quality criteria). Vacations, too, need to be morally. The industry has responded and continues to invent new eco-labels. Meanwhile, it is precisely the "environmentally conscious" milieus that travel particularly far and frequently, leaving behind the largest ecological footprint.[54]

To see pure hypocrisy in this would be premature in most cases. Rather, most people act unconsciously within the framework of the acquired habitus or the internalized value system. That is the case here too. In order to avoid cognitive dissonances, "conscious purchasing decisions"—of which one is so proud in educational circles—are then brought into line with this framework *post festum* if necessary. Emblematic for this is Tesla's luxury cars—a must-have for the rich and beautiful in California and elsewhere. In contrast to Mercedes-Benz's luxury cars, they glide over the asphalt without exhaust fumes. And yet they are just as fast. However, they run out of power on the way from Los Angeles to San Francisco. Worse still, if you take into account not only driving but also production, power generation, and disposal, the environmental balance is less than flattering. But

---

Breakfast at Tiffany's filled the cinemas. Today, luxury brands should ensure that the image of their main customers does not rub off on them.

[50]So Bourdieu (1992).

[51]See Torp (2012).

[52]See Spode (2008).

[53]See also Schulze (1993); Enzensberger (1996).

[54]See Hallerbach (1993).

rational aspects of purchasing are at best secondary. What could better demonstrate the wealth of money *and* morals than this functionally meaningless car? A green potlach with quite the "archaic features."

Concluding Remark

Theorists such as Sombart, Veblen, and Bourdieu have clairvoyantly revealed the hidden social functions of luxury consumption. Here, too, I was concerned with the functional achievements of luxury in its sociocultural environment and the construction of luxury by this very environment. As a scarce good, it proved to be *the* engine of history. However, we still do not know everything about it. Our analyses are too one-dimensional. For Pierre Bourdieu, human existence—and thus also luxury—is nothing other than a permanent struggle; every action is subject to the postulate of positioning in the "social space." A strange social Darwinian thinking for a "left" sociologist.

On the other hand, I would like to point out that people do things just because they enjoy them, and that they also find things beautiful just because they *are* beautiful. In this, the human being probably does not differ from other higher animals. There is obviously, as suggested in the introduction, besides the socially constructed, also a silent, "natural" aesthetics and joy. Magpies are fascinated by silver rings, chimpanzees play obliviously with a ball. In this sense, luxury is nothing but a beautiful toy. And you want to own it. In this sense, the Tesla stands above all criticism.

Schiller said that humans are "only human beings where they play" and Johan Huizinga even saw the "origin of culture in the game." In any case, the approach presented here covers only part of the fascination of luxury. There is still more research to be done here.

2   Behavioral Explanations for Luxury Consumption

Hannes Gurzki and David M. Woisetschläger

Introduction

"Luxury travel is booming," the ITB announced last year in a press statement (ITB 2018). Growth rates for high-priced trips, that are priced at least 500 € per night, have been nearly around 18% since 2014 compared to 9% for international travel in general (ITB 2018).

Luxury is everywhere. No longer are the boundaries between physical luxury goods and luxury services clear-cut, if they have ever been. Luxury brands such as Armani or Bulgari, originally focused on fashion and jewelry, have extended their brands into the luxury hospitality sector together with partners to offer a range of services from cafés and restaurants to hotels (Albrecht et al. 2013a). Luxury brands are establishing their service ecosystem to orchestrate luxury experiences from door to door. And growth is set to continue as more and more consumers desire luxury (see Sect. 3.3).

But what is driving consumer desire for luxury tourism? And what can luxury tourism leaders earn from it? The objective of this chapter is to highlight the key motivations for luxury consumption and luxury tourism in particular. Studies about travel motivations abound in both practitioner-oriented as well as academic litera-ture. What is considered a motivation thereby also widely differs from more abstract concepts such as self-fulfillment to very concrete ones, such as relaxing or visiting a specific touristic site. The challenge is thus to find a unifying framework, which requires a certain level of abstractness, yet considers the differing views on motiva-tions. Such a framework can be helpful to reflect on current practices for managers active in the luxury tourism area.

This section first provides a view on luxury in order to better understand the concept. It then goes on to discuss luxury consumption motivations and motives. Understanding the needs, wants, and desires luxury satisfies helps luxury brands to devise strategies targeting these core motivations. Lastly, it provides a practical framework with guidelines for luxury brand managers in the tourism sector.

The Essence of Luxury: It Is All About an Extraordinary Experience

The idea of luxury is as old as humankind and has been an integral part of the history of civilizations from ancient Egypt, Rome, and China (Kapferer and Bastien 2009). The meaning of the term luxury can be derived from the Latin words "luxus" and "luxuria" meaning a deviation from normal and ordinary standards (Grugel-Pannier 1996; Valtin 2005). Thus, in its basic sense, luxury is the extraordinary (Gurzki 2018).[55] The downside of this view is that the extraordinary depends on the ordinary and is thus dependent on the context and individual consumer. Luxury is a complex phenomenon that covers multiple dimensions. The luxury consumer is embedded in a material, social, economic, cultural, and historical environment in which marketers influence and create different realizations of luxury (Gurzki and Woisetschläger 2017). From each dimension, luxury is something extraordinary. Only the

---

[55]This chapter is based on the unpublished thesis of Gurzki, H. (2018). The Creation of the Extraordinary – Principles of luxury. Dissertation. Braunschweig: TU Braunschweig.

extraordinary has the potential to create the desire and lure of luxury and create the strong motivational forces for luxury consumers to fulfill this desire (Gurzki 2018).

The upside of this perspective is that anything that is extraordinary can become a luxury. This is true for abstract concepts such as love and time, but also for concrete concepts such as luxury goods and services. Luxury products and services often also embed these fundamental values of love and time, for example through the link to the brands heritage and history, the time spent by craftsman and artists to design the product or the experience or the passion and attention to every detail that gives the luxury object a soul and feeling of contained love (Fuchs et al. 2015).

So why do consumers crave for the extraordinary? A key to understanding luxury consumer motivations lies in the concept of desire. Something is desired because an individual intensely desired or because others desire it and thus there is social agreement on the desirability (Berry 1994). Desires can be defined as "ever-changing, infinitely renewable wishes inflamed by imagination, fantasy, and a longing for transcendent pleasure" (Belk et al. 1996, p. 368). Desire has many dimensions and layers, mainly the desire for pleasure, identity-play and self-enhancement, social relationships, and self-transcendence (Gurzki 2018). Desire thus exists in the interplay between the collective social and the individual sphere, as well as the mind and the body.

In particular, the desired luxury object or experience has the power to make the consumer feel special and extraordinary (Gurzki 2018). This special feeling can have different sources such as the pleasure and intensity of the experience itself, its sensory and mental stimulation, the possibility to experience profound meaning and opportunity for the consumer to engage in identity-play, or the social status and relational consumer benefits.

The Drivers of Luxury Consumption

This section briefly illustrates different lenses on luxury motivations that are commonly used to explain luxury consumption. The aim is to illuminate the motivational processes that drive luxury consumer decisions while highlighting the complexity and interdependency of the different dimensions. Thus, rather than merely showing different motivational dimensions, this section outlines four concepts that are illustrating the motivations of luxury consumers. The concepts are thus interlinked, yet each perspective provides a different view on luxury consumption.

First, motivational theories that outline how fundamental human motivations influence luxury consumption ("Luxury consumption satisfies needs"). Second, self-concept theories highlight the role of identity in symbolic purchase decisions ("Luxury consumption affirms and enhances the consumer's identity"). Third, an experiential view focuses on the emotions experienced through luxury consumption ("Luxury consumption offers positive emotions and pleasure"). Here the realization of imagined pleasure or the reinforcement through the actual experience are core drivers of luxury consumption. Fourth, a consumer value perspective looks at how luxury consumption enables consumers to create consumer value ("Luxury consumption creates value for the consumer").

Motivational Theories: Luxury and Needs

Research on consumption motivations has proposed conceptual models with different layers of needs (Kenrick et al. 2010; Maslow 1943). In his popular motivational theory, Maslow proposed that human motivations drive certain behaviors. Motivations are derived from individual needs, whereby these needs can be arranged in a hierarchy. Most basic is the physiological needs necessary for survival, such as the need for food and water. If these needs are not satisfied, all other needs become less important. The need further up the hierarchy is safety, which includes physical safety as the absence of violence and threats, but also the absence of illnesses. Once these needs are satisfied, love, belonging, and social relationships become important. As the fourth category of needs, he proposes esteem needs, which includes a positive evaluation and respect from others and having a sense of achievement. The highest level of needs according to Maslow (1943) is the need for self-actualization and expressing one's identity, for example through creativity such as arts or music. While the approach has been criticized and never been empirically validated, it provides some useful insights into motivation: There are human needs, which are organized in a hierarchy that drive human actions. Typically, multiple needs and motivations to fulfill the needs are present at the same time (Maslow 1943). Kenrick et al. (2010) proposed a revised hierarchy of needs, based on evolutionary psychology, which has spurred attention in luxury research. According to evolutionary theory, the main motivation of any organism is survival and reproduction (Kenrick et al. 2010). The needs of human beings are thus in ascending order from basic to advanced: immediate physiological needs, self-protection, affiliation, status and esteem, mate acquisition, mate retention, and parenting (Kenrick et al. 2010; Sundie et al. 2011).

Luxury goods and experiences are able to satisfy a broad range of needs. Berry (1994) proposes that luxury provides a more refined version to fulfill basic needs, and thereby rather than being a necessity for survival, being a choice and thus a want or desire. He proposes that four different categories of luxury goods exist, for which this choice becomes apparent: sustenance, shelter, clothing, and leisure. While any food would satisfy the hunger need, luxury foods such as caviar or champagne provide a more refined version to satisfy them, making clear the distinction between necessity and luxury. Luxury goods and experiences can also provide protection, such as in the case of identity-threats (Braun and Wicklund 1989). For example, luxury products can restore feelings of powerlessness for the consumer (Madzharov et al. 2015; Rucker et al. 2012). This explains the behavior of consumers driven by self-protection motives to engage in compensatory consumption to restore their identity, for example, an increase in the consumption of visible and conspicuous luxury goods for individuals with recent changes in socioeconomic status (Carr and Vignoles 2011). Feelings of powerlessness such as rejection by sales personnel in store can lead to a higher willingness to pay for luxury products (Madzharov et al. 2015).

The key to luxury goods and experiences is their ability to provide social status, recognition, and self-esteem to their consumers (Drèze and Nunes 2009; McFerran 2013). This ranges from the visible display of conspicuous goods (Han et al. 2010) to the experience of exclusive luxury experiences or the participation in loyalty programs (Arbore and Estes 2013). For example, the membership of a higher and aspired loyalty status tier can create a feeling of superiority for its members (Drèze and Nunes 2009). For this motive, particularly soft benefits such as recognition play a strong role members (Drèze and Nunes 2009). This prestige consumption is mainly driven by the desire to belong to a certain aspirational target group to which the luxury good or experience promises access, such as in the case of an exclusive vacation (Bearden and Etzel 1982). Particularly for public luxuries the consumer's social environment and reference group plays a strong role for the purchase decision (Bearden and Etzel 1982). McFerran (2013) has shown that the feeling of status is even heightened if the consumer receives a preferential treatment in the presence of others, even if the benefit itself is lower, which they term the "entourage-effect."

Luxury purchases are related to higher-level needs such as self-expression (Albrecht et al. 2013a), self-definition (Kisabaka 2001), self-actualization (Wong and Ahuvia 1998), self-realization (Lasslop 2002), self-enhancement (Mandel et al. 2006), self-transformation (Llamas and Thomsen 2016), and even self-transcendence (Arndt et al. 2004b; Rindfleisch and Burroughs 2004) through their high symbolic content and identity-relevance (Csikszentmihalyi 2000). For example, the choice of aesthetic products can lead to self-enhancement (Townsend and Sood 2012). The self-enhancement motive through luxury purchases can be triggered by showing successful persons similar to oneself (Mandel et al. 2006). Particularly luxury experiences are relevant for self-transformation (Llamas and Thomsen 2016). This includes altruistic acts or the experience of authentic places which are experienced as a luxury, creating feelings of pleasure, meaning, and human connectedness (Llamas and Thomsen 2016). This also highlights how broad the concept of luxury is and where the potential for triggering these motivational sources lies for luxury tourism. Lastly, self-transcendence can be a strong motivational driver to engage in luxury consumption (Arndt et al. 2004b; Rindfleisch and Burroughs 2004). This relates back to one of the oldest theories about the motivations to consume luxury: the human awareness of death and the human desire for immortality (Arndt et al. 2004a, b; Kapferer and Bastien 2009; Maheswaran and Agrawal 2004; Mandel and Heine 1999; Rindfleisch and Burroughs 2004). Research has proposed that luxuries and desired objects can provide a way to cope with these fundamental fears (Arndt et al. 2004b). Timeless and enduring luxury possessions propose something eternal and they offer stable resources to construct and enhance one's identity which helps to cope with these existential threats (Kapferer and Bastien 2009).

Self-Concept: Luxury and Identity

One of the most powerful drivers of luxury consumption is the ability of luxury to provide meaning and that it offers a way to express a lifestyle or construct one's self through its symbolic content (Hogg and Michell 1996; McCracken 1990). Possessions, experiences, and ideas can become a part of the identity, provide meaning, reflect who we are and thereby symbolically extend the self (Belk 1988, 2013; Rochberg-Halton 1984; Sirgy 1982). Extraordinary possessions or experiences offer a chance for identity play and provide means to build and experience one's ideal and aspirational self (McCracken 1990). Through the individual's investment of resources to gain control (high prices), creativity (personalization), and knowledge (consumption codes and rituals) these experiences and objects become part of the consumer's self (Belk 1988). The motives to protect and enhance one's identity are powerful motivational drivers, particularly for luxury consumption.

The self has two dimensions and two states: the actual self (current state) and the ideal self (desired state) as well as the personal self (own perception) and the social self (perceptions of how the individual is seen by others) (Sirgy and Su 2000). Self-concept theory proposes that individuals aim to create a positive self-concept and enhance their self-esteem while striving for self-consistency (Sirgy 1982). If the consumer perceives a mismatch between the actual and the ideal desired self, the consumer will act to make them congruent (Sirgy 1982). For self-consistency, they strive to match the symbolic properties of their consumption choice with their actual self, whereas for self-enhancement they match the symbolic properties with their ideal self (Sirgy and Su 2000). This can be driven by both external motives of the social self, such as status and recognition or internal motives of the private self, such as self-fulfillment (Sirgy 1982). The social recognition and even an individual's belief about social recognition can enhance the self-image (Grubb and Grathwohl 1967). Whereas the personal (or private) self is more strongly related to luxury consumption for self-realization motives, the social (or public) self is more strongly related to affiliation and distinction (Meffert and Lasslop 2004).

In a purchase decision, such as the choice of a travel destination, consumers form an image of their destination based on the atmospheric, service, price, location, and promotion cues (see Sirgy and Su 2000). Then they look for functional congruity, that is a match of the service quality, price, and other characteristics with their expectations as well as self-congruity, that is the match of the symbolic characteristics to their identity (Sirgy and Su 2000). The stronger the congruity is, the higher the motivation to visit the destination (Sirgy and Su 2000). For luxury, particularly the ideal self seems to play a strong role in consumption choices, triggered by self-enhancement motives. Factors such as the conspicuousness of the destination, the presence of others (contouring), the consumer's age, prior experience and knowledge, the involvement of the consumer or time pressure also influence which self-concept is activated (Sirgy and Su 2000). For example, conspicuous destinations are more likely to trigger the activation of the social self, whereas inconspicuous destinations are more likely to trigger the private self (Sirgy and Su 2000). Previous research has shown that consumers that for consumption decision that have a high personal relevance for the consumer, consumers are more likely to trade up to luxury goods (Silverstein and Fiske 2003). When relevant to the identity, they are also more likely to make distinct choices compared to other consumers, which further underscores the identity-relevance of consumption and the consumers need for uniqueness which can be satisfied through the consumption of luxury goods (Berger and Heath 2007; Snyder 1992). Particularly, experiences have a strong impact on self-definition as they facilitate social relationships (Bhattacharjee and Mogilner 2014).

Experiential Consumption: Luxury and Emotions

Consumers have both existential and experiential needs (Csikszentmihalyi 2000). Luxury consumption creates pleasure and intense emotions that make the consumer feel special and unique, which distinguishes them from ordinary experiences (Gurzki 2018). Hagtvedt and Patrick (2009, p. 609) proposed that a luxury brand "[. . .] has premium products, provides pleasure as a central benefit, and connects with consumers on an emotional level." Similarly, Kapferer and Bastien (2009) asserted that luxury is pleasurable and personalized. Thus, the experience, fantasy, feelings, aesthetics, and pleasure are crucial for luxury (Holbrook and Hirschman 1982). The ability to create positive emotions, states of comfort and pleasure is essential for luxury goods and distinguishes luxury from non-luxury (Berry 1994; Nia and Zaichkowsky 2000). Through the immersion into luxury, consumers feel special and experience intense bodily reactions (Venkatesh and Meamber 2006). While some studies have also investigated the negative emotions that arise from luxury (e.g., Dubois and Laurent 1994), the large part confirms the strong and positive emotions (such as Drèze and Nunes 2009, Hagtvedt and Patrick 2009). For example, the product design and aesthetics can cause positive emotions, enjoyment, pleasure, and arousal, often as a subconscious emotional reaction (Kumar and Garg 2010). Using EEG (electroencephalogram) recordings to investigating the neural processes of luxury consumption, Pozharliev et al. (2015) have shown that luxury brands elicit a stronger emotional response compared to products with a basic branding.

Extraordinary experiences provide pleasure and can also lead personal growth, self-transformation, and experience of flow (Arnould and Price 1993). They have a high emotional intensity and require a strong integration, awareness, and involvement of the consumer who plays an integral part in co-creating the experiences (Arnould and Price 1993). The pleasure can come from different sources, such as the individual engagement with the brand, the social environment and recognition, the consumer–brand relationship, the novelty and curiosity of the consumption, the meaning and symbolism, or the pleasure of the experience itself (Hirschman and Holbrook 1982; Holbrook and Hirschman 1982, Vigneron and Johnson 2004).

Extraordinary experiences, such as travel and culture provide a greater happiness compared to ordinary experiences, especially if they are identity-relevant (Bhattacharjee and Mogilner 2014). While experiences can be scripted through the culturally or socially learned rules that guide the interactions, often consumers have vague expectations and just need to be ready to immerse themselves into the experience to create the spontaneous moments of delight that occur from the dynamic interpersonal interactions (Arnould and Price 1993). Through the sum of interactions with the brand and its touchpoints, the consumer can form a relationship with the brand that even extends to other products (Hagtvedt and Patrick 2009).

Consumer Value: Luxury and Dimensions of Consumer Value

Through their ability to satisfy needs, create meaning, and express one's identity and experience emotions, luxury brands create consumers value. Consumer value has long been a core concept to explain consumer behavior (Hennigs and Wiedmann 2012; Tynan et al. 2010; Valtin 2005; Vigneron and Johnson 1999; Wiedmann et al. 2007, 2009). While early studies have equated consumer value with price or focused on a single value dimension to explain consumer behavior, more recent research has started to consider the multidimensional structure and complexity of consumer value (see Sanchez-Fernandez and Iniesta-Bonillo 2007). Particularly to understand luxury consumer behavior, this multidimensional understanding is important in understanding the different value drivers and providing specific directions for managers on how to improve them (Sanchez-Fernandez and Iniesta-Bonillo 2007). One of the most comprehensive value typologies is the value typology of Holbrook (1999). For Holbrook (1999, p. 5), consumer value is an "interactive relativistic preference experience" forming an interconnected system. Consumer value is created through the interaction of a consumer (subject) and a product or experience (object). It is relativistic in that the value is personal (differs between individual), situational (depends on the context of evaluation), and comparative (is judged in comparison to other objects). Moreover, it is preferential in that consumer value carries a preference judgment. Most importantly, consumer value lies in the experience rather than in the brand or object itself and depends on the co-creation of value by the consumer (see Holbrook 1999). Based on this conceptualization, Holbrook (1999) proposed a conceptual framework of consumer value along three dimensions: the motivation for the experience (extrinsic: experience is instrumental as means to an end versus intrinsic: experience is appreciated in itself), the main source of value (self-oriented: value from the individual versus other-oriented: value from others), and the level of consumer agency (active: consumer has the agency versus reactive: object has the agency). This leads to eight different types of value: efficiency (such as convenience or a favorable input/output ratio), excellence (such as a high quality of the experience), status (such as impression management and creating a successful image), esteem (such as self-esteem and self-protection through reputation and possessions), play (such as fun, entertainment, mastery, curiosity), esthetics (such as beauty), ethics (such as justice, virtue, and morality), and spirituality (such as faith, ecstasy, sacredness, or magic). Most crucially: the value perceived and created depends on the individual who plays an active role in the value creation. Thus, the same experience might create different types of value for different consumers. An experience can create different types of value for a consumer and all types might even be created at the same time. The structuring dimensions are thereby extremes at the two ends of a continuum (see Holbrook 1999). This framework has also been empirically tested in tourism research, whereby status and esteem have been combined to form a social dimension, and ethics and spirituality have been excluded (Gallarza and Gil 2008).

Luxury promises a high consumer value across value dimensions through its symbolism (Levy 1959; Solomon 1983), experiential character (Hirschman and Holbrook 1982; Holbrook and Hirschman 1982), and high functional quality (Berthon et al. 2009). Most differentiating compared to non-luxury brands is their high ratio of non-functional to functional consumer value (Nueno and Quelch 1998). For luxury brands, Vigneron and Johnson (1999) proposed that luxury brands differ to non-luxury brands in that they propose higher conspicuousness (Veblenian motivation), uniqueness (snob), social (bandwagon), emotional (hedonist), and quality (perfectionist) value. More recent studies identified four consumer value dimensions for luxury brands: functional (usability, uniqueness, quality), financial (price), individual (self-identity, hedonic, materialistic), and social (conspicuousness, prestige) (Wiedmann et al. 2007, 2009). The functional dimension captures the main product-related benefits such as quality, durability, usability, and reliability. The financial dimension captures aspects such as price and investment needed, but also factors such as resale cost. The individual dimension relates to the value that relates to the consumer's identity value, materialism or hedonism, and experiential orientation. Social value captures the value that the consumer receives from social interactions, such as recognition and prestige. These factors are showed that this structure is similar across cultures (Hennigs and Wiedmann 2012). Similarly, Tynan et al. (2010) identified five sources of consumer value: utilitarian, symbolic/expressive, experiential/hedonic, relational, and cost/sacrifice. In their view particularly the symbolic/expressive, experiential/hedonic, and relational values are the key differentiators for luxury brands. Choo et al. (2012) proposed that fashion brands possess utilitarian (excellence, functional), hedonic (aesthetics, pleasure, experience), symbolic (self-expressive, social), and economic value as the key value dimensions, yet they also highlighted the importance of feelings and affect as well as the epistemic value that stems from novelty or knowledge. The true consumer value comes from the value co-creation in designing luxury experiences, a view that is likely to also receive more prominence in academic research in the future (Tynan et al. 2010).

Discussion and Implications for Luxury Tourism

Luxury is an extraordinary experience. It promises the fulfillment of needs, identity enhancement, pleasure and creates value for the consumer. As an experience is co-created and depends strongly on the consumer's involvement and integration.

The experience of pleasure, sensuality, and immersion thereby plays a strong role for the consumer and helps to build an emotional relationship with the brand. What is most distinguishing comparing extraordinary luxury experiences to ordinary non-luxury experiences are the strong emotions they elicit, making the consumer feel special (Gurzki 2018). Luxury experiences need to be extraordinary across dimensions to fuel the consumer desire. And if executed well, they are able to satisfy a broad range of needs at the same time or throughout different moments of the customer experience. Integrating the concepts outlined above, we propose four different categories of needs and motivations that could serve as inspiration to think about the core consumer motivations luxury tourism experiences and provide guidelines in their design.

The luxury basics cover physiological and safety needs, but also convenience needs such as efficiency or service excellence. Mastering the luxury basics means providing the highest level of service quality, flawlessly across touchpoints. The consumer experience is simple, effortless, and personal no matter whether the guest chooses to interact digitally with high tech or analogous with high touch. Convenience also means the consumer can interact with the brand and purchase whenever and wherever they want in a way that suits their needs (e.g., rental platforms for fashion products that are available on-demand[56]). And for many, this increasingly also means asking for on-demand services. With a further evolution of the market, in which luxury service innovations become adopted quickly by premium and mass market players, being at the forefront of innovation is key to deliver on the luxury basics and remain in the luxury game.

The social connectors cover the needs related to social relationships, belonging, and affiliation, but also distinction and status, which are essentially two sides of the same coin (see Sect. 3). Hence, creating social value for an aspirational community is crucial. To satisfy these needs, luxury brands need to think about both exclusivity (whom to keep out of the club) and inclusivity (whom to admit to the club). Yet social connection also means, understanding the special needs of diverse target groups such as families, seniors, or millennials. Bonding with the right community is an increasing trend, particularly with Millennials. An example of a brand strongly targeting this motivation is W Hotels by organizing their own music festival series with exclusive benefits for loyal members.[57]

The special personal moments are a third category integrating a range of individual needs. *Private escapes* promise comfort, relaxation, and privacy (such as secluded islands or nature). *Adventures* promise activity, thrill, and discovery (such as safaris). *Feel-good moments* promise wellbeing that contributes to the individuals physical, mental, emotional health (such as spa resorts). *Engaging indulgences* that offer aesthetic, fun, playful, engaging, and immersive multisensory experiences (such as fine dining). And *learning opportunities* that cater to the need

---

[56]https://www.renttherunway.com/

[57]https://www.wakeupcallfest.com/en/

for esteem, recognition (and self-recognition), personal freedom, curiosity, creativity, and self-expression such as through connoisseurship (such as cultural trips).

Lastly, the transformative experiences that go beyond individual moments and strongly cater to more profound self-actualization needs and identity transformation. While also being an individual need it goes beyond. While the desire to become a better self and personal growth is at the center, the motivations are also often spiritual or ethical in nature. Meaning and authenticity are core values that consumers are increasingly looking for. Mindfulness, being in harmony with oneself and the environment, sustainability, and having a sense of place and time. The desire for this profound transformation could help explaining the demand for authentic and immersive local experiences luxury tourists are seeking for. And the popularity of offerings such as AirBnB's Luxury Retreats[58] or the sustainable and culturally rich luxury such as Fogo island inn.[59] Many of the special moments and social connections can build up to this transformation and let consumers experience the magical and profound that makes the luxury experience extraordinary for them.

Designing extraordinary experiences requires a deep understanding of the target consumers, with their needs, identity, and expectations. At heart, this requires the art of listening, personal interactions, and relationships that have shaped the high touch culture of service excellence in luxury tourism for decades. Authentic and meaningful experiences can satisfy needs and trigger motivations that consumers hardly find outside of luxury. In addition, new technologies such as big data and artificial intelligence provide ways to complement the high touch service culture with high tech to create seamless personalized extraordinary experiences across all touchpoints. This offers new ways to create and deliver consumer value to the luxury market.

3   The Purchase Decision and Service Consumption Process for Luxury Goods

The Creation of Desire

Hannes Gurzki and David M. Woisetschläger

Introduction

How do luxury consumers decide? And what can luxury tourism managers and marketers do to increase their chances of success with this diverse target group?

---

[58]https://www.luxuryretreats.com
[59]https://www.fogoislandinn.ca

Often, the lure of luxury cannot easily be grasped rationally, which has challenged economic theory: Why do consumers have such intense feelings toward luxury products and services, personify luxury objects and experiences, have a high desire for them regardless of price, do not consider substitutes, are unwilling to sell special possessions for market value, are unwilling to discard them, and have feelings of elation or depression regarding them (Belk 1991a)?

Luxury is based on desire (Gurzki 2018).[60] The more an object is desired, the higher its luxury status (Berry 1994). Desire thereby is created in two ways: Through social agreement on the desirability of the experience or object and through the intensity by which the individual desires it (Berry 1994). Thus, both the emotional intensity of the imagined experience by the individual as well as the social environment and the imagined benefits such as recognition and esteem play a role (Belk et al. 2003). For luxury brand managers, it is crucial to understand the principles of desire in order to be able to manage them systematically.

This section will briefly outline the three stages of service consumption and propose how they differ between luxury and non-luxury services. We propose that this approach best serves to illustrate commonalities between non-luxury services consumption and the consumption of luxury services, while highlighting the key differences that managers need to be aware of. This chapter builds on traditional psychological decision-making processes and complements them with findings from a cultural view of consumption. Particularly for luxury goods as cultural products, where meaning, symbolism, and the experience are essential characteristics, we believe this combination enriches the understanding of the process and is helpful to provide spaces for reflection and insight to luxury brand managers (Albrecht et al. 2013a).

Luxury and the Creation of Desire

Luxury is about the creation of desire. This desire creation underlines all stages of the consumption process. Desire is central to luxury as it deals with the passionate aspects of consumption and goes beyond high involvement and hedonic, aesthetic, or symbolic consumption (Belk et al. 2003). Desire originates from the Latin word desiderium, meaning a "grief for the absence or loss of a person or thing," particularly a "yearning for the unreachable stars" (Belk et al. 2000, p. 100). This conceptualization highlights the need or lack of the desired experience as a key driver behind luxury consumption (Illouz 2009). A more positive view however could also be that the desire and motivation to reach the unreachable stars is an energy that fuels the consumer's creativity and innovation capabilities (Kozinets et al. 2016). This desire creates the motivational drive which is projected onto consumption objects and experiences (Belk et al. 2003). Desires are strongly emotional and can even be

---

[60]This chapter is based on the unpublished thesis of Gurzki, H. (2018). The Creation of the Extraordinary – Principles of luxury. Dissertation. Braunschweig: TU Braunschweig.

partly irrational (Belk et al. 2000). Already hundreds of years ago thinkers such as Augustine and Hobbes proposed that humans have an unlimited desire for sensual pleasures, wealth, and power, all of which luxury objects promise (Graeber 2011).

Thus, desire can be defined as "ever-changing, infinitely renewable wishes inflamed by imagination, fantasy, and a longing for transcendent pleasure" (Belk et al. 1996, p. 368). This conceptualization highlights several important aspects. First, desire is a dynamic process in which the object of desire (the luxury experience) can change. This means that potentially anything can become an object of desire, if the longing it creates is large enough for a large enough number of consumers. Yet any luxury can also lose its luxury status and specificity, which carries a danger to luxury brands (Berry 1994). Second, the process is infinite as it is driven by the consumer's innovation and there is no limit to it. Any luxury experience can always be surpassed by a higher level of refinement (Illouz 2009). Imagination is thus central to create desire (Belk et al. 2003). Often, fantasizing about the idealized experience can be as pleasurable, if not even more pleasurable than the actual experience (Belk et al. 1996, 2000; Illouz 2009). The experience is thereby through to promise pleasure, recognition, an escape from reality, or even profound bodily or spiritual transformations that cannot be found otherwise (Belk et al. 2003). The fantasizing and longing are longer and more intense the more difficult the object or experience is to access or possess, and as long as there is still hope that it can be reached (Belk et al. 2003). Thus, in the process of desire creation, it is crucial to keep the dream alive and preserve the symbolic distance between the experience and the consumer (Belk et al. 2003). This can also include the requirement of personal sacrifices such as time to find the best and most authentic places or financial sacrifices (Belk et al. 2003).

This creates a cycle of desire through imagination and fantasizing, longing, working toward the desired experience, frustration, and the sustaining of hope (Illouz 2009), which is essential in the pre-purchase phase for luxury goods. Recent empirical studies confirm that non-consumption can increase the desire for the object (Dai and Fishbach 2014). The process of desiring also creates pleasure in itself and can create physiological states similar to sexual arousals, such as increased heart rate and blood pressure (Belk 2001; Belk et al. 1996). Desire can thus be a source of happiness and the fear of being without desire can lead to negative feelings of emptiness and disappointment (Belk et al. 2003). The role of culture is crucial in this process, as the locations of desire and aspired fantasies often have a deeply rooted and shared cultural basis from which the collective imagination emerges (Belk et al. 2003; Illouz 2009). In a global consumer culture, tastes and preferences such as for food are shaped by the collective tastes of the aspired target groups and act as a marker of group membership (Graeber 2011). Particularly myths play a great role in shaping the desire of extraordinary experiences (Arnould and Price 1993). They provide narratives and stories that guide social life and help to make sense of the world (Stern 1995). Through myth, even contradictions and paradoxes of everyday life can be overcome, such as the myth of eternal youth or attraction, which provides potent culturally shared and often even unconscious blueprints for the creation of desire (Levy 1981). The existence of the luxury experience or object provides a reassurance that the hopes, ideals, and desires can become a reality (McCracken 1990). Particularly luxury as being out-of-reach acts as a powerful symbolic container for these aspired lifestyles, as it hardly faces the risk of empirical tests that might destroy the dream (McCracken 1990). Thus, maintaining and enacting the myth and thereby seducing luxury consumers has become a potent marketing strategy to create consumer desire (Kapferer and Valette-Florence 2016). Today, maybe more than ever before in an increasingly globalized, commodified, and rationalized world, the extraordinary lies in authentic, meaningful, and even magical experiences to which luxury promises an access (Llamas and Belk 2011; Tian and Belk 2006, p. 272). Some authors have even gone as far as to propose that a driving force of contemporary consumption is a fundamental human desire for miracles, the magical, mystical, and spectacle (Tian and Belk 2006). The myth and imagination of an extraordinary and transformative experience can thus become a powerful object and driver for consumer's "thoughts, energies, hopes, desires, and expenditures of time and money" (Belk and Costa 1998, p. 219). What might sound somewhat abstract in theory works well in practice: Research has shown that advertisements based on myths and the underlying cultural patterns can create more appealing and successful advertisements and experiences (Brown et al. 2013; Johar et al. 2001).

The luxury brand is of utmost importance as it embodies and nurtures the brand myth through symbolic narratives, a sense of authenticity, and an idealized community (Brown et al. 2003). The consumer's identification and active involvement with the brand myth increase desirability and brand liking (Holt 2004). Moreover, brands with a credible and strong identity and heritage have a higher cognitive, effective, and intentional brand strength (Wiedmann et al. 2011). Often the personality and story of the creator, the brands traditions and practices, the difficulty in accessing the brand, the history and authenticity of the brand, the quality and service excellence of the brand, stories about place and time or individual stories about user experiences provide starting points to shape the brand story and fuel the consumer imagination (Kapferer 2012; Napoli et al. 2014). Thereby, a sincere communication, commitment to quality and historical evidence, formal and informal classifications (such as awards), a strong link to the product's origin and place together with the use of historical and cultural referents of production methods, stylistic consistency, and the emphasis of non-commercial motives are helpful in rendering this communication authentic (Beverland 2005, 2006). Thereby different forms of authenticity can be distinguished (Leigh 2006): Objective authenticity that is created in reference and comparison to a specific place of origin. Subjective or constructive authenticity that is based on the experiences, dreams, or fantasies. Existential authenticity that is based on the consumer's motivation for pleasure and fun and be both intrapersonal (physical or psychological) or interpersonal in emphasizing a collective sense of the self (Alexander 2009; Leigh 2006).

Desire and the Service Consumption Process

Empirical studies comparing the differences in the different stages purchase process between luxury and non-luxury goods or services are sparse (Albrecht et al. 2013b; Gurzki and Woisetschläger 2017). Yet conceptually there are several pragmatic guidelines that help luxury managers. While many of the insights from the existing often product-focused literature can be transferred to luxury services, luxury experiences exhibit several characteristics that should briefly be mentioned (Wirtz and Lovelock 2018): For services, intangible elements dominate the value creation, even more so for luxury services. Moreover, the service quality varies widely as services are hardly standardized across providers (heterogeneity) and often require the presence and even active involvement of the consumer (inseparability of production and consumption). Moreover, services cannot be stored (perishability) (Wirtz and Lovelock 2018). These service characteristics also influence the marketing mix: Besides product, price, placement, and promotion, the dimensions of people, processes, and physical environment receive a higher importance. This has implications on the consumer decision marking, the service consumption process, and its management which will be outlined in this section with a focus on the service consumption process. The service consumption process has three stages: the pre-purchase stage, the service encounter stage, and the post-encounter stage (Lemon and Verhoef 2016; Wirtz and Lovelock 2018).

Pre-purchase Stage

The pre-purchase stage typically is concerned with the consumer becoming aware of the need (problem recognition), information search, the evaluation of alternatives, and the purchase decision. Service research distinguishes between need recognition, meaning a strive to restore the original state before the need arose and opportunity recognition, meaning the opportunity to achieve an ideal state (Wirtz and Lovelock 2018). For luxury brands, both types of need recognition play a role, although opportunity recognition seems to be more present. One example of need recognition would be compensatory consumption of luxury objects through which individuals try to restore feelings of powerlessness (e.g., Rucker and Galinsky 2008). Opportunity recognition in a broader sense could be seen as consumption related to the enjoyment, pleasure, or even self-transformation (e.g., Llamas and Thomsen 2016). Whereas for functionally oriented products, the need recognition is usually quite clear (such as hunger), the need recognition for luxury experiences is more complex for different reasons. First, typically luxury caters to higher level needs, which can be satisfied in many different ways (see previous chapter). Second, dimensions needs are typically not independent for luxury objects, as luxury objects are able to satisfy a broad range of needs at the same time. For example, a Michelin-star dinner consisting of different stages is likely to satisfy the entirety of needs (although in a refined way) from basic physiological needs (such as the intake of nutritious and delicious food), safety needs (such as the feeling of a safe environment or food safety), social needs (such as the possibility to new similar-minded food enthusiasts), esteem (such as through learning about new ways of preparing food or discovering new ingredients) or even self-actualization (such as feelings of morality through the consumption of sustainably produced local food). Second, these needs are influenced by many factors such as the social environment, cultural values, or even the subconscious. For luxury, in particular, we propose that the subconscious plays a large role and is strongly influenced by the cultural myths we carry around. Again, fine dining could serve an example here. The types of food we eat are strongly influenced by cultural beliefs. Sometimes fine-dining plays with these boundaries and tries to stretch them. Such cases are L'Arpège with chef Alain Passard or Plaza Athénée of Chef Alain Ducasse who have challenged the notion that meat is a necessary ingredient of French cuisine by changing to vegetarian cooking.[61] Third, the needs are dynamic and change over time. For example, throughout a single luxury trip, there are all different types of needs depending on the situation, daytime, and so forth. Thus, to play a role in the decision-making process, the needs have to be salient to the consumer. We propose that the concept of desire as briefly introduced above provides a useful conception to simplify these complex and dynamic processes and is particularly helpful in understanding the role that myths and cultural narrative play in shaping subconscious drives.

In the pre-purchase stage, inspiration (such as through storytelling and visual imagery) is key to fuel the consumer's desire (Böttger et al. 2017). Once the consumer recognizes the need or desire, he or she starts to search for information about how to satisfy it. For luxury, the consumer is looking for excellence and perfection and given its high price, likely to engage in extensive problem solving (Wirtz and Lovelock 2018). Yet, one of the challenges consumers face particularly for luxury consumption decisions, that they have a high share of experience and credence attributes due to their highly experiential and symbolic nature. Particularly the evaluation of the symbolic properties requires high consumer expertise and literacy (Bengtsson and Firat 2006). Given the complexity and price of the decision and the fact that services cannot be returned, perceived risk could be expected to be higher. However, research has found, that the perceived risk for luxury brands compared to other brands in their category is lower as their strong reputation and their ability to generate positive emotions dominates (Chaudhuri 1998; Yu et al. 2018).

Typically, in the information search phase, the consumer also identifies suitable solutions, compares them, assesses their fit to the needs, and forms initial service expectations. As luxury is able to satisfy particularly higher-level and more abstract needs, this can lead to a broad choice set. For example, a motive such as "I want to reward myself" could lead to the choice of a luxury vacation, spa retreat, or dinner, but also extend to renting or even purchasing a luxury car, buying a handbag, or even going outside of the world of consumption such as spending time with the family or

---

[61]https://www.vogue.com/article/best-vegetarian-plant-based-vegan-restaurants-paris

exploring nature. Looking at the range of service alternatives within the choice set, factors such as explicit (such as advertising) and implicit (such as price) service promises, word-of-mouth, past experiences, personal needs, and service philosophies impact the desired service level (Zeithaml et al. 1993). Based on situational factors, the own perceived role in the service encounter, the perceived service alternatives, and the predicted service, consumers derive what they consider as an adequate service (Zeithaml et al. 1993). The difference between the desired and an adequate service is their zone of tolerance (Zeithaml et al. 1993). And this zone of tolerance can be expected to be rather narrow for luxury experiences. Thus, luxury marketers need to consider that image they shape from the first point of contact, as this determines how the customer evaluates the service experience. Lastly, the purchase decision and reservation are made.

While we propose that these phases with the above-mentioned specifications work for luxury, we believe there is more to it. Desire is fueled by imagination that builds the motivational drive of the consumer. If a luxury experience is perceived as extraordinary, unique, and authentic, there is little choice for the consumer other than to fulfill the desire and find ways to rationalize the decision afterward. While there is so far little empirical research on this topic, we believe this is a fruitful avenue for academic researchers in order to provide a deeper understanding of the mechanisms behind what is long common practice in the industry (Kapferer and Bastien 2009). And there is already some support from cognitive psychology. For example, Hansen and Wänke (2011) have shown that luxury brands have a more abstract and higher-level mental construal. This means that they are described with more abstract language by both advertisers and consumers. In higher-level construal, information is represented in a more schematic, decontextualized, and abstract manner ("such as feeling special"), whereas in low-level construal information is often unstructured, contextualized and concrete (such as "eating a sandwich") (Liberman and Trope 2008). What might sound like an unimportant linguistic detail has profound implications on decision-making. For example, for complex decisions, such as the choice of a vacation that relies on the judgment of multiple attributes, a higher distance and abstract construal leads to a stronger focus on desirability as compared to feasibility characteristics (see also Fiedler 2007). These construal levels are influenced by perceived psychological distance such as social (aspirational consumer group), temporal (link to traditions and past), hypothetical (link to fiction, stories and myths surrounding the destination), or spatial (link to exotic or other-worldly destinations) distance of luxury brands (Trope and Liberman 2000; Trope et al. 2007). In a choice situation, an option that is perceived as distant that has different desirable characteristics has great chances of being chosen. Emotions also affect construal levels and have a cognitive basis that influences consumer decision-making. Hence, the importance of seduction and storytelling rather than pure selling for luxury brands (Kapferer and Valette-Florence 2016). And more tactically, how to narrate them across the brand owned touchpoints from media (e.g., websites, social media, advertising, loyalty programs) to the service marketing mix (e.g., price, product attributes, service, sales force), partner-owned touchpoints such as distribution partners, and even customer-owned touchpoints such as payment methods, or social/external touchpoints such as other customers or third-party information sources (such as Tripadvisor) (Lemon and Verhoef 2016).

Service Encounter Stage

The service encounter lies at the heart of every service experience. Particularly in luxury, service experiences can be characterized by a high degree of personalization and human interaction. While traditional models for the service encounter stage emphasize the customer need for behavioral, decisional, and cognitive control during service experience, luxury goes beyond this (Wirtz and Lovelock 2018). The consumer needs to play an active part in the experience—up to the point to be completely immersed by the experience. Lemon and Verhoef (2016, p. 71) define customer experience as "a multidimensional construct focusing on a customer's cognitive, emotional, behavioral, sensorial, and social responses to a firm's offerings during the customer's entire purchase journey." Brakus et al. (2009) suggest that there are four different types of experience: sensory (senses), affective (feelings, emotions), behavioral (bodily experiences), and intellectual (curiosity, thinking). Schmitt (1999) proposes a different typology based on five types of experiences: sensory (sense), affective (feel), cognitive (think), physical (act), and social-identity (relate). A great experience has been shown to influence satisfaction and loyalty directly and indirectly by strengthening the brand personality associations (Brakus et al. 2009).

We propose that extraordinary luxury experiences can be conceptualized as ways to experience the magical, spiritual, or even sacred (Dion and Arnould 2011; Dion and Borraz 2015). Through the processes of what the consumer behavior literature calls sacralization, luxury objects and experiences, places, times, tangibles, intangibles can be endowed with intangible meaning such as a magical aura (Belk 1991b; Dion and Arnould 2011). They can express a relationship with time such as a sense of the past and nostalgia (Belk 1991b), life and death (Roux and Lipovetsky 2003), or become vehicles for transcendent experiences (Belk et al. 1989). Belk et al. (1989) describe several characteristics of what they term sacredness, that have a high relevance to luxury experiences and objects: They are a quintessence in that they are exactly what they ought to be, have an effective presence, and are authentic, beyond commerce. The sacred or extraordinary is the opposition to the profane and ordinary of everyday life. For some extraordinary experience, something unique, different or supernatural can reveal itself. Similarly, some experiences have ambivalent attitudes, such as strong approach and avoidance tendencies at the same time. These experiences require sacrifice (such as time or money invested) and commitment to build an intense emotional connection. Through the performed rituals artifacts become endowed with the magical powers. The experiences are surrounded by myths and mystery, thereby often beyond rational understanding. The different elements of the experience are objectified and subsumed under a higher transcendental meaning and frame of reference. Often the experiences involve the feeling of communitas, which is a transcendence of normal social roles and structures. As a result, they lead to ecstasy and flow, and a transcendence of life and matter (see Belk et al. 1989).

The special status of the experience or luxury objects is sustained through enacting the myth and the performance of rituals (Belk et al. 1989). Rituals thereby help to render the experiences meaningful, charge them with symbolic values, and thereby establish a social communication system (Lury 2011; McCracken 1986). Of particular importance for luxury experiences is that they remain extraordinary, which requires certain rituals to separate them from the profane and ordinary of everyday life (Belk et al. 1989). Stern (1995, p. 165) proposed that rituals are "the repetitive sequence of behaviors referenced in mythic narratives." Rook (1985, p. 252) defined ritual as "a type of expressive, symbolic activity constructed of multiple behaviors that occur in a fixed, episodic sequence and that tend to be repeated over time."

Previous researchers have used the metaphor of the theater to describe service encounters with consumers and company employees performing and staging their roles and scripts (Wirtz and Lovelock 2018). Rituals include ritual performance roles, ritual artifacts such as consumer products, ritual scripts on the use of the artifacts, and a ritual audience that goes beyond the participants in the ritualistic performance (Rook 1985). While some rituals and social exchanges are extensively scripted, others provide less guidance and more flexibility in their performance (Rook 1985). Rituals can be categorized into five main areas based on their main meaning and source of behavior (Rook 1985): Cosmology with religious (such as meditation), magic (such as gambling), and esthetic rituals (such as arts performances); cultural values with cultural rituals (such as festivities, e.g., Valentine's day) or rituals marking the rites of passage (such as graduation or marriage); group learning rituals with civic (such as elections), small group (such as business negotiations), or family (such as birthday) rituals; individual and personal rituals (such as grooming rituals); and biological rituals which include basic interactional rituals (such as greeting and mating)

Regarding consumption, the main types of individual and social rituals performed are related to the exchange, possession, grooming, and divestment of the consumption object (McCracken 1986). Exchanges such as the purchase or gifting are often ritualized (McCracken 1986). In purchase situations, money can acquire a sacred meaning depending on the source of the money and the nature of the exchange and thereby distance the exchange from the ordinary sphere of commerce (Belk and Wallendorf 1990). Particularly in luxury exchanges, it is important to separate the artistic and commercial worlds to preserve the extraordinary and magical aura of the luxury product and experience by separating it from the profane and rational world of money (Belk and Wallendorf 1990). This separation of the artistic and commercial world becomes particularly obvious in gifting rituals in which money is transformed into a means to develop a social connection (Belk and Wallendorf 1990). This symbolic relationship between the gift giver and gift receiver, which proceeds in several stages (Sherry et al. 1993). In the gestation stage, the gift giver selects and appropriate gift with the desirable symbolic properties wished to be transferred to the gift receiver. In the presentation stage, the gift is presented to the receiver. In the reformulation stage, the receiver evaluates the gift and updates the relationship to the gift-giver, which also dictates the conditions for future gift exchanges (Sherry et al. 1993; Sherry Jr 1983). The gifting process thus has economic (such as the value of the gift and its economic and symbolic properties), personal (such as the influence of one's self-concept on the choice of the gift), and social dimensions (such as the relationship between the gift giver and receiver) (Sherry Jr 1983). For example, social norms prescribe that gifts can hardly be denied or thrown away and need to be reciprocated with a similar value, which can even lead to competitive altruism in which gift-givers try to outperform each other (Boone 1998; Giesler 2006; Mauss 2002; Sherry Jr 1983). For luxury brands, self-gifts also play an important role particularly in fostering self-esteem, justifying indulgences, or allowing oneself to discovery, escapism, or perfectionism (Kauppinen-Räisänen et al. 2014; Mick and DeMoss 1990). Self-gifts also promise an experience in itself and are not limited to the traditional purchasing situation, but occur throughout the consumer experience (Kauppinen-Räisänen et al. 2014).

Possession rituals are performed to claim the symbolic ownership, for example through personalization (McCracken 1986). The symbolic transfer of meaning needs to be continually renewed through grooming rituals, such as the protection and cleaning of the luxury object to preserve it (McCracken 1986). This also extends to ensuring the appropriateness of the object to perform social roles through numinous elements (related to transformation and even magical experiences), judicious elements (related to the appropriateness and its evaluations as right or wrong), dramatic elements (related to playfulness), formal elements (the performance to certain standards), and ideological elements (the formation of identity) (Rook and Levy 1983). Lastly, divestment rituals that mark the symbolic departure from the object and owner, which can include removing traces of the owner to symbolically purify the object for resale (McCracken 1986, p. 80). These rituals highlight the importance of objects in identity formation. They are used to perform certain social roles and their loss can even lead to identity confusion (McCracken 1986). While these rituals are predominantly relevant to luxury objects, they also extend to the artifacts used in the ritual performances during service encounters. Other rituals with high relevance to consumption are collecting (combination singularizes whole), inheritance (singularizing through age), and external sanction (sanction by external authority) (Belk et al. 1989). Moreover, previous research has highlighted the role of rituals as heuristics in decision-making (Dion and Borraz 2015; Rook 1985). For example, Sherry et al. (1993, p. 237) proposed that "ritual behavior generally serves a positive function as an automatic decision maker, and traditional ritual helps to give meaning to contemporary personal and social life."

Post-encounter Stage

In the post-encounter stage, the experiences are evaluated against expectations (Wirtz and Lovelock 2018). The challenge for luxury experiences: Consumer expectations could not be higher! Customers expect a flawless delivery of the basic needs, the highest standard of quality for performance needs, and an exceptional performance on the delighters which are the essence of luxury (based on the model of Kano 1984). A positive service experience is created if the experience exceeds the expectations and in turn influences behavioral outcomes such as word-of-mouth or loyalty (Anderson and Sullivan 1993; Oliver 1980, 1993; Wirtz and Lovelock 2018). Particularly for luxury brands the zone of tolerance is low and exceeding the demanding expectations is hard (Zeithaml et al. 1993). In their SERVQUAL framework, Parasuraman et al. (1988) propose that service quality has five main dimensions (Wirtz and Lovelock 2018): Tangibles such as the physical environment, facilities, equipment, appearance of personnel; Reliability such as the ability to perform service dependably and accurately; Responsiveness such as the willingness to help customers and provide prompt service); Assurance such as the credibility ad trustworthiness of the service provider, the competence of the service provider and its employees, the courtesy, respect, and friendliness of the personnel as well as the feeling of security; and Empathy such as access and the approachability of the service provider, communication in an active and understandable form as well as understanding the customer including caring and the individualized attention the firm provides to its customers

These dimensions have been shown to be important drivers of customer satisfaction for tourism experiences (Lai and Hitchcock 2017). For the context of luxury hotels, Lai and Hitchcock (2016) proposed to add environment (for example, security, comfort, variety of services offered), technology (presence of in-room and hotel technologies), and entertainment (such as tourist attractions, recreation facilities or shopping centers) to the list of criteria. Using a different approach, Kim et al. (2011) proposed that hedonism (such as indulgences or excitement), novelty (uniqueness of experience), local culture, refreshment (sense of freedom), meaningfulness, involvement, and knowledge gained are important factors by which consumers judge tourism experiences. Moreover, consumer characteristics play an important role. For example, research has shown that consumers with a strong desire for exclusivity evaluate experiences more favorably compared to consumers with weak desire (Kim 2018).

Rituals also play an important role in the post-encounter stage. Now the expectations and dreams have been put to the empirical test of reality. And in the infinite cycle of desire, once the object of desire is attained, the desire must be reformulated and the object re-enchanted or a new location of desire must be found, either in a new object or the displacement to the object's entirety (Belk et al. 2003). Fascinatingly, even if the luxury experience fails to meet the standards of the imagined myth, the myth itself is often kept alive and transferred to the next luxury object, experience or their collection (McCracken 1990). The cycle of luxury needs to be fueled by further refinements and innovation to add something on top. This cycle services to enlarge and refine the individual's taste and prevents a sufficiency (McCracken 1990). Through the collection of things or experiences the dream can be maintained (Belk 1995a, b). Collecting itself has become seen as a form of luxury consumption as the entirety of objects and experiences and their completeness make the collection extraordinary and set it apart from the ordinary world (Belk 1995a, b). This also goes for souvenirs as mnemonic devices and materializations of the experience that have been infused with the magic and provide a reminder and good marketing strategy for luxury experience managers (Fernandez and Lastovicka 2011).

Discussion and Implications for Luxury Tourism

These observations have several implications for luxury tourism marketers and managers. Luxury players need to ask themselves what makes their experience extraordinary and what creates desire for their guests. This also means decoding the symbolic myths and rituals that shape the experience and enriching them with the luxury brand. As the creation of desire and the extraordinariness of the experience are crucial throughout the entire consumption journey, companies need to deeply understand the experiential world of their consumers and build an experiential platform to cater to their needs and desires (Schmitt 2003). To realize this, continuous innovation is essential (Schmitt 2003). One example from outside the world of luxury comes from a company that has very well understood the role of dreams and desires, such as playfulness and fun: Disney. Guests can personalize their Disney experience online before visiting Walt Disney World and manage their entire vacation, including the booking of fast passes, dining reservations, and entertainment times (Abtan et al. 2016).[62] Using the technology around their "MagicBands" wristband, guests can get access to special interactive features or will be greeted personally when they dine in a restaurant as the staff knows in advance who will be arriving and what their likely preferences are.[63] This moves the delivery of a seamless service experience to the next level, even including predictive models and anticipating guest needs before they might be aware of them. Why this experiential innovation might raise privacy concerns among more critical travelers, it illustrates the importance of innovation and pushing to be at the forefront of consumer trends to "wow" consumers. While this does not suggest that luxury operators should follow this blueprint blindly, it shows how innovation can be done. What has always worked for luxury, the creation of desire, works even better in a digital world. While in the offline age, marketers were predominantly occupied reacting to consumer journeys. Now in a digital world, they have new means to guide consumers and thereby manage the entire customer journey through automation, proactive personalization, contextual interaction, journey innovation (Edelman and Singer 2015). While this might not be a cheap endeavor (for example, Disney has invested more than $1 billion and employs a team of more than 1000 people to manage their customer experience), it shows what is required to remain competitive in the field of providing magical experiences. Without making this effort to invest in designing the extraordinary experiences of the future, even luxury brands (maybe with few exceptions for the ultra-high-end), risk losing their aura and will be facing hard times (Abtan et al. 2016). But in the best case, they can create a loyalty loop in

---

[62]https://disneyworld.disney.go.com/en-eu/plan/

[63]https://www.wired.com/2015/03/disney-magicband/

which the consumers enjoy, the experience, advocate it to others, and bond with the brand (Edelman and Singer 2015).

4   Luxury Consumption as a Process of Overcoming Fear

When Overcoming Fear Provokes a False Understanding of Luxury

Marc Aeberhard

> The world has enough for everyone's needs, but not everyone's *greed.*
> Mahatma Gandhi

"Fear not!" (Isaiah 41:10)—this brief invitation, which could not be more universal and which reverberates equally across all continents, cultures, ways of thinking, or religions back to the earliest beginnings of mankind, is equally the programmatic answer to the ultimate driving force of all being: the overcoming of fear. Reasoned by the existential threat to life and limb in an inhospitable system, to be eaten by saber-toothed tigers, trampled by mammoths, or swallowed by natural forces. Even today, this instinct slumbering deep within us is still as topical and alive as it was tens of thousands of years ago. Only that the saber-toothed tiger of that time gave way to today's power-hungry bosses, unscrupulous investors, treacherous terrorists, or headline-hungry paparazzi.

The question underlying this book about the definition of luxury and its manifestations thus does not begin with a succinct display of material trophies such as expensive watches, fast sports cars, nobly equipped yachts, pompous estates, or glittering jewels, but rather with a philosophical and developmental sociological consideration. It is a journey of the mind which, like a spiral, begins with the threatening nature and ends in the threatened nature.

Hunger as a Driving Force for Storage

With the settlement of mankind and the transition from hunters and gatherers to simple horticultural societies thousands of years ago, people no longer had to travel from herd to herd and from pastureland to pastureland with all their possessions but could begin to store and accumulate goods at a stationary place (cf. Lenski 1977, p. 164 et seq.). Whether it is a coincidence or it is in the primordial nature of man to want to own more than he could carry with him—originally in accordance with the circumstances—is difficult to determine. The fact remains, however, that in geographical/climatic zones, which were not characterized by abundance but by deprivation, storage was used for the first time and later perfected. The result was an economic surplus (Cf. Lenski 1977, p. 75 ff.). The foundation for economic stock management was laid and the need to search for food on a daily basis was thus abolished. In addition to the economic surplus, another dimension emerged: free time. Time that was not implicitly and explicitly needed to make a living. So, the restless striving of mankind demanded an occupation in this free time: It was the birth of art, culture, religion, but also power and celebration of power (cf. Lenski 1977, p. 62 et seq.). As a result of the specialization of skills in the craft and intellectual fields, but also in the development of personality profiles (charisma), the first stratifications were formed within communities, which over the coming millennia developed into social structures. The social diversity experienced today is the result of a complex historical, cultural, economic, geographical, climatic, technical, topographical interdependence and development that has been (and still is) shaped by disputes within and outside individual communities (cf. Lenski 1977, p. 25 ff.). But the goal of all human communities is the same: the maximization of well-being, however differently it may be defined. Maximize well-being and not only legitimize it in a social context, but also create so much incentive that imitation is stimulated. Power and privilege thus become the driving force of a unique dynamization in humanity. After "wanting-to-have" and "wanting-to-have-more" it is the birth of "wanting-to-have-it-all" (Cf. Epicurus 1991, p. 2). And thus inevitably the creation of unequal distribution, which leads to envy. This primeval instinct of "wanting to have and not losing it again," which resulted from the primeval fear of losing one's prey to wild animals or other tribes, has remained deep within us even today. Even today, we are afraid of losing, betting, speculating, or getting our belongings stolen. And so it is not surprising that over thousands of years man became exceedingly creative to protect his belongings, all this with only one goal in mind, the "creation of security" (cf. Lenski 1977, p. 308), the overcoming of the fear of having to hunt/collect to make a living without any means, fearing for survival. But security is relative, by no means an absolute quantity, and as long as human beings continue to hog possessions in ever more comprehensive scenarios, the greater the danger and consequently the fear of losing them again. The (subjective) need for security is increasing more and more, especially in today's societies. Modern man feels threatened in many ways—consciously or unconsciously, justified or exaggerated. This includes:

– The threat to life and limb
– The threat of losing property through robbery and theft
– The threat of losing privacy through data leaks and violations of personal rights, and thus becoming blackmailable
– The threat of losing one's livelihood and therefore one's health through poisoning as a result of environmental sins of all kinds
– etc.

It is therefore not surprising that security has become a very important pillar of the new understanding of luxury and will become even more important in the future. This includes not only hiring bodyguards or security companies, but also and especially the preservation of social structures (well-functioning states, but then also respect, decency, politeness, etiquette, and style) and natural resources (healthy soils, clean water, fresh air, intact nature, etc.).

Renunciation in Response to Fear of Loss

All those life plans and concepts that see themselves as a complete anti-program of what has just been said are therefore exciting. Reference is made here to asceticism, as well as to eremitic forms of life or the founding, development, and management of monastic structures and philosophies (Cf. Berg 1977, p. 15 ff.). But they remain the exceptions, despite the fact that in the current epoch they are of remarkably growing interest in Western societies, which for some decades have been characterized by the overcoming of hunger (and thus belong to the post-material societies). Is it the search for a higher experience? Is it the overtaxing with the complexity of today's societies or is it the decoupling from one's own primeval human roots, which were/are shaped by the primeval fear of being eaten and starving, or is it—exactly the opposite—the potentiated fear and thus the search for the (ultimate) security? Or is it simply the desire to break out of a society that is characterized by an ever-increasing pace (cf. Kersting 2005, p. 67)—and thus exposed to an ever-faster material rapture—and find its own pace of life? Deceleration becomes the ultimate buzzword, but also the maxim of life.

Fear of Social Isolation

But man is and remains in his (primordial) being a herd animal (cf. Weinschenk 2009, p. 11 ff.) and needs social interaction and integration. This integration takes place according to socialized, traditional patterns, that begins in early childhood and develops deep into adulthood. Belonging, being appreciated, and loved in its most intimate form are existential needs that are indispensable for the healthy being of man (Cf. Maslow 1943, p. 370 ff.). But also in the emotional world, in the perception, "am I accepted or not," there is always the fear of rejection, the fear of being expelled from the pack, of not satisfying the boss, of being insufficient in the team, or of not being loved by the other person (see Katz and Kahn 1966, p. 526 f.).

The Two Primeval Fears: Being Eaten and Not Being Loved

These two deep-seated primeval fears of being eaten and not being loved are the most important stimulants for performance. John Calvin's church doctrine has made use of these fears just as much as the first capitalists and they continue to swing in perverted form in reality TV shows like "The Apprentice" until today: What do I have to do to earn God's grace? The theory of predestination secularized at the beginning of the nineteenth century in the course of the incipient industrialization and served as an argument that happiness and grace could only be achieved by maximizing economic creativity (Cf. Weber 1904, p. 17 f., 75 ff.). Or—put in modern terms—how do I make my boss happy? Consequently, the luxury discussion has a very special connotation under this aspect when suddenly the overcoming of primeval fear is: It does not matter whether I am liked. Because I am who I am. The credo of hedonism (cf. Bachmann 2013, p. 35 ff.) as an answer to fear and thus a recipe for luxury?

But all these motives have one thing in common: the deep, deep desire to overcome an economic, mental, social, political, etc. provoked fear. Overcoming fear becomes a remarkably current need, super need, luxury need, luxury factor.

Overcoming Fear: A Questionable Luxury

But woe betide the system that extrapolates hedonism and egocentric fulfillment of needs as an attitude to an entire society. It would be the unbridled mass self-realization of entire societies. Regulative and normative orders would be leveraged and social structures would disintegrate into chaos. Here the limits of individualization have been reached. With increasing social density—above all in urban areas and megacities—the personally anticipated desire for development, reproduction of the workforce (cf. Ulich 1991, p. 305 ff.) and the individually manifested living out of all those needs suppressed by socialization and social etiquette become an important counter-experience in a restricted everyday life. In short: leisure time development becomes an oasis in which non-conformity and accumulated frustration potential can be lived out (like survival camps, extreme sports that reach the limit of life-threatening (extreme climbing, wingsuit flights, apnea diving), paintball tournaments, etc.). In these examples, fear is overcome by adrenaline stimulation (cf. Wolf 2018, p. 112). Whereby the basic psychological question lies in the motive of such activity: Is the overcome fear experienced as catharsis and thus as happiness (Tolle p. 95 f, p. 114 ff.) or is it the need to experience a "Back-to-the-instincts-roots"-experience in a sensually blunted world characterized by control and domination?

No matter how the question is turned around, one answer always results from the developments described and the exemplary behavior of people in post-material societies: Basic instincts stimulation in a deserted world, secularized societies, demystified cultures and disrespectful encounters with nature should suggest feelings of happiness and thus complete satisfaction as a result of overcome fear. Marketing specialists and product developers define the outstanding significance of (even faster, higher, wilder, crazier, more dangerous) experiences and lust (Epicurus 1991, p. 1 ff.) as the new dimension of a modern understanding of luxury in the last refuges of the planet.

In Humility Lies the Soul of True Luxury

Indeed, the striking demand "Make the world your subject" seems to have been reached in such actions, and yes, all fears seem to have been overcome and Rousseau's motto "Back to Nature" has been fulfilled (cf. Stangroom and Garvey 2006, p. 72 ff.). However, respect for creation remains forgotten. The unbalanced trampling around in highly complex ecosystems has completely distorted the signs of centuries of human evolution: it is no longer man who must be afraid of the saber-tooth tiger, but today it is the tiger who must be afraid of man.

Fear, oh man! Fear yourself! For true luxury lies only in knowledge, insight (Epicurus 1991, p. 2), respect, and humility.

5   Luxury Consumption as a Healing Process

Luxury Is Unavailable!

Stephan Hagenow

"Dislocations" as a Starting Point for a March Stop: A New Definition of Luxury

If it is true that luxury tourism has entered a phase of global saturation, in which offers have become interchangeable and comparable, and luxury seekers are predominantly looking for the immaterial, then it is probably time for a march stop. The desire to outdo each other leads to an inner emptiness in an artificial world for time. This can be neither healing nor health-promoting—these two points are extremely important to most people. Let us remember the designation of origin of the word luxury: it comes from Latin and originally means "dislocated" in the sense of deviation from the norm. Luxations are painful, require professional help, and are associated with suffering and patience. In common usage, luxury is often understood to mean wastefulness, unnecessary enjoyment, or exaggerated pleasure. In this chapter, however, I would like to use a concept of luxury that is closer to origin. *Luxury is the process of healing individual, biographical and cultural injuries to body and soul.*

Taking this definition, it becomes clear that each healing process is individual. The sufferer wants to be cared for, he or she wants to be perceived in his or her wholeness and not reduced to his or her dislocated limbs.

Before one thinks up any more antics, technical refinements, or other "contortions" in order to stand out among the mass of providers, it probably needs a fundamental rethinking of what one actually wants or what can actually be achieved. Everyone agrees that the feeling of luxury is highly individual and can hardly be unified. For some, it is a good book, for others a noble ambience, for others an adventure trip through unknown areas, or a self-retreat and the focusing on oneself.

In my opinion, the different perception of luxury is mainly due to the fact that all people have different coping strategies in their different "contortions." If luxury only serves to forget the contortions for a moment, this is not sustainable. When people are forced to get too far out of their comfort zone, they are afraid. When people are suddenly left on their own with a lot of space without social interaction, they become lonely. If one's own dislocations and possibilities are neglected, and if adventurous luxury offers are to provide a kick never before achieved, then in the medium term this is a depletion of body and soul. If luxury becomes a promise of salvation that cannot be kept, aggression can even occur. The quintessence is therefore: Only those who take the trouble to perceive the individual dislocations and the associated coping strategies of their customers will receive satisfied feedback.

People Are Looking for the "Unavailable"

And with the keyword "unavailability" we are right in the middle of the core of all religions. All religions know that true perfection and complete wholeness are eschatological desires, that is, they are a foretaste of Paradise. We can make a difference by eating healthy, by doing business sustainably to secure our livelihoods in the long term, by working fairly and not enjoying our luxury at the expense of others. But perfection only exists as an appetizer—not as a cheap consolation, mind you, but as a foretaste. And here lies the theological direct connection to the already mentioned "contortions": Every human being is vulnerable in his more or less pronounced self-confidence, carries injuries and scars around with him, lives with great joys and inner abysses, has to deal with his family of origin, his traditions and the culture in which he grew up, has to deal with resistance against other "Is," has to learn to deal with successes and disappointments. It is, therefore, no coincidence that the Bible, and in its wake the Judeo-Christian tradition, largely renounces to imagine Paradise. Paradise is a concept of longing that creation and with it man is as perfect as he was originally intended. There are mostly collective categories, justice will be served, creation will live in harmony with each other, there will be no more violence and no more death. In this life and in this world, there will not and cannot be an ideal world. Anyone in the luxury segment who tries to convince us of the opposite and promises heaven on earth will fail. Moreover, people will be disappointed and even further away from enduring their own contortions. It does not matter how many hotels and resorts are using the name "Paradise," they can only intensify the longing, but not fulfill it.

Luxury as a Foretaste of Paradise

But religion is a killjoy—even if all religions have a healthy skepticism against any kind of wastefulness, hedonism, and wealth. The Old Testament prophets and also

Jesus address an unmistakable message of judgment to the rich who live at the expense of others and do not share. They will have as much trouble as a camel who has to go through the eye of a needle. But most religions also know how to enjoy non-everyday things, even Jesus is accused by his opponents of being a friend of "gluttons and wine drinkers" and his first miracle in John's Gospel is to turn water into wine. And Jesus enjoys being anointed and touched by a woman with precious anointing oil, an outrageous luxury at that time.

All religions want to be coping strategies to deal with their own contortions and also to make clear that being human always happens in a larger context than we can grasp with our modest possibilities. And especially Christianity relieves man of the compulsion to want or have to establish salvation himself—it does not work at all. Salvation and healing are always grace, a gift, and therefore unavailable. And Protestantism attaches particular importance to the fact that every human being is perceived in his ambivalence, "as righteous and sinner/dislocated." No one is just righteous or just dislocated. The Bible actually only describes salvation to the effect that each and every one must answer with his or her individual history before God and mankind and that God's grace will be comprehensive. True luxury in the sense of healing cannot be earned either, the most ostentatious Trump Tower is of no use, it remains transient. Even the super-rich cannot afford true luxury—because they cannot heal themselves.

What does this basic idea mean in the context of this book? A false conclusion as tourism manager would now be to put one's hands in one's lap and leave paradise to God. Most people have now realized that genuine fulfillment is linked to immaterial values and only becomes fruitful in the social context. Successful luxury in the sense of experiences of freedom from one's own contortions will be the ambience in which a foretaste of paradisiacal conditions can be tasted. But just with foretaste, an amuse-gueule, not mixed with the big ladle, individually made with devotion and refinement and served with personal attention. But even here the border must always be clear. Of course, a person in the hotel business needs a high degree of social competence and empathy, but it is always a temporary relationship marked by professionalism. If you fall into the trap of getting too close to the guest, you will either lose yourself—because you can never get enough attention—or you will inevitably be disappointed, because it is not a real relationship, but a paid and temporary one.

Luxury as Support in the Search for Meaning

If the luxury segment has recognized these limits and has attuned itself to the unavailability of true luxury, then there are certainly new possibilities. Luxury then means that there can be a framework which is given to the guest, in which he feels safe and protected and in which he has the leisure to deal with himself and his possibilities. The latter is not taken from you by anyone—except God himself. In Psalm 139 a desperate worshipper searches for the meaning of life: "Explore me, O God, and know the real me! Dig deeply and discover who I am! Put me to the test and watch how I handle the strain. Examine me to see if there is an evil bone in me, and guide me down Your path forever." The praying person notices that he is overwhelmed with the search for meaning. He seeks stability in the unavailable, which for him is God. Only in the interaction with God does the praying person feel that he or she is with himself or herself—with all joy, pride, and self-confidence, but also with all despair and abysses. A luxury resort must therefore offer a room or rooms in which opportunities for retreat are opened up. It must be a protected and, above all, a public space, a space in which people of all shapes and sizes feel comfortable, a space of silence and encounter—and which has only this purpose and is not used for animation and fitness in between or constantly plays music in the background. The hotel suite or bungalow on the island does not fulfill this purpose. It must be a space where stories are told and heard, where people are open to the unavailable, where people can complain about their contortions. Where people can learn again to take responsibility for themselves and those entrusted to them (relatives, companies, employees) by giving a piece of responsibility to the unavailable. This experience, this pausing, can be incredibly enriching, but sometimes it is not so pleasant to meet and endure oneself. No wonder, then, that the one who is praying in Psalm 139, 5 ff also has flight thoughts: "Where can I go from your Spirit? Where can I flee from your presence? If I go up to the heavens, you are there; if I make my bed in the depths, you are there. If I rise on the wings of the dawn, if I settle on the far side of the sea, even there your hand will guide me, your right hand will hold me fast. You hem me in behind and before, and you lay your hand upon me. Such knowledge is too wonderful for me, too lofty for me to attain."

Translated into our theme: Even the most creative leisure activities, the most intensive wellness treatments or enchanted places of this world do not by themselves allow experiences of healing. Interaction is necessary, space for encounters with oneself and the unavailable.

It would be really innovative to have trained people on staff that are unobtrusive, under the obligation of secrecy, and are not paid by results. Professional companions could show the dislocated wrong ways, open up new ways, or—in the words of the 139th Psalm—open up a perspective for the future. Such companions can be, but do not have to be, clergymen. However, it is essential that they are trained and that they are able to perceive and meet people professionally. These encounters will be more lasting than if the service employee knows whether the guest would rather have the breakfast egg cooked for 3 or 6 minutes. The professional care must be right in a luxury area, it must pamper body and soul—but it must be purposeful and must not become an end in itself. Otherwise, it just becomes bling-bling, hollow, transient, and frustrating. By the way, many churches have recognized this a long time ago. They specifically offer holiday pastoral care at holiday resorts, accompany people on pilgrimages who are in search of meaning and healing. Spirituals travel on cruise ships and are there for all passengers regardless of their culture and religion. And all over the world people visit holy places and rooms, probably not only for cultural-historical reasons. They light candles in foreign churches, celebrate Pujas, put their shoes down in front of mosques or put on a hat in the synagogue. They look for the unavailable, the encounter with themselves, for new ways and life perspectives, but also for a place to complain and pause.

Luxury as an Art of Living and Ethical Responsibility

Ultimately, travel is always a journey to oneself—the luxury segment only differs in standards and cost. In the distance, far away from everyday life, one is closer to oneself, in the encounter with foreign values and cultures one examines one's own basic values and in the best case one integrates these experiences into one's own value system and new perspectives are integrated. But this process is never complete. In the luxury segment, there is a wider variety of ways to welcome travelers individually, to ask them about their contortions and desires, and to open up new perspectives for them. People like to return to such places; they are not interchangeable, they are places of longing because they keep the longing for the unavailable awake and offer comfort and security when the road is too arduous and the contortions plague. Luxury is the art of enjoying life in all its fullness—luxury refers to what is coming, to the fullness in which all people can enjoy it. Therefore, fair, ecological, and sustainable tourism must not be a mere marketing instrument. Fullness can only arise if the trips are accompanied by people who are paid decently and treated with respect. If over-exploitation of nature is carried out by tourism, this prolongs the time to perfection. Nature itself is only cipher, the most beautiful beach, the most romantic sunset, and the most breathtaking view are harbingers, amuse-gueule of paradise, not paradise itself.

The Ten Commandments of Luxury Tourism
1. Those responsible are aware of the unavailability of true luxury.
2. People are seekers who have to deal with their "dislocations," their luxations of life.
3. Travelers are in search of meaning and distance from their everyday lives.
4. Luxury tourism opens up space and provides public spaces that are purposeless, inviting, interdenominational, and interreligious—not interchangeable, but interwoven with the destination.
5. At a luxury location, professional, independent companions are available to help people integrate their new experiences.
6. The offers do not harass the travelers, but they do encourage them to leave the comfort zone through the offers. All service employees maintain a professional service distance and do not fake interpersonal connections.
7. Individual services are offered at a high level. The services address the real needs, perhaps even the dislocations, the desires for healing.
8. The luxury of care serves the soul, spirit, and body in a balanced relationship.

(continued)

9. Service providers must be paid fairly and be treated with respect. Tourism must be lived ecologically and sustainably, because luxury cannot be beneficial at the expense of others or nature.

10. Luxury recognizes its limitations and refers to the future perfection where people are in peace and justice with themselves, nature, and the Unavailable/God. Luxury offers a taste of paradise. It gives us an idea of the fullness with which God thought of creation and of the perfection to which we are called.

Literature

Abtan O, Barton C, Bonelli F, Gurzki H, Mei-Pochtler A, Pianon N, Tsusaka M (2016) Digital or die: The choice for luxury brands. https://www.bcg.com/de-de/publications/2016/digital-or-die-choice-luxury-brands.aspx. Accessed 25 Feb 2019

Albrecht C-M, Backhaus C, Gurzki H, Woisetschläger DM (2013a) Drivers of brand extension success: what really matters for luxury brands. Psychol Market 30(8):647–659

Albrecht C-M, Backhaus C, Gurzki H, Woisetschläger DM (2013b) Value creation for luxury brands through brand extensions: an investigation of forward and reciprocal effects. Market ZFP 35(2):91–108

Alexander N (2009) Brand authentication: creating and maintaining brand auras. Eur J Market 43 (3/4):551–562

Anderson EW, Sullivan MW (1993) The antecedents and consequences of customer satisfaction for firms. Market Sci 12(2):125–143

Antweiler C (2008) Rational use and productive luxury? In Jäckel M, Schößler F (eds) Luxury. President of the University, Trier

Arbore A, Estes Z (2013) Loyalty program structure and consumers' perceptions of status: Feeling special in a grocery store? J Retail Consum Serv 20(5):439–444

Arndt J, Solomon S, Kasser T, Sheldon KM (2004a) The urge to splurge revisited: further reflections on applying terror management theory to materialism and consumer behavior. J Consum Psychol 14(3):225–229

Arndt J, Solomon S, Kasser T, Sheldon KM (2004b) The urge to splurge: a terror management account of materialism and consumer behavior. J Consum Psychol 14(3):198–212

Arnould EJ, Price LL (1993) River magic: extraordinary experience and the extended service encounter. J Consum Res 20(1):24–45

Bachmann A (2013) Hedonism and the good life. Mentis, Münster

Barlösius E (1999) Sociology of food. Juventa, Weinheim/Munich

Bearden W, Etzel M (1982) Reference group influence on product and brand purchase decisions. J Consum Res 9(2):183–194

Belk RW (1988) Possessions and the extended self. J Consum Res 15(2):139–168

Belk RW (1991a) The ineluctable mysteries of possessions. J Soc Behav Pers 6(6):17–55

Belk RW (1991b) Possessions and the sense of past. In: Belk RW (ed) SV—highways and buyways: naturalistic research from the consumer behavior odyssey. Association for Consumer Research, Provo, UT, pp 114–130

Belk RW (1995a) Collecting as luxury consumption: Effects on individuals and households. J Econ Psychol 16(3):477–490

Belk RW (1995b) Collecting in a consumer society. Routledge, London

Belk RW (2001) Specialty magazines and flights of fancy: feeding the desire to desire. Eur Adv Consum Res 5:197–202

Belk RW (2013) Extended self in a digital world. J Consum Res 40(3):477–500

Belk RW, Costa JA (1998) The mountain man myth: a contemporary consuming fantasy. J Consum Res 25(3):218–240

Belk RW, Wallendorf M (1990) The sacred meanings of money. J Econ Psychol 11(1):35–67

Belk RW, Wallendorf M, Sherry J Jr (1989) The sacred and the profane in consumer behavior: Theodicy on the odyssey. J Consum Res 16(1):1–38

Belk RW, Ger G, Askegaard S (1996) Metaphors of consumer desire. Adv Consum Res 23:368–373

Belk RW, Ger G, Askegaard S (2000) The missing streetcar named desire. In: Ratneshwar S, Mick DG, Huffman C (eds) The why of consumption: contemporary perspectives on consumer motives, goals and desires. Routledge, London, pp 98–119

Belk RW, Ger G, Skegaard S (2003) The fire of desire: a multisited inquiry into consumer passion. J Consum Res 30(3):326–351

Bengtsson A, Firat AF (2006) Brand literacy: consumers' sense-making of brand management. Adv Consum Res 33:375–380

Berg D (1977) Poverty and science. Contributions to the history of the study of mendicant orders in the 13th century. Schwann, Düsseldorf

Berger J, Heath C (2007) Where consumers diverge from others: Identity signaling and product domains. J Consum Res 34(2):121–134

Berry CJ (1994) The idea of luxury: a conceptual and historical investigation. Cambridge University Press, Cambridge

Berthon P, Pitt L, Parent M, Berthon J-P (2009) Aesthetics and ephemerality: observing and preserving the luxury brand. Calif Manag Rev 52(1):45–66

Beverland MB (2005) Crafting brand authenticity: the case of luxury wines. J Manag Stud 42 (5):1003–1029

Beverland MB (2006) The 'real thing': branding authenticity in the luxury wine trade. J Bus Res 59 (2):251–258

Bhattacharjee A, Mogilner C (2014) Happiness from ordinary and extraordinary experiences. J Consum Res 41(1):1–17

Boone JL (1998) The evolution of magnanimity. Hum Nat 9(1):1–21

Böttger T, Rudolph T, Evanschitzky H, Pfrang T (2017) Customer inspiration: conceptualization, scale development, and validation. J Market 81(6):116–131

Bourdieu P (1992) The subtle differences. Suhrkamp, Frankfurt a. M.

Brakus JJ, Schmitt BH, Zarantonello L (2009) Brand experience: what is it? How is it measured? Does it affect loyalty? J Market 73(3):52–68

Braun OL, Wicklund RA (1989) Psychological antecedents of conspicuous consumption. J Econ Psychol 10(2):161–187

Brockhaus (1894) Conversation encyclopedia. 17 vol. Brockhaus, Vienna

Brockhaus (1966) Encyclopedia. Twenty books. Brockhaus, Wiesbaden

Brown S, Kozinets RV, Sherry JF Jr (2003) Teaching old brands new tricks: retro branding and the revival of brand meaning. J Market 67(3):19–33

Brown S, McDonagh P, Shultz II, J C (2013) Titanic: consuming the myths and meanings of an ambiguous brand. J Consum Res 40(4):595–614

Carr HL, Vignoles VL (2011) Keeping up with the Joneses: Status projection as symbolic self-completion. Eur J Soc Psychol 41(4):518–527

Chaudhuri A (1998) Product class effects on perceived risk: the role of emotion. Int J Res Market 15 (2):157–168

Choo HJ, Moon H, Kim H, Yoon N (2012) Luxury customer value. J Fashion Market Manag Int J 16(1):81–101

Csikszentmihalyi M (2000) The costs and benefits of consuming. J Consum Res 27(2):267–272

Dai X, Fishbach A (2014) How nonconsumption shapes desire. J Consum Res 41(4):936–952

Dion D, Arnould E (2011) Retail luxury strategy: assembling charisma through art and magic. J Retail 87(4):502–520

Dion D, Borraz S (2015) Managing heritage brands: A study of the sacralization of heritage stores in the luxury industry. J Retail Consum Serv 22:77–84

Drèze X, Nunes JC (2009) Feeling superior: the impact of loyalty program structure on consumers' perceptions of status. J Consum Res 35(6):890–905

Dubois B, Laurent G (1994) Attitudes toward the concept of luxury: An exploratory analysis. Asia Pac Adv Consum Res 1:273–278

Edelman DC, Singer M (2015) Competing on customer journeys. Harv Bus Rev 93(11):88–100

Elias N (1969) Courtly society. Luchterhand, Neuwied

Enzensberger HM (1996) Reminiscences of abundance. The mirror, 51

Epicurus (1991) From overcoming fear. dtv, Munich

Fernandez KV, Lastovicka JL (2011) Making magic: fetishes in contemporary consumption. J Consum Res 38(2):278–299

Fiedler K (2007) Construal level theory as an integrative framework for behavioral decision-making research and consumer psychology. J Consum Psychol 17(2):101–106

Fuchs C, Schreier M, van Osselaer SMJ (2015) The handmade effect: what's love got to do with it? J Market 79(2):98–110

Gallarza MG, Gil I (2008) The concept of value and its dimensions: a tool for analysing tourism experiences. Tourism Rev 63(3):4–20

Giesler M (2006) Consumer gift systems. J Consum Res 33(2):283–290

Gorgeous J (2002) From a mass holiday to a luxury holiday. In: Fifth C-B-R tourism symposium, Munich Trade Fair, Munich

Graeber D (2011) Consumption. Curr Anthropol 52(4):489–511

Great E (2005) A new earth. Michael Joseph, London

Grubb E, Grathwohl H (1967) Consumer self-concept, symbolism and market behavior: a theoretical approach. J Market 31(4):22–27

Grugel-Pannier D (1996) Luxury: An examination of the history of concepts and ideas with special consideration of Bernard Mandeville. Peter Lang, Frankfurt a. M.

Gurzki H (2018) The creation of the extraordinary—principles of luxury. Dissertation. TU Braunschweig, Braunschweig

Gurzki H, Woisetschläger DM (2017) Mapping the luxury research landscape: A bibliometric citation analysis. J Bus Res 77:147–166

Habermas J (1975) Structural change of the public. Luchterhand, Neuwied

Hagtvedt H, Patrick VM (2009) The broad embrace of luxury: Hedonic potential as a driver of brand extendibility. J Consum Psychol 19(4):608–618

Hallerbach B (1993) The travel behaviour of environmentally conscious people. Diploma thesis, Trier

Han YJ, Nunes JC, Drèze X (2010) Signaling status with luxury goods: the role of brand prominence. J Market 74(4):15–30

Hansen J, Wänke M (2011) The abstractness of luxury. J Econ Psychol 32(5):789–796

Haupt H-G, Torp C (eds) (2009) The consumer society in Germany 1890-1990. Campus, Frankfurt a. M.

Hennigs N, Wiedmann K-P (2012) Consumer value perception of luxury goods: a cross-cultural and cross-industry comparison. In: Wiedmann K-P, Hennigs N (eds) Luxury marketing. Gabler, Wiesbaden, pp 77–99

Hillmann K-H (2007) Dictionary of sociology, 5th edn. Kröner, Stuttgart

Hirschman EC, Holbrook MB (1982) Hedonic consumption: emerging concepts, methods and propositions. J Market 46(3):92–101

Hogg MK, Michell P (1996) Identity, self and consumption: a conceptual framework. J Market Manag 12(7):629–644

Holbrook MB (1999) Consumer value: a framework for analysis and research. Routledge, London

Holbrook MB, Hirschman EC (1982) The experiential aspects of consumption: consumer fantasies, feelings, and fun. J Consum Res 9(2):132–140

Holt DB (2004) How brands become icons: The principles of cultural branding. Harvard Business School Press, Boston, MA

Illouz E (2009) Emotions, imagination and consumption a new research agenda. J Consum Cult 9 (3):377–413

ITB (2018) Luxury travel continues to grow. October 25. https://www.itb-berlin.com/Press/PressReleases/News_47502.html. Accessed 24 Feb 2019

Jäckel M (2008) How demonstrative was and is consumption? In Jäckel M, Schößler F (eds) Luxury. President of the University, Trier

Johar GV, Holbrook MB, Stern BB (2001) The role of myth in creative advertising design: theory, process and outcome. J Advert 30(2):1–25

Kano N (1984) Attractive quality and must-be quality. J Jpn Soc Qual Control 14(2):39–48

Kapferer J-N (2012) The luxury strategy: break the rules of marketing to build luxury brands. Kogan Page, London

Kapferer J-N, Bastien V (2009) The luxury strategy: break the rules of marketing to build luxury brands. Kogan Page, London

Kapferer J-N, Valette-Florence P (2016) Beyond rarity: the paths of luxury desire. How luxury brands grow yet remain desirable. J Product Manag 25(2):120–133

Kapferer J-N, Vincent B (2009) The luxury strategy: break the rules of marketing to build luxury brands. Kogan Page, London

Katz D, Kahn RL (1966) The social psychology of organizations, 2nd edn. Wiley, New York

Kauppinen-Räisänen H, Gummerus J, von Koskull C, Finne Å, Helkkula A, Kowalkowski C, Rindell A (2014) Am I worth it? Gifting myself with luxury. J Fashion Market Manag 18 (2):112–132

Kenrick DT, Griskevicius V, Neuberg SL, Schaller M (2010) Renovating the pyramid of needs: contemporary extensions built upon ancient foundations. Perspect Psychol Sci 5(3):292–314

Kersting W (2005) Justice and the art of living. Philosophical side issues. Mentis, Paderborn

Kim Y (2018) Power moderates the impact of desire for exclusivity on luxury experiential consumption. Psychol Market 35(4):283–293

Kim J-H, Ritchie JRB, McCormick B (2011) Development of a scale to measure memorable tourism experiences. J Travel Res 51(1):12–25

Kisabaka L (2001) Marketing for luxury products. Product Marketing Promotion Agency, Cologne

Knebel H-J (1958) Sociological structural changes in modern tourism. Dissertation, Hamburg

Kocka J (ed) (1987) Citizenship and bourgeoisie in the 19th century. Vandenhoeck & Ruprecht, Göttingen

Kozinets R, Patterson A, Ashman R (2016) Networks of desire: how technology increases our passion to consume. J Consum Res 43(5):659–682

Kriedtke P, Medick H, Schlumbohm J (1977) Industrialization before industrialization. Vandenhoeck & Ruprecht, Göttingen

Krünitz JG (1773) Economic encyclopedia. 242 Bde. Pauli, Berlin

Kumar M, Garg N (2010) Aesthetic principles and cognitive emotion appraisals: How much of the beauty lies in the eye of the beholder? J Consum Psychol 20(4):485–494

Lai IKW, Hitchcock M (2016) A comparison of service quality attributes for stand-alone and resort-based luxury hotels in Macau: 3-Dimensional importance-performance analysis. Tourism Manag 55(C):139–159

Lai IKW, Hitchcock M (2017) Sources of satisfaction with luxury hotels for new, repeat, and frequent travelers: A PLS impact-asymmetry analysis. Tourism Manag 60:107–129

Lasslop I (2002) Identity-oriented management of luxury brands. In: Burmann C, Meffert H, Koers M (eds) Brand management. Gabler, Wiesbaden, pp 327–351

Leigh TW (2006) The consumer quest for authenticity: the multiplicity of meanings within the MG subculture of consumption. J Acad Market Sci 34(4):481–493

Lemon KN, Verhoef PC (2016) Understanding customer experience throughout the customer journey. J Market 80(6):69–96

Lenski G (1977) Power and privilege. Suhrkamp, Frankfurt a. M.

Lepenies W (1981) Melancholy and company. Suhrkamp, Frankfurt a. M.

Levy SJ (1959) Symbols for sale. Harv Bus Rev (July-August), 117–124

Levy S (1981) Interpreting consumer mythology: a structural approach to consumer behavior. J Market 45(3):49–61

Liberman N, Trope Y (2008) Still here, still now. University of Chicago Press, Chicago

Llamas R, Belk R (2011) Shangri-La: messing with a myth. J Macromarket 31(3):257–275

Llamas R, Thomsen TU (2016) The luxury of igniting change by giving: Transforming yourself while transforming others' lives. J Bus Res 69(1):166–176

Lury C (2011) Consumer culture. Polity, Cambridge

Maase K (1997) Unlimited pleasure. Fischer, Frankfurt a. M.

Madzharov AV, Block LG, Morrin M (2015) The cool scent of power: effects of ambient scent on consumer preferences and choice behavior. J Market 79(1):83–96

Maheswaran D, Agrawal N (2004) Motivational and cultural variations in mortality salience effects: contemplations on terror management theory and consumer behavior. J Consum Psychol 14 (3):213–218

Mandel N, Heine SJ (1999) Terror management and marketing: He who dies with the most toys wins. Adv Consum Res 26:527–532

Mandel N, Petrova PK, Cialdini RB (2006) Images of success and the preference for luxury brands. J Consum Psychol 16(1):57–69

Maslow AH (1943) A theory of human motivation. Psychol Rev 50(4):370–396

Mauss M (1978) Sociology and anthropology II. Ullstein, Berlin

Mauss M (2002) The gift: The form and reason for exchange in archaic societies. Routledge, London

McCracken G (1986) Culture and consumption: A theoretical account of the structure and movement of the cultural meaning of consumer goods. J Consum Res 13(1):71–84

McCracken G (1990) Culture and consumption: new approaches to the symbolic character of consumer goods and activities. Indiana University Press, Bloomington

McFerran B (2013) The entourage effect. J Consum Res 40(5):871–884

Meffert H, Lasslop I (2004) Luxury brand strategy. In: Bruhn IM (ed) Handbook of brand management. Gabler, Wiesbaden, pp 927–947

Merkel I (2009) Contrary to the ideal: consumer policy in the GDR. In: Haupt H-G, Torp C (eds) Die Konsumgesellschaft in Deutschland 1890-1990. Campus, Frankfurt a.M., pp 289–204

Mick D, DeMoss M (1990) Self-gifts: phenomenological insights from four contexts. J Consum Res 17(3):322–332

Montanari M (1993) The hunger and the abundance. Beck, Munich

Moraw P (1988) The court feasts of Emperor Friedrich Barbarossa of 1184 and 1188. In: Schultz U (ed) Das Fest. Beck, Munich, pp 70–83

Napoli J, Dickinson SJ, Beverland MB, Farrelly F (2014) Measuring consumer-based brand authenticity. J Bus Res 67(6):1090–1098

Nia A, Zaichkowsky J (2000) Do counterfeits devalue the ownership of luxury brands? J Product Brand Manag 9(7):485–497

Nueno JL, Quelch JA (1998) The mass marketing of luxury. Bus Horiz 41(6):61–68

Oliver RL (1980) A cognitive model of the antecedents and consequences of satisfaction decisions. J Market Res 17(4):460–469

Oliver RL (1993) Cognitive, affective, and attribute bases of the satisfaction response. J Consum Res 20(3):418–430

o. V. (1897) Mirror 34:125

o. V. (2011) The relevant one. Reconsider Knowl Ethics 22(1)

o. V. (2018) Luxury. http://de.wikipedia.org/wiki/Luxus . Accessed 25 Aug 2018

Parasuraman AJ, Zeithaml VA, Berry LL (1988) SERVQUAL: a multiple-item scale for measuring consumer perceptions of service qualitative. J Retail 64(1):12–40

Pozharliev R, Verbeke WJMI, Van Strien JW, Bagozzi RP (2015) Merely being with you increases my attention to luxury products: using EEG to understand consumers' emotional experience with luxury branded products. J Market Res 52(4):546–558

Reith R, Meyer T (eds) (2003) Luxury and consumption. Waxmann, Münster

Rindfleisch A, Burroughs J (2004) Terrifying thoughts, terrible materialism? Contemplations on a terror management account of materialism and consumer behavior. J Consum Psychol 14 (3):219–224

Rochberg-Halton E (1984) Object relations, role models, and cultivation of the self. Environ Behav 16(3):335–368

Rook DW (1985) The ritual dimension of consumer behavior. J Consum Res 12(3):251–264

Rook DW, Levy SJ (1983) Psychosocial themes in consumer grooming rituals. NA Adv Consum Res 10:329–333

Roscher W (1861) Views of the economy from the historical point of view. Winter, Leipzig

Roux E, Lipovetsky G (2003) Eternal luxury: From the age of the sacred to the time of brands. Gallimard, Paris

Rucker D, Galinsky A (2008) Desire to acquire: Powerlessness and compensatory consumption. J Consum Res 35(2):257–267

Rucker DD, Galinsky AD, Dubois D (2012) Power and consumer behavior: How power shapes who and what consumers value. J Consum Psychol 22(3):352–368

Sanchez-Fernandez R, Iniesta-Bonillo MA (2007) The concept of perceived value: a systematic review of the research. Market Theor 7(4):427–451

Schildt A (1995) Modern times. Christians, Hamburg

Schmitt B (1999) Experiential marketing: how to get customers to sense, feel, think, act, and relate to your company and brands. Free, New York

Schmitt B (2003) Customer experience management: a revolutionary approach to connecting with your customers. Wiley, Hoboken

Schulze G (1993) The adventure society. Campus, Frankfurt a. M.

Sherry JF Jr (1983) Gift giving in anthropological perspective. J Consum Res 10(2):157–168

Sherry JF, McGrath MA, Levy SJ (1993) The dark side of the gift. J Bus Res 28(3):225–244

Silverstein MJ, Fiske N (2003) Luxury for the masses. Harv Bus Rev 81(4):48–57

Sirgy MJ (1982) Self-concept in consumer behavior: A critical review. J Consum Res 9(3):287–300

Sirgy MJ, Su C (2000) Destination image, self-congruity, and travel behavior: Toward an integrative model. J Travel Res 38(4):340

Snyder CR (1992) Product scarcity by need for uniqueness interaction: a consumer catch-22 carousel? Basic Appl Soc Psychol 13(1):9–24

Solomon M (1983) The role of products as social stimuli: A symbolic interactionism perspective. J Consum Res 10(3):319–329

Sombart W (1922) Luxury and capitalism, 2nd edn. Duncker & Humblot, Munich

Spode H (1993) The power of drunkenness. Leske and Budrich, Opladen

Spode H (2008) Resource future. Budrich, Opladen

Spode H (2009) How the Germans became "travel world champions". VS Publishing House for Social Sciences, Wiesbaden

Stangroom J, Garvey J (2006) The most famous philosophers. Premio, London

Stern BB (1995) Consumer myths: Frye's taxonomy and the structural analysis of consumption text. J Consum Res 22(2):165–185

Stradner J (1917) Tourism, 2nd edn. Leykam, Graz

Sundie JM, Kenrick DT, Griskevicius V, Tybur JM, Vohs KD, Beal DJ (2011) Peacocks, porsches, and thorstein veblen: conspicuous consumption as a sexual signaling system. J Pers Soc Psychol 100(4):664–680

Tian K, Belk R (2006) Consumption and the meaning of life. In: Belk RW (ed) Research in consumer behavior, vol 10. Emerald, Bingley, pp 249–274

Toothed UM (1985) Lecture notes. Unpublished document. Bern

Torp C (2012) Growth, security, morale. Wallstein, Göttingen

Townsend C, Sood S (2012) Self-affirmation through the choice of highly aesthetic products. J Consum Res 39(2):415–428

Trautwein S (1919) Society and conviviality in the past and present. Teubner, Berlin

Triebel A (1991) Two classes and the diversity of consumption. Dissertation, FU Berlin, Berlin

Trope Y, Liberman N (2000) Temporal construal and time-dependent changes in preference. J Pers Soc Psychol 79(6):876–889

Trope Y, Liberman N, Wakslak C (2007) Construal levels and psychological distance: effects on representation, prediction, evaluation, and behavior. J Consum Psychol 17(2):83–95

Tynan C, McKechnie S, Chhuon C (2010) Co-creating value for luxury brands. J Bus Res 63 (11):1156–1163

Ulich E (1991) Industrial psychology. Publisher of the Professional Associations, Zurich

Valtin A (2005) The value of luxury brands: Determinants of consumer-oriented brand value and implications for luxury brand management. German University Publisher, Wiesbaden

Veblen T (1958) Theory of fine people. Kiepenheuer & Witsch, Cologne

Venkatesh A, Meamber L (2006) Arts and aesthetics: Marketing and cultural production. Market Theor 6(1):11–39

Vigneron F, Johnson LW (1999) A review and a conceptual framework of prestige-seeking consumer behavior. Acad Market Sci Rev 1999(1):1–15

Vigneron F, Johnson LW (2004) Measuring perceptions of brand luxury. J Brand Manag 11 (6):484–506

von Rotteck C, Welcker C (1834) State encyclopedia. 12 volumes. Hammerich, Altona

Weber M (1904) Die Protestantische Ethik und der Geist des Kapitalismus, vol 1. Verlag Wirtschaft und Finanzen, Leipzig

Weber M (2009) Protestant ethics and the spirit of capitalism. Anaconda, Cologne

Weinschenk S (2009) Neuro web design. New Riders, Berkeley

Wiedmann K-P, Hennigs N, Siebels A (2007) Measuring consumers' luxury value perception: a cross-cultural framework. Acad Market Sci Rev 2007(7):1–21

Wiedmann K-P, Hennigs N, Siebels A (2009) Value-based segmentation of luxury consumption behavior. Psychol Market 26(7):625–651

Wiedmann K-P, Hennigs N, Schmidt S, Wuestefeld T (2011) The importance of brand heritage as a key performance driver in marketing management. J Brand Manag 19(3):182–194

Wirtz J, Lovelock CH (2018) Essentials of services marketing, 3rd edn. Pearson, Essex

Wolf D (2018) Psychology dictionary, PAL. Mannheim

Wong N, Ahuvia A (1998) Personal taste and family face: Luxury consumption in Confucian and Western societies. Psychol Market 15(5):423–441

Yu S, Hudders L, Cauberghe V (2018) Selling luxury products online: The effect of a quality label on risk perception, purchase intention and attitude toward the brand. J Electron Commerce Res 19(1):16–35

Zeithaml VA, Berry LL, Parasuraman A (1993) The nature and determinants of customer expectations of service. J Acad Market Sci 21(1):1–12

Prof. Dr. Hasso Spode studied history and sociology as well as religious studies and philosophy; doctorate on the change of dietary behavior and habilitation on the history of alcohol. Teaching and research activities mainly at the Technical University of Berlin and the Leibniz University Hannover; since 1999 also director of the Historical Archive for Tourism at the Technical University of Berlin. Numerous publications and media appearances, e.g., on the history of tourism.

Hannes Gurzki is a consultant at The Boston Consulting Group. He is an expert in luxury, marketing, branding, and consumer behavior. He studied business administration and cultural management at the University of Mannheim and holds an MBA from ESSEC Business School. His research appeared in leading academic journals such as the Journal of Business Research or Psychology & Marketing.

Prof. Dr. David M. Woisetschläger is Professor of Services Management and Director of the Institute of Automotive Management and Industrial Production at TU Braunschweig University. His research interests are in the fields of Brand Management, Customer Relationship Management, Sales, and Sponsorship. He is also a consultant for new service development and the analysis of customer survey and behavioral data with an industry focus in the automobile and telecommunications industries. His research is published in leading international journals such as the Journal of Marketing, Journal of the Academy of Marketing Science, Journal of Business Research, and Journal of Retailing.

Marc Aeberhard founded Luxury Hotel & Spa Management Ltd. in Zurich in 2004 and has been acting as Managing Director ever since. The company has access to a global network of travel trade partners, lifestyle and travel media and (U)HNWI and works closely with public relations and sales and marketing agencies in Frankfurt, Munich, Paris, Dubai, Milan, New York, Hong Kong, and London. Furthermore, the native Swiss is a member of the consulting networks of the Gerson Lehmann Group, USA and Hotellerie Suisse, Bern. He also takes on an active role in the "Luxury" task force of the management of ITB Berlin, Germany. As author and co-author of various specialist publications, his name can be found regularly. He also holds guest lectures in Berlin, Istanbul, Lausanne, Lucerne, Munich, Singapore, Stuttgart, Thun, Vienna, Worms, Zurich, etc. The graduate hotelier graduated with distinction from the Ecole Hôtelière de Lausanne (EHL) and previously completed his studies in business administration as lic.rer.pol. (MBA) at the University of Bern. The luxury hotelier has more than 20 years of experience in the fields of hotel opening, management, and renovation/refurbishment in Abu Dhabi, Germany, France, Maldives, Morocco, Seychelles, Sri Lanka, Switzerland, Thailand, Ukraine, and Cyprus of small hotels in high and top end. Many of the hotels have been awarded international prizes. All projects are based on the definition of New Luxury and work according to the principles of the Triple Bottomline.

Stephan Hagenow doctor of theology and pastor, is the Head of the Department of Personnel Development of the Parish of Bern, Jura, Solothurn (Switzerland). He is responsible for accompanying and advising pastors and parishes, checking and approving job descriptions for the parish office, home pastoral care, catechetics and deaconry, health and risk prevention, as well as individual promotion and technical management of the regional parish offices for the inner-church area, theological and spiritual foundation work in the personnel development of the parish, participation in the acceptance into the Bernese church service, representation of the leadership in the field of theology and development of foundations and statements on church-theologically relevant basic questions.

Marketing Management of Luxury Providers

Marc Aeberhard, Magda Antonioli Corigliano, Sara Bricchi, Juliet Kinsman, and Keiko Kirihara

Like the mass consumer goods business, the luxury sector is also dominated by providers of *physical* luxury goods, who have contributed significantly to the steady growth of the premium segments worldwide with their professional marketing know-how. Therefore, examples from the physical consumer goods sector are used in this chapter in order to elaborate on the special features of luxury marketing—in particular with regard to the tourism industry.

M. Aeberhard
Luxury Hotel & Spa Management Ltd., Zürich, Switzerland

M. Antonioli Corigliano (✉)
ACME Management School, Bocconi University, Milan, Italy
e-mail: magda.antonioli@unibocconi.it

S. Bricchi
SDA Bocconi School of Management, Bocconi University, Milan, Italy
e-mail: sara.bricchi@unibocconi.it

J. Kinsman (✉)
BOUTECO, London, UK

K. Kirihara (✉)
Worms University of Applied Sciences, Worms, Germany
e-mail: kirihara@hs-worms.de

© Springer Nature Switzerland AG 2020
R. Conrady et al. (eds.), *Luxury Tourism*, Tourism, Hospitality & Event
Management, https://doi.org/10.1007/978-3-030-59893-8_5

1   Marketing Strategies

The Marketing Strategy as a Derivation of the Corporate Strategy: Global Versus Multinational

Keiko Kirihara and Marc Aeberhard

The marketing strategy is derived from the corporate strategy. It can vary depending on the industry, life cycle of the company, ownership structure, etc. For example, a company's growth strategy may result in an expansive brand/product program strategy that is more focused on innovative new products, while a consolidation strategy tends to have a restrictive effect on the brand/product portfolio, and so on.

However, if we look at the basic business strategy of recent decades, which has led to the geographic expansion of the business world (with dramatic sociocultural and political changes in the world), we must look at the globalization strategy. Almost every company, independent of the industrial sector, has dealt with the question of the extent to which globalization and thus the standardization of business can be realized (or to what extent concessions have to be made with regard to regional–local market conditions) in order to operate successfully in the markets of the world.

Because somewhere between the theoretically contrary strategies—the globalization strategy versus the multinational strategy—lies in practice the "right" or successful *glocal* strategic implementation. Depending on the individual case, a more globally standardized or locally differentiated strategy is pursued; but there is no universal solution for business success.

International marketing is also derived from the context of these strategic corporate decisions. Since the 1990s at the latest, international marketing has been concerned with the question of the extent to which globally standardized brands and the marketing mix are assertive in the world's markets in order to generate synergies, in particular economies of scale effects. In the world of mass consumer goods, the acid test is almost part of everyday business, and the aim is always to achieve the highest possible degree of standardization on the basis of rather low margins. Regional–local compromises are therefore only realized in cases where standardization measures significantly impair competitiveness in the respective markets.

Only China's economic rise to become the largest and thus strategically most important market for almost all industries in the future has promoted a glocal strategic approach on a broad scale. The Chinese market and its huge business potential with more than 1.3 billion consumers often have a very independent sociocultural as well as political-legal profile compared to the western home markets. In the case of China, classic globally oriented mass-market companies such as Starbucks or KFC have implemented an extremely cautious marketing mix that takes account of regional and cultural peculiarities, while at the same time paying attention to the development and maintenance of a global brand concept.

Corporate and Marketing Strategies in the Luxury Sector

But what about the luxury sector? In general, most companies in the world's luxury markets use a globally standardized strategy. For example, on various journeys to distant continents and cities, one can see that more or less the same brands and products are offered everywhere. It seems to be the case that all people in the world desire the same luxury brands. While the French (LVMH, Kering, etc.) dominate by far in luxury-oriented consumption (cosmetics, fashion, luxury foods, etc.), suppliers from Germany (Mercedes, BMW, Porsche, Miele) and Italy (Ferrari, Boffi, etc.) have emerged as strong players in investment goods such as automobiles or sanitary and kitchen equipment. The Western European luxury style thus seems—at least so far—to set the global standard.

In addition to the globally established Western luxury style, another explanation for the successful implementation of global standardization could be that luxury customers worldwide have a much more homogeneous profile than consumers in the mass markets of the world. This is understandable when one compares the income and financial circumstances, level of education, or lifestyle of luxury customers. Undoubtedly, the lifestyle of a wealthy Chinese will show parallels with his counterpart from Germany or America, while the buying behavior of a rather simple consumer from India differs clearly from that of a simple Frenchman.

Marketing Strategies in Tourism

Are these phenomena of the physical consumer goods luxury markets transferable to tourism?

There is no general answer to this question. Basically, it can be stated that the conventional tourism/tour operator business has been and still is rather local, at best regionally oriented. If, for example, one compares the needs and requirements of a German with those of a Chinese tourist, there are still significant differences in terms of destinations, leisure preferences, and available leisure time.

With regard to the hotel business in the high-end sector, it can be assumed that companies are targeting global target groups, as comparable financial circumstances mean that the demands and sometimes also lifestyles of very wealthy hotel guests are similar. With regard to product policy, however, globalized standardization in the high-end sector of the hotel industry is now being vehemently denied. Standardization was the recipe for success of Hilton, Intercontinental, and Sheraton about 50 years ago. Today, even companies such as Four Seasons and Ritz Carlton have realized that although they can appear with a global brand name, they must formulate the product regionally/local (for example, Four Seasons at Sayan or Four Seasons in Chiang Mai). Comparable to the shopping malls that offer the same thing all over the world, it can be assumed that the demanding luxury customers are tired of the globally standardized standard hotels. The increasing demand for regional–local products and experiences has obviously led to this strategic turn.

In contrast to the physical consumer goods business, the pressure to generate economies of scale is not particularly high. Cost effects in production are largely eliminated in the service sector. In addition, due to the sometimes-large distances between individual hotels, there are only limited synergies via purchasing cooperations.

The focus of the global strategy is therefore primarily on the brand and the communication policy (creation of a UAP) through which access to global luxury customers is to be ensured. For example, access to global communication and information structures as well as ranking tools that help potential customers assess a tourism service provider (star category, membership in hotel consortia, international chain hotels, trip advisors, etc.) is of great importance for the classification of a destination.

New digital business models such as Airbnb, on the other hand, are pursuing a global strategy from the outset. However, Airbnb is also taking a glocal approach in the operational implementation of the strategy on the basis of marketing mix factors. While a uniform global product portfolio with standardized prices is valid for customers worldwide, communication with target groups is adapted glocal taking into account regional–local characteristics (see also section "Communication Policy").

What first started as an online provider of "bed and breakfast" with "airbed" 10 years ago, Airbnb's current business is not based in the luxury sector. However, for some time now it has been unmistakable that the company is expanding its portfolio with more profitable quality products (Airbnb Plus) and is differentiating its product range more specifically for target groups (Airbnb Collection). With the launch of Beyond by Airbnb in 2018 (following the acquisition of Luxury Retreat in 2017), the company has irrevocably entered the luxury market (Cf. Airbnb 2018). It remains to be seen how Airbnb will work the top-end segment and what impact the market entry will have on luxury tourism.

Components of the Marketing Strategy

If the basic orientation has been derived from the corporate strategy, the following questions need to be answered in order to concretize the marketing strategy:

- Where? Which target market?
- Who? Which target group?
- What? What are the benefits?

With regard to the target market, the "Where?" may include a geographical definition of the market, but also the question of the product category and the price segment. In the case of a global corporate strategy, all markets in the world would then be relevant in principle, so that there would be no geographical limitation (but possibly geographical prioritization). In the meantime, the pursuit of a multinational strategy would result in the assessment and selection of certain regions or countries. With a strategic focus on luxury markets, it is clear that global top-end price segments represent the target market, which are far above the price segments of the mass business.

The definition of the target market also outlines the target group, although further differentiation or segmentation of the market according to specific customer groups generally takes place depending on the individual strategic orientation (brand positioning) (see also Sect. 2).

In order to answer the question as to what benefits are to be offered in order to arouse desire in the target group, a "unique" functional and/or psychological benefit should be formulated which clearly differentiates the product from the competitive environment and creates purchasing preferences.

The questions about "Who?" and "What?" are summarized in a brand concept that outlines the positioning and benefits of the brand and the product in the form of a functional–rational as well as a psychological–emotional benefit and thus comprises an essential component of the marketing/brand strategy.

Special Features of Luxury Brands and Products

The management of luxury brands is subject to somewhat different laws than that of a mass brand or a consumer goods product for the mass market, whereby the following aspects are to be emphasized:

– **Focus on psychological–emotional benefit dimensions**

While the benefit of a mass consumer product always consists of a rational–functional as well as psychological–emotional benefit dimension (with different focal points depending on brand, product category, etc.), the focus of a luxury product is clearly in the psychological–emotional service spectrum.

This does not necessarily mean that the functional benefit of a luxury product is meaningless, but it is usually not decisive for the purchase. This aspect becomes clear when, for example, one looks at high-quality products with outstanding functional performance, which nevertheless cannot be assigned to the luxury category. A typical example is the Samsung Galaxy brand for the smartphone market. Certainly, a quality product that surpasses the functional performance of its main competitor, the Apple iPhone, but is still not perceived as a luxury product.

So what does it take for a product to be perceived as a luxury product? There are numerous attempts and approaches to outline the concept of luxury. The difficulty in clearly defining the concept of luxury could be attributed on the one hand to the fact that the understanding can vary greatly depending on the viewer. While they see luxury consumption as a means of demonstrating their social status, for others it may be a waste of money.

Increasingly there is also talk of "time as the greatest luxury." In this case, luxury would be an intangible good. This post-materialistic need, which is becoming more and more established as a luxury trend, especially in the established industrial nations, has a high relevance for the service sector, especially for tourism, which already offers immaterial dimensions of use. In luxury tourism in particular, apart from functional services such as transportation or sightseeing, which are regarded more as hygiene factors, psychological–emotional benefit dimensions/service packages are also at stake. The time, which is so short and therefore precious, should be filled with enjoyment in a meaningful way in order to "reward" oneself with it.

– **The brand/product as a reward**

A brand (or product) delivers a relevant benefit to the consumer if it satisfies his needs or serves his motive. In neuromarketing one speaks of "reward" of the brain, if a need, an important motive of the consumer, could be successfully addressed and satisfied by the benefits of a brand. The physiological effect of a "reward stimulus" can be observed via brain scans when the amygdala, the so-called "pleasure center" of the brain, has been activated by increased oxygen consumption. According to the explanatory approach of neuromarketing, dopamines (also popularly referred to as "happiness hormones") are distributed as harbingers of the expected reward or the successful satisfaction of needs, so that the brand/product is perceived as a reward from the consumer's point of view. It is well known that dopamine is also released when drugs are consumed, which promotes addictive behavior. In the case of brand consumption, one speaks of increased brand loyalty and customer loyalty.

Taking the example of Samsung, the brand would therefore be perceived less as a reward, while Apple would be able to offer exactly this reward effect to loyal buyers. This also explains the still long queues in front of the world's Apple stores when a new, innovative model comes onto the market and "Apple fans" even spend the night in front of the stores to be among the first buyers to hold the new product (at a high pioneer price) in their hands.

In various neuromarketing studies, mass consumer goods brands such as Coca-Cola were also identified as "reward brands" that were able to bring about clear differentiation and preference building among their users. Nevertheless, it can be assumed that the fascination of reward brands is rather located in the luxury sector, whereas the purchase of consumer products for daily use does not normally result in any or at least no excessive distribution of dopamine, despite all brand preferences.

– **Extrinsic and intrinsic "rewards"**

In principle, needs or motives can be divided into extrinsic and intrinsic characteristics. If one differentiates the needs for luxury goods according to these motif categories, one could distinguish between extrinsically influenced status consumption and intrinsically motivated, rather hedonistically oriented luxury consumption.

Kapferer and Bastien (2012, p. 18 ff.) describe luxury consumption as a "social marker," pointing out that the essence of luxury is always based on symbolic evidence of social advancement or belonging to the upper class. In addition, both authors admit that luxury consumption has an intrinsic character that they classify as "substantial" and thus as a more sustainable form of luxury consumption. While extrinsically motivated status consumption is subject to a certain fast-paced lifestyle due to its dependence on zeitgeist and trends, intrinsic hedonistic luxury has a higher durability.

This permanence/sustainability is explained by the identity of the luxury brand, which is based on origin, authenticity, and cultural values that have grown with it and are shared and internalized by luxury consumers. Such an identity-related connection between luxury brand and consumer can represent a long-term, even lifelong relationship that is not easily affected by changes in the market environment. Langer and Heil (2015, p. 22) therefore even speak of an "extended self" to underline the symbiotic character of the relationship between loyal consumer and luxury brand.

– **Conceptual cornerstones of luxury brands**

Luxury brands and products focus on psychological–emotional dimensions of use and are perceived as a reward by their consumers. A fundamental distinction can be made between a more extrinsically pronounced or intrinsically controlled motif or benefit performance. The intrinsic utility dimension of a luxury brand can be attributed to greater sustainability, since it is based on a substantial and thus more consistent correspondence of identity (values, culture) between brand and consumer.

If these premises for the definition of luxury brands and products are supplemented by the criteria exclusivity (in the sense of high-end price, scarcity/limited availability) and uniqueness (in the sense of unrivaled, non-substitutable), a clearly defined profile emerges that differs significantly from quality products (See Kapferer and Bastien 2012, p. 66 et seq.). So from the consumer's perspective, when the three cornerstones of luxury consumption—identity, exclusivity, and uniqueness—come together in a brand, the purchase and use are perceived as an extraordinary (luxury) experience, as a reward.

Referring once again to the example of Apple vs. Samsung, one could summarize that the iPhone is apparently always able to arouse loyal customers' psychological–emotional desires in order to satisfy status-oriented needs as well as to influence customer psychology by creating identity. Despite the high intensity of competition in the smartphone markets (including Samsung and Huawei, which offer comparable or even better functional benefits at an attractive price), Apple's iPhone remains unique for most customers and therefore not easily replaceable. When introducing the new generation model, Apple always makes sure that access to the product is expensive and time-consuming, so that a certain degree of exclusivity is maintained.

Of course, this condition is not carved in stone. If Apple does not manage to further nourish the brand's fascination by introducing "rewarding" product components (for example, innovations that better serve the motifs), the iPhone will eventually lose its luxury status.

Consequences for the Strategy of Luxury Brands

Starting from the claim of uniqueness based on an incomparable range of identities offered by luxury brands, one could conclude that a positioning—as is the norm in the conceptual determination of consumer goods brands—is unnecessary (See Kapferer and Bastien 2012, p. 65 f.). Because luxury brands with a unique selling proposition have no real competitors, since the (historically grown) brand identity is authentic, timeless, and thus characterized by consistency. The brand concept thus results "from within," i.e., from the origin and identity of the brand and not from external market dynamics and current trends.

The classic approach of brand positioning, on the other hand, is based on consumer insight, the identification of the needs and wishes of the target group. Taking into account the competition (and its offers of benefits), a positioning is then formulated which places the own offer of benefits in the market as relevant as possible (for the target group) and differentiating (from the competition).

So, do luxury brands really need no positioning? Langer and Heil (2015, p. 24), with their concept of "Category Potential Analysis," show how customer and competition analyses can be carried out for the luxury sector in order to ultimately determine one's own brand concept on the basis of a positioning analysis.

In order to discuss this question, two luxury brands will now be examined, each of which is regarded as the epitome of luxury in the markets in which they operate successfully.

– **Hermès**

Hermès is a French fashion and lifestyle brand that has repeatedly been described as one of the most valuable luxury brands in the world. This uncompromising luxury company is one of the few that does not belong to a luxury group (LVMH or Kering etc.) but has been family-run since the company was founded in Paris in 1837. Based on its brand history as a supplier of equestrian equipment for the nobility, the brand has expanded globally in various market segments in order to participate in the lucrative volume business (cosmetics, accessories, etc.). However, care has always been taken to ensure that the brand perception focuses on the most exclusive core products (Hermès bags, etc.) and that all initiatives are implemented in harmony with the brand identity, which is based on origin (French luxury culture), craftsmanship (established luxury competence), and the claim to the very highest quality and exclusivity.

In the market environment in which Hermès operates significantly, there are indeed no relevant competitors. Which leather handbag can compete with a Kelly or Birkin bag, which usually cost more than 10,000 euros? And above all: Which company is in a position to permanently enforce such prices in the market? Hermès represents the highest level of exclusivity, there is nothing comparable with regard to the timeless appeal of brand products in an otherwise fast-moving industry.

According to Axel Dumas, currently CEO of the family business, Hèrmes does not maintain a marketing department and emphasizes that products and initiatives are never developed as an immediate response to market demands (Cf. Roll 2018). Dumas confirms in this context that Hermès does not carry out a competitive positioning analysis. The company extracts its competence and market relevance from its identity "from within" and develops further to maintain it.

– **Porsche**

The Porsche brand can also look back on a long history. Founded in 1931 by Ferdinand Porsche, the ingredients for the so-called Porsche DNA, the brand identity, consist of a successful racing history, the superior competence of German engineering, the unique esthetics, and, ultimately, the brand core product in which this DNA manifests itself, the 911.

Since the introduction of the 911, the design concept has remained virtually unchanged, representing the materialization of Porsche's core brand values (superiority, dynamics, and esthetics) for decades. Technically, the 911 is continuously updated while maintaining the design concept in order to always meet the quality standards of the state of the art.

In contrast to Hermès, the 911, as a sub-brand and product, plays a far more fundamental role in Porsche's brand identity than, for example, the Kelly or Birkin Bag. Long into the 1970s, Porsche pursued a monobrand strategy with the 911 being the only product (and this includes the original Porsche 356 and its successor models). It was not until 1975 that attempts were made to address lower price segments in the sports car market with the 924, 944 and later with the Boxter and Cayman. However, Porsche only made its breakthrough into the large (but still exclusive) volume business with its entry into the SUV sector, initially with the Cayenne, followed by the Macan, the highest-volume model in the portfolio to date.

Porsche is also pursuing a global strategy with an umbrella brand approach based on its core brand product, the 911. The premium sports car segment also includes brands such as Ferrari and Lamborghini, as well as top Mercedes and BMW models. Nevertheless, a Porsche driver (or someone who plans to become one) would never consider these brands as an alternative to "his dream car." There is no substitute for Porsche buyers and prospective buyers, since nothing comparable exists from their point of view.

The case studies of Hèrmes and Porsche show that they derive their persuasiveness and fascination from their brand identity. For their buyers and potential buyers, there are no alternatives. A positioning in the classical sense—as a strategic differentiation effort vs. competitive environment—is therefore obsolete. However, this statement only applies to "genuine" luxury brands. Not all brands in the luxury segment have such a stable and unique brand identity and can be classified as luxury brands in the narrower sense. The less prominent brands, which are still in the process of development or are trying to establish themselves in the high-end sector through revitalization, are usually also subjected to a competitive and positioning comparison.

Luxury Brands in Tourism

After looking at brand icons from the world of physical luxury consumer goods, the question arises as to whether comparably successful brands also exist in luxury tourism. As already briefly mentioned, one is most likely to find what one is looking for in the hotel business: Brands such as Ritz Paris, Plaza New York, Raffles Singapore, as monobrands with a strong brand identity, still fascinate global luxury customers today. Here, too, it seems that the brand's radiance is based on history, origin, and thus identity. Behind the brand name of a Hotel Ritz Paris stand legends and stories (such as those of Coco Chanel) as well as at The Plaza New York (meeting place of the upper ten thousand) or Raffles Singapore (lifestyle of the great colonial period). However, while these "historical brand icons" are to be regarded as unique solitaires in the hotel landscape, umbrella brand concepts such as Four Seasons, One & Only, and Ritz Carlton have succeeded in building brand substance and exploiting the potential to expand successfully globally.

If identity and origin are so decisive, the question arises whether it is even possible to establish a new brand in the luxury business in a relatively short time. In principle, this is possible but extremely difficult. Complete new launches of luxury hotel brands often only seem to succeed with extraordinary concepts and their sensational designs, which have to be communicated with great effort via global marketing activities and penetrated into the market (for example, Soneva in the Maldives, North Island in the Seychelles). In times of multimedia and digitalization, the development of new brands in the tourism sector is therefore also complex and very cost-intensive, as is the case in the physical consumer goods sector.

How can we successfully meet the increasing demand from the growing global luxury clientele? In order to avoid the risky and cost-intensive development of new brands, it has become increasingly evident that since the first launch of the "Designer Hotel" from Versace (on the Gold Coast of Australia in 2000) brands from outside the industry—preferably from the fashion and lifestyle industry—are being transferred to the luxury hotel business.

The objective of these brand transfers is always to use existing brand awareness and images for the new business segment. The global "luxury community" ought to be addressed via the brands they are familiar with and appreciate, with the promise of being able to immerse themselves in the respective brand worlds and thus turn their hotel stay into a unique (brand) experience. From the point of view of fashion companies, the business potential of a brand can thus be expanded and further exploited through its conceptual expansion as a lifestyle brand. The expansion already achieved by most fashion brands in the interior and accessories markets (almost every luxury fashion brand offers a home collection) could thus experience a stronger upswing and level the entry into new markets that contribute to increasing synergy effects between the divisions.

For the hotel business, the initiators hope that these lifestyle brands will open up a new strategic orientation in order to differentiate themselves in the increasingly intense competition with an individual identity offering. The transfer of a brand from the physical product area to the intangible service sector can also be seen as an attempt to make the range of services even more concrete and to make them visually and tactilely perceptible for customers. In addition, the hotel industry hopes to develop new service and experience dimensions for its business segment in collaboration with marketing experts from outside the industry. It is about more than just design and interior, namely the development and realization of unique experiences during the hotel stay. Hotel restaurants, services, but also the selection of hotel guests are intended to transport the customer into a physical and sociocultural world with a far higher degree of identity and individualization.

However, this type of brand transfer to a new business segment is not always successful. The central success factor is first and foremost a consistent, brand conceptually clean transfer to the hotel business in all its facets in order to convince the discerning brand connoisseurs. In addition, it should always be checked to what extent a fashion brand is suitable for a transfer outside the industry. It should be borne in mind, for example, that the fashion sector is always subject to fast-paced change, while the hotel business stands for consistency. A designer hotel concept could make it necessary to regularly adapt the "look" of a hotel under a fashion brand to trends, which could drive up the already high costs for design and maintenance (Cf. Shankman 2018).

The alliance between fashion label and hotel business (Armani, Missoni, Bulgari hotels, etc.) does not always seem to be crowned with success. Particularly in the established industrialized countries, where an intrinsically motivated and thus discreet luxury style is increasingly establishing itself, extrinsic consumer concepts are less popular. The market situation is different in the emerging markets. Here, demonstrative luxury consumption plays a central role, so that a stay in a well-known designer hotel meets the status-oriented consumer need (see also Sect. 2).

The "instagramization" of experience consumption and thus the media aspect of luxury consumption should not be underestimated. In times of social media, a global luxury lifestyle brand as a hotel address tends to generate more attention and followers than conventional hotel brands. This aspect can be of decisive importance for young, brand-conscious luxury customers who are growing up in China and other emerging markets (Cf. Ipsos 2017a, b). A gap between the regional luxury markets seems to be developing, as a countertrend is emerging in established industrialized countries. Luxury customers, especially the high-net-worth individuals, are increasingly shying away from media presence and attach importance to discretion, data protection, and media-free stays.

In conclusion, it can be stated that the increasing intensity of competition has also led to initiatives in the luxury segment to exploit brand potential from other business areas. In addition to brand transfers, co-branding approaches are increasingly being observed in the hotel industry. In addition to working with star chefs, a classic co-branding activity (for example Ducasse in Plaza Athénée, Paris), the Beverly Hills Hotel cooperates with Hollywood A celebrities as brand ambassadors, while the Loews-Hotel Group has contracted well-known personal trainers (cf. Festa 2015). Decisive for a successful cooperation, however, are the coherence and the conceptually comprehensible complement of both brands as well as the authenticity of the benefits. A star chef who only sells his (brand) name without actually being present in the hotel restaurant offers no added value, at least for the connoisseur. The fact that even in extrinsically influenced Dubai, customers are increasingly turning away from luxury restaurants with "star chef licenses" could possibly be seen as the first sign of changes in attitudes in emerging markets.

2 Market Segmentation

Keiko Kirihara and Marc Aeberhard

An essential success factor for a marketing strategy approach is the identification and processing of relevant markets or market segments that are as homogeneous as possible in terms of demand structure and profile. This is the only way to ensure that the marketing mix measures are accepted effectively and can thus be success-fully implemented. Only products and marketing measures (such as communication, distribution, pricing, etc.) that are aimed at a specific target group and therefore satisfy its specific needs will have a high chance of success in the markets. Different segmentation approaches, which are presented in the following, are used to define relevant markets.

Segmentation Approaches as a Result of the Corporate and Marketing Strategy

The fundamental question of segmentation depends initially on the general corporate strategy. A company that is committed to a global business strategy is unlikely to undertake geographic segmentation of the market. A different starting situation arises with a multinational strategy: geographical criteria are used to identify and concretize the relevant market segment. For example, markets can be segmented by region (Europe, Asia, etc.), state, federal state, etc.

However, since the geographical location of target groups can be associated not only with economic and political but also with other criteria such as psychographic, behavioral, or sociocultural characteristics, globally oriented companies have long tended to segment strategic market regions accordingly.

Psychographic segmentation criteria are basically lifestyles, individual personal-ity traits, and belonging to a social class, while behavioral criteria are largely related to the consequences of different attitudes, values, and norms.

In the age of globalization, Europe is often seen as one region. However, reality shows that, for example, there is a North-South divide in Western Europe, which means that attitudes and values (and thus purchasing behavior) can vary greatly depending on whether the consumer lives in North/Central Europe or in Southern

Europe/Mediterranean regions of Europe. In the cosmetics sector, for example, it is well known that attitudes toward women's beauty vary greatly from region to region in Europe. While in the Mediterranean region the attitude that women should be feminine dominates the majority, this attitude is rather underrepresented among women in Central and Northern Europe. Gender-specific understanding of roles as well as other differences in values and norms thus characterize the North–South relationship even within Western Europe. Furthermore, there are also differentiated behavioral patterns due to different economic conditions (purchasing power), climate, hygiene behavior, eating habits, etc., factors that have developed evolutionarily over a long period of time.

The East-West divide in Europe is even more drastic. Almost 30 years after the collapse of the Eastern bloc, there is still a considerable economic gap, apart from the already large differences in socio-political and cultural backgrounds. In 2016, for example, the purchasing power of Ukrainians reached just 80% of the average purchasing power of a Lichtenstein citizen (Cf. GfK 2016).

Taking in Account the Sociocultural Aspects

If such differences in attitudes and behavior already exist within Europe, which in the age of globalization is more or less regarded as one cultural area, then it is understandable what is happening in the rest of the world and in the regions. It is therefore all the more astonishing that the strategic departments of global parent companies continue to strive to implement a marketing strategy that is as globally standardized as possible, while a systemic integration of regional–local approaches is rarely found. However, sociocultural peculiarities of strategic markets (such as China) are increasingly taken into account in global-glocal marketing, especially since market research results often suggest local adjustments.

Demographic Segmentation

Finally, reference is made to demographic segmentation, which includes criteria such as age, gender, marital status, education, occupation, and income. In the age of global migration, demographic characteristics such as ethnicity, religion, and nationality are also relevant within national borders.

Compared to psychographic or behavioral criteria, the advantage of demographic characteristics is the availability of statistics that allow concrete quantification of segments. Demographic criteria thus represent the first opportunity to quantitatively outline a still little-known market and assess its segments in terms of business potential. In developing and emerging countries, in particular, purchasing power and income classifications are an initial basis for market entry decisions for Western consumer goods companies. If there is insufficient purchasing power for Western products, all further research efforts are obsolete.

In our saturated and secularized economies, the aging of society and thus the shift in target groups to the so-called "silver generation" of baby boomers, the largest and most affluent consumer group, also play a key role in the segmentation and identification of target markets. While our society still cannot say goodbye to the youth cult of the 1970s, the relevant target groups for almost all consumer goods are getting older and older. How industry deals with this contradiction is reflected in the diversity of measures. For example, companies try to adapt their product offerings better to the changing needs of older consumers (such as better legibility or operation of technical products) without explicitly addressing age, while the desire to preserve youth is translated into advertising statements such as "vitality" and "joie de vivre." In the leisure and tourism industry, the dominance of older target groups has been reflected as a boom in the spa and wellness sector. The dream of eternal youth is as old as mankind, but never before has so much energy been spent as in this day and age to fulfill this wish. In the markets of the established industrial nations, the preservation of health and vital appearance is a mega-trend with high sustainability.

In global terms, however, the growing and large age group of young people in developing and emerging countries represents the relevant target group. While in China, due to the rigorous one-child policy of the 1980s (which was finally abolished in 2015), a mushroom-like structure of the age pyramid is discernible, India remains on a growth course with regard to the population. For our Western industries, this could mean that in younger growth markets other age-related product requirements have to be met than in our domestic markets, which continue to be characterized by an aging society.

Basically, it can be stated that several criteria are used to identify relevant target segments. Demographic criteria are suitable for an initial assessment of the potential of markets and market segments. Sales potential analyses thus allow an initial preselection of markets. Psychographic and behavioral criteria as well as sociocultural aspects require the use of market research studies in order to obtain relevant data. They form the basis for brand and product concepts and play a central role in determining positioning, especially in markets where psychological–emotional dimensions are decisive for purchasing preferences.

Market Segmentation in the Global Luxury Sector

The trend is continuing worldwide: Various "Global Wealth" studies (BCG, Cap Gemini, etc.) confirm a disproportionate increase in the assets of high-net-worth individuals (HNWI/High Net Worth Individuals) in 2017 compared with global GDP growth (Cf. BCG 2018).

There is no doubt that this development will continue to fuel growth in the global luxury markets of various consumer goods. The development of assets by country and region is also interesting: The USA continues to be the leader in the category of rich individuals, followed by Japan, China, and Germany, which together represent over 60% of the world's high net-worth/HNWIs. Since 2015, however, the Asia-Pacific region has already outperformed North America in terms of asset size with China being the key growth driver. Over the next 10 years, this region is expected to account for approximately 40% of the world's HNWI assets (Cf. BCG 2018).

These tectonic shifts in global wealth distribution have also been noted by the luxury industries. In the development of luxury cars, automobile manufacturers such as Mercedes and BMW are less and less targeting the domestic market and Western European customers, but are primarily concerned with the needs and wishes of the "new rich" in China and other emerging markets. The resurrection of the Maybach brand by Mercedes was conceived against the background of this strategy. The production of the luxury limousine, which has been discontinued in the meantime, was intended to meet exactly the extreme luxury requirements of the wealthy elites of these regions. Globally oriented luxury brands such as Hermès, which otherwise consistently pursue their globally standardized strategy, also make concessions when it comes to the Chinese market. For example, flagship stores with an integrated Hermès Museum and in-house craftsmen have been set up especially for China, to convey the origins and brand identity of the house up close.

Up to now, the luxury segments have predominantly been handled in a globally standardized way, often with a psychographic segmentation of the global customer groups. However, the examples of Mercedes and Hermès show that against the background of new strategic target markets (primarily China, but partly also Russia, India) and luxury customer types, a regionally differentiated view of the previously comparatively homogeneous target group is necessary.

Because as already mentioned in Sect. 1, there are fundamental differences in the understanding of luxury between the established industrial countries (the USA, Western Europe, Japan) on the one hand and the emerging countries or the rising new economic powers (China, Russia, India, etc.) on the other. While in the established industrial countries luxury is rather an expression of an individual lifestyle (with a tendency toward intrinsic hedonistic motives), in the growth markets such as China luxury consumption is seen as a status and demonstration of economic success and social advancement (Cf. Wittig et al. 2014, p. 66 ff.).

The reasons for these differences are, on the one hand, the different life cycle phases in which the respective luxury markets find themselves and the different levels of development and maturity of the luxury customers. On the other hand, however, sociocultural differences such as different views of esthetics can also be referred to. While eye-catching luxury goods ("bling bling" luxury), for example, are very popular in the Arab Emirates, Russia, and also in collectivist China, this luxury style has always been regarded as vulgar and less desirable in Japan and in Western European regions (Fig. 1).

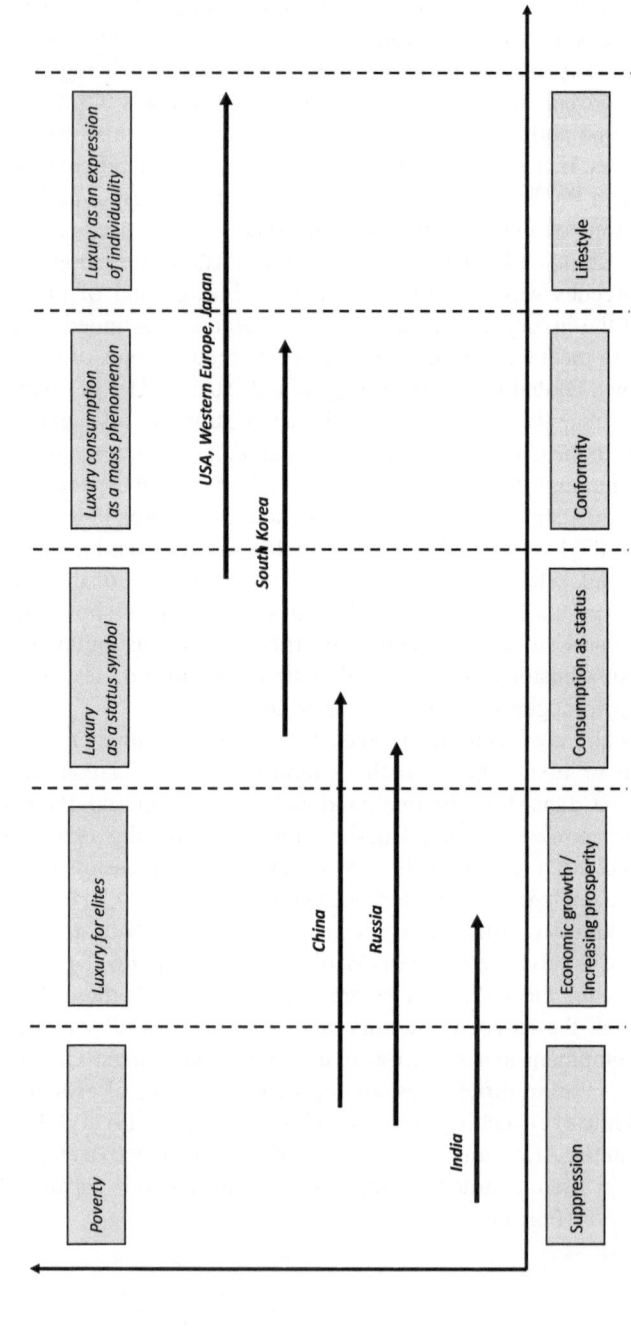

Fig. 1 Luxury segments in tourism (source: Roland Berger Master Circle 2005)

Luxury Segments in Tourism

Transferred to the tourism sector, it is the status-oriented customers who practically buy their social affiliation by staying in hotels with brand names and the "right" destinations, ordering expensive wines with well-known labels and visiting Michelin-starred restaurants, while the intrinsically shaped, discreet luxury customers avoid precisely this attention and rather distinguish themselves through connoisseurship and insider contacts within their peer group.

The high degree of maturity of tourism in the established industrial countries is also reflected in a further differentiation of the luxury hotel industry into the top-end and high-end segments. While the latter is more likely to be classified as a conventional luxury hotel (average daily rate—ADR up to approx. 500 euros, including large hotel brands, etc.), the top-end segment (ADR > 1000 euros) is characterized by exclusive, more "private" accommodation with "retreat and hideaway character." This "New Luxury" approach represents a dramatic paradigm shift in the saturated Western markets toward a new understanding of luxury stays that encompasses aspects such as space, time, service, safety, etc. and introduces a new dimension of individualization in the hotel business.

In summary, a new order of magnitude of diversity is emerging, above all regionally and socioculturally, when one considers the global target groups of the luxury industries. In the past, the customer profile was mainly western, but the luxury customers of the future from countries such as China, India, Russia, etc. show a high degree of diversity and cultural otherness. The knowledge and consideration of the different views, attitudes, and needs/motives on the subject of luxury and luxury consumption are of significant importance for the successful development of these markets.

"Millennials" in China

Even though the introduction of anti-corruption legislation and currency devaluation in 2015 have significantly dampened luxury consumption in China, the luxury segments remain on a growth course. Since then, the tendency to pursue luxury consumption abroad (80% of luxury consumption takes place abroad) has been conspicuous in order to probably evade public control and possible repression. This is currently accompanied by the fact that the particularly travel-friendly "millennials" (25–34 years) are striving for luxury consumption not only as an acquisition of physical products but also as an experience consumption (Cf. Ipsos 2017a, b).

At first glance, Millennials in China thus seem to differentiate themselves less and less from young consumers in established industrial nations. The decisive difference, however, lies in the fact that the desire for demonstration and "proof" of the luxury experience is overly pronounced for the Chinese millennials. Social media play a central role here, underlining the extrinsic focus for luxury experiences as well.

3   Marketing Tools

Keiko Kirihara and Marc Aeberhard

Once the strategy derived from the corporate and marketing objectives has been defined, it is time to implement it operationally using the four marketing mix instruments. Now we are entering the market-effective action level, i.e., marketing activities that are aimed directly at the consumer, the customer. The decisive factor for the successful implementation of the strategy in practice is first of all to coordinate the various marketing mix instruments (product, price, communication, and distribution policy) inclusively and to ensure strict coherence. Because, if the individual measures and their signals to the target group are only insufficiently coordinated and not conclusive in themselves, the brand and benefits could be perceived as diffuse and thus only inadequately address the target group.

Product and Service Policy

The core of the marketing instrument is the product. On the basis of a defined brand strategy, the target group approach is concretized in the form of a product offering consisting of a bundle of (functional–rational/psychological–emotional) benefits. Thus, the product can be physical (consumer and capital goods, etc.) or nonphysical (services, travel, experiences, ideas, digitized films, etc.).

Due to their nonphysical nature, intangible service products or services have special features such as the inability to verify the benefits before purchase and the associated increased purchase risk. The involvement of the user in the provision of services and the inability to store services are also typical of service products. A particularly big challenge is also the control and assurance of quality standards, since services are usually provided by personnel, by people, often in complex combination with other service components of a physical nature. In addition, in tourism in general and in the hotel industry in particular, many factors influencing the product cannot be controlled, but are decisive for the purchase. These include in particular weather and climate influences, sociopolitical and economic developments/instabilities, trends and fads, etc.

Service products can be categorized from the proportion between nonphysical and physical product components (or personnel vs. equipment load) (Cf. Kotler and Keller 2016, p. 423). While in restaurants, for example, the tangible part of the product predominates, in legal advice the intangible benefit takes up the majority of the product. In the case of a holiday trip, the ratio may vary in individual cases, but the proportion of nonphysical product components tends to increase. The higher the proportion of services, the more likely it is that the special features or challenges mentioned will become apparent in the case of a service product. In connection with the increased purchase risk, one also speaks of experience or even trust goods, in which the consumer relies on his previous user experience at the time of the purchase decision or—if this is not possible—must trust the supplier and his promise of benefits.

The consequences for the marketing mix instruments can be considerable. The product ratings of other consumers (word of mouth, rating portals, etc.) play an even more important role for the final purchase decision than for physical goods, which can be checked out in advance by the prospective buyer. Pricing and the image of the corporate and product brand, as well as the physical product part, can also provide decisive indications of product quality from the customer's perspective.

The Luxury Product in Tourism

As already mentioned several times, the focus of a luxury brand and the corresponding product is on the psychological–emotional benefit dimension, which—successfully addressed to the luxury consumer—is perceived as a reward.

According to Wittig et al. (2014, p. 159), (luxury) travel comes closest to the idea of a luxury experience. The affluent spend almost half more on holiday travel than the average and, according to a study by VISA, the discrepancy is expected to increase even more to around 80% in the near future. While cost-conscious consumers are increasingly booking their holidays online, 70% of luxury travelers have their holidays organized by travel agencies. Also against the background of the mega-trend of individualization, it can be assumed that the need for unique experiences, for undiscovered, exotic places in the world will continue to increase; especially among those who can afford to satisfy these needs.

According to the wants/needs of the luxury clientele, the place of stay is less the global hotels of well-known companies, but extraordinary, individual hotels, which—as witnesses of local history—often make the stay a unique experience (Cf. Wittig et al. 2014, p. 160). Here, too, the cornerstones of luxury consumption are discernible: luxury journeys that convey unique experiences are in demand. Places and residences of authentic origin/identity that are still untouched (or at least less known) by global tourism enable luxury travelers to immerse themselves in a still undiscovered world.

Especially via the online channel, luxury niche products are increasingly being offered and lucratively communicated to the global, individualistically oriented luxury clientele:

National Geographic offers expeditions with an ecological–intellectual touch with *National Geographic Expeditions*. As a mission, the spin-off of the venerable Royal Geographical Society has set itself the goal of strengthening ecological awareness of the environment and nature. A wide range of different destinations and travel formats (from simple train journeys to private jets, from family trips to photo explorations, etc.) promise a very high degree of individualization of the offer for unique and authentic experiences. This promise is to be secured through the experience and competence of the organization with the "guarantee seal" of the

National Geographic brand, supported by many years of cooperation with international institutes and experts (www.nationalgeographic.com).

Tourism experts who have gathered their knowledge and experience in the luxury sector offer extraordinary and authentic luxury holidays beyond mainstream tourism through their websites. The operators of the Portal *Design Hotels, for* example, recognized the need for individual travel early on and have been successfully offering individual design hotels worldwide since the early 1990s. Another provider, *Mr. & Mrs. Smith,* gives individual travelers the choice between boutique hotels and villas, which are rated and recommended by prominent guests such as Stella McCartney or Cate Blanchett (Cf. Spiegel Online 2017).

Another alternative form of exclusive stay is the "clubhouses" of exclusive private clubs such as *Soho House*, which have established themselves in selected metropolises around the world. Founded more than 20 years ago as an alternative to the traditional and stiff upper-class clubs in London, Soho House offers a social platform for the global elite from various industries. Preference is given to club members from the creative industries such as film, music, and art (Cf. Wittig et al. 2014, p. 168). The less commercial Soho House has since attracted a number of imitators, especially as the ingredients of a unique, very exclusive location have a strong appeal to many wealthy people without the prominent network. There are now similar private exclusive clubs in almost every global metropolis, such as the Silencio in Paris or the China Club in Berlin (Cf. Fitzmaurice 2017). Some providers of luxury brands (e.g., Dunhill) have also recognized this trend and are establishing "brand clubs and houses" in the centers of the world in order to offer their luxury customers a comprehensive brand experience and to bind them to themselves in the long term (e.g., Dunhill operates clubs in London, Tokyo, Shanghai, Hong Kong) (Cf. Wittig et al. 2014, p. 163).

The top-end customer is also "submerging" more and more. While 40 years ago the top hotels of this world were still sufficient for the discerning clientele, today they react to the changed perception of luxury by building their own infrastructure. Thus, the guest does not go to the hotel anymore, but lets the hotel come to him. The Haute Volée maintains its châlets, villas, yachts, or lofts in the hottest destinations, cared for by its own team, from butler, housekeeping, cook to driver. The luxury villas outperform most hotels many times over in terms of furnishings, fittings, and comfort. In addition, they are individually furnished and passed from guest to guest under the hand, so that this clientele can remain among themselves, under the radar, in paparazzi-free zones.

Airbnb, the digital giant among the providers of individual stays, has currently expanded into the higher-priced segment with Airbnb Plus. The entry into the luxury segment with *Beyond by Airbnb,* which is even higher in price and has so far been unoccupied by Airbnb, was completed in 2018 (Cf. Airbnb 2018). It can be assumed that the luxury trend in tourism will be brought to a broader market base with such new launches.

Pricing Policy

Pricing policy as an integral part of marketing is derived from the positioning of the brand and the product. If it is a quality product, the price positioning will be in the upper mid-price segment or in the upper price segment—depending on the brand concept. The direct impact on sales gives pricing policy an important position in the marketing mix, although from a marketing point of view it is seen less as a tactical instrument than as a brand strategy instrument. Pricing directly influences buying behavior in most markets, which are more or less elastic. Before a consumer's needs and desires stands a budget, which usually sets limits to his or her consumption needs.

The price positioning should be perceived by the relevant target group as balanced between the required absolute price setting and the benefits offered. In the event of an imbalance (e.g., perception of an excessive price), the product offer could be perceived as "usury," while a very low price setting could raise doubts, e.g., with regard to product quality. Conceptual and positioning considerations as well as subjective customer perceptions play a decisive role in pricing policy.

Role of Pricing Policy in the Luxury Sector

In the luxury segment, pricing policy plays a less dominant role than in the price segments of the mass markets. In the case of target groups that have large consumption budgets and therefore have above-average purchasing willingness and purchasing power, one can rather assume that purchasing behavior is price-inelastic.

Pricing policy in the luxury sector serves primarily to exclude those who have less purchasing power and thus to underline the exclusivity of luxury consumption. In emerging markets, where extrinsic demonstrative luxury consumption often dominates, the announcement of an extremely high and thus exclusive pricing (most expensive luxury property in the world or similar) can have a motivating effect on the willingness of rich customers to buy. In such cases, one could even assume a positive price elasticity.

Wealthy or even rich people, however, are not considered ignorant without any price or cost awareness. The perception of a balanced and thus "right" price–performance ratio of the product range also applies to luxury customers in order to support the purchase decision. With reference to the concept of the psychological–emotional dimension (luxury consumption as a reward), it is particularly true for the luxury markets that the subjectively perceived or expected "reward" should correspond to the premium price of luxury consumption so that a purchase decision can be made. However, it should be noted that the psychological–emotional benefit is more difficult to quantify than the functional–rational product performance.

At least as important as the "right" price–performance ratio in the luxury sector is the consistency of pricing and thus the stability of price positioning. Luxury and its rewarding experiences cannot be offered at a reduced price as a bargain. The credibility of the highest quality standards, the authenticity of the luxury brand, and its products prohibit any seasonal price adjustments. For example, it is part of

Ferrari's policy that prices should be fixed regardless of all influences and should not be negotiable.

For intangible products such as luxury travel, which are not storable, the exclusion of seasonal price reduction opportunities represents a considerable challenge. For example, the hotel industry is struggling with the phenomenon that every room that is not sold irretrievably means potentially lost turnover. Therefore, the temptation is always great to plunge into the adventure of supply and demand and to expose oneself to the sometimes extremely volatile elasticity of demand. In principle, larger hotels are more willing to adjust their pricing policies. However, the smaller and more exclusive a house is, the more sensitively price reductions are assessed. Experience in dealing with both direct customers and travel agents shows that once prices have been adjusted, Pandora's box has been opened. Tough price negotiations become unavoidable and as a rule, the hotel usually loses in the end.

Novel business models such as Secret Escapes, which have specialized in "bargains for luxury travelers," are trying to use this dilemma of luxury providers for themselves. Top-end vendors penalize such business models, but they pose a challenge to top-of-mass vendors by undermining the maximized price target. In the luxury hotel industry, the following applies: Those who participate in such portals are desperately struggling with massive sales problems.

Communication Policy

After brand and product positioning has been translated into products and value propositions, brand and product messages must be conveyed correctly and effectively to the target group. It is always crucial for successful communication that the intended messages are coded accordingly in signals (speech, images, music, etc.), which in turn must be correctly decoded and understood by the target group.

Everyone knows that in a conversation, in communication between people, there are plenty of reasons for misunderstandings. Now one can imagine that in a global context, in communication with a variety of target groups (with different languages, sociocultural backgrounds, etc.), the danger of misunderstandings is enormous. That is why global digital companies like Airbnb work with a variety of local marketing experts and translators to transfer their global communications (primarily via websites) and messages so that they always "feel right" to their customers. This includes, for example, not only correct translation but also the consideration of regional dialects and local expressions.

In addition to the complexity of global communication, a paradigm shift in communication policy has taken place against the background of digitalization. Whereas before the Internet age the distribution of roles between sender (industry/advertiser) and receiver (consumer/customer) was clearly defined, the digitalization of communication led to the dissolution of this strict distribution of roles. Consumers and customers are not only recipients of messages, but can also be found on blogs and vlogs as part of a community, or as influencers, etc. creating content and sending messages over the web. In this context, there is also talk of a democratization of opinion. In any case, the fact is that in times of the Internet, companies no longer have the possibility to comprehensively control and steer the communication flows that affect their brands and products.

In principle, the communication policy consists of advertising, public relations, sales promotion, personal sales, and direct marketing. But here, too, the boundaries between the communication media are increasingly blurring due to the Internet. In the past, advertising and PR were strictly separate communication instruments, now the transitions are smooth. In a highly complex and competitive media landscape, pure push advertising alone is not enough to build brand awareness and image, while the promotion of pull effects via content marketing is becoming increasingly important. Here, companies try to offer information relevant to target groups (only indirectly related to the brand/product) online in order to direct potential customers to their own communication media/websites and ultimately to generate attention and interest for their own offers.

Against the background of the advance of digital media, the following categories of communication forms can be defined: Paid–Owned–Earned Media. While Paid Media covers the classic area of "purchased" advertising, Owned Media represents the information provided by the company (such as the website). Earned Media, on the other hand, represents the information/content that customers or potential customers produce by, for example, looking at the brand/product, reporting, and evaluating their experiences. This type of content, which is generated by users and potential customers, represents the most valuable form of communication from the company's point of view, as long as it reports positively about the brand/product. Earned Media is comparable to word of mouth, which is known to be very credible and therefore extremely effective for image building. The importance of social media for the industry as a communication channel is therefore largely due to the fact that Earned Media is primarily generated in social networks.

Special Features in the Luxury Sector

Mass media are not and have never been important for luxury markets. First of all, it should be noted that the dispersion losses would be far too great. Why should a product be advertised heavily to consumers if most of them cannot afford the product in the first place? In addition, it should be noted that exclusivity and thus difficult access to a genuine luxury brand is an important cornerstone for the appeal of luxury consumption. Against this background, mass media advertising of an exclusive product that excludes many consumers is a contradiction in terms.

This should clarify the relationship of the luxury industry to mass media, especially to TV. In certain glossy magazines, one will nevertheless still find print advertisements of High-End brands such as Hermès or Porsche. The reason for this is not so much the disproportionately well represented wealthy readers as the intention of the companies to always maintain a certain level of brand awareness and topicality. Exclusivity and status consumption, which most people do not know (and therefore do not envy), lose their appeal, at least for extrinsically motivated luxury customers.

There are also, of course, the intrinsically motivated customers of the luxury sector, or at least those who are reluctant to be identified in public as privileged wealthy people. In a society of envy, such as Germany, many luxury customers prefer the intrinsic pleasure of luxury as so-called connoisseurs or extrinsic recognition exclusively from one's peers. For example, more discreet initiatives such as art and culture sponsoring, charity parties and events are preferred communication measures in which luxury customers can remain among themselves and are not subject to social criticism.

In principle, personal contact continues to be of central importance for target groups in the luxury segment. There is no doubt that there is a more individual and therefore tailor-made form of dialog with the customer. As already explained in connection with the product policy, personal relationship management, and competent individual advice will also be increasingly decisive for the success of supplier companies in luxury tourism. Especially since countless brand promises in advertising communication, which usually only promise superlatives in the hotel industry, have lost their credibility. Even the well-intentioned beginnings of evaluation portals such as Tripadvisor or Holidaycheck etc. are now largely undermined in their effect, as these instruments are increasingly being misused to harm competitors or even to demand additional services from tourism providers. The importance of word-of-mouth recommendations is therefore the most important and most reliable measure today to spread a message precisely, especially in the top-end segment. This results in an interesting inner circle consisting of selected travel agents, hoteliers, and customers such as "The Net" or the Aman-Junkies.

Against the background of increasing digitalization, the sensible and complementary addition of digital media to personal communication will nevertheless be able to give luxury providers a decisive competitive edge. Presence in social media, use of new visualization technologies (virtual and augmented reality, etc.), digital information provision and digital dialog to complete the individual communication mix are increasingly becoming standard requirements that must function smoothly and efficiently, especially in the luxury segment. However, the highest priority is and remains the maintenance of a personal relationship of trust and the assurance of the loyalty of discerning and demanding customers.

Distribution Policy

The product is delivered to the end consumer via the distribution and sales channels. It is only here, via the sale, that the company's value-added process ends, in that the product produced is converted into sales and (minus the costs) into profit. It is often the trade or the intermediary organizations that not only transport, store, and distribute the product, but also carry out sales, promotion, and liability tasks. The knowledge generated by their proximity to customers and their experience create sales advantages for retailers or intermediaries, which they demand from the manufacturing companies through margins or commissions.

The balance of power in the service sector/tourism is quite different. Here, intermediary organizations (brokers/agencies) predominate, which are increasingly being pushed back by the online channels. In the case of a nonphysical product, essential logistical tasks as well as physical distribution are omitted, so that direct online sales are suitable for the suppliers (tour operators and airlines, etc.). At the same time, the companies are also using sales portals of intermediaries (e.g., www.booking.com) or online travel agencies (e.g., Expedia) as part of a multi-channel approach in order to offer their products to target groups.

Distribution in the Luxury Sector

The more complex, the more consulting intensive and thus the more individual the relationship between brand or product and customer is, the more crucial it is for the manufacturer to be close to the customer and to gain a specific insight into his or her sensitivities. In most luxury industries, for example, very close contact is maintained with retailers and intermediaries, which is achieved through regular and intensive brand and product training as an integrated marketing mix instrument.

The exclusivity of luxury consumption is also cultivated in distribution policy—at least by a few companies. For example, a potential buyer of Hermès' Kelly Bag will not find the bag in the store. A rather inexperienced prospective customer will probably even be rejected for the time being with reference to a long waiting list. Another example of deliberately managed distribution bottlenecks is the long waiting times in the automotive industry when ordering a new luxury model. Here, too, distribution policy is intended to further fuel the desires of customers through artificial scarcity.

However, the focus of most luxury suppliers lies more on the efforts to use the variety of available distribution channels in order to achieve the most effective brand presence possible. The distribution focus of (physical) luxury brands often lies in direct sales with their own flagship stores with the aim of presenting as comprehensive and fascinating a brand world as possible. The focus here is obviously not only on sales and thus direct sales but also on the communicative effect of an offline business as a sensual experience of the brand world.

As a rule, flagship stores are generally accompanied by an online presence and sales as an increasingly important complement to the distribution spectrum (e.g., Burberry, Fendi, etc.). However, the online sales channel had long been neglected by some companies in the luxury goods industry on the assumption that luxury goods can only be sold through personal contact. But companies like Prada, which have underestimated the increasing dominance of online sales in the luxury sector (and thus also the "digital natives" as the next generation of luxury customers), have recently paid the price bitterly by stagnating sales in a market that continues to grow (Cf. Schroder 2018). Particularly in key markets such as China, e-commerce is progressing at an incredibly high pace and it is imperative for market success that the aforementioned millennials as an increasingly central customer group be handled with the latest mobile marketing initiatives.

Special Features of Luxury Tourism

Holiday trips in the high-end and top-end segment often cause considerable expenses (the price of a trip can easily reach the equivalent of a mid-range car). Therefore, the importance of the travel agent of trust is increasing again. It is about real advice (with on-site experience) and ultimately about trust. The discerning customer wants to make sure that he does not experience any unpleasant surprises on site, that his ideas are perfectly implemented, fulfilled, or even exceeded, and that he does not lose any time himself for travel planning and booking. It is about the concept of the one-stop shop. Seen in this light, the specialized segment of exclusive tour operators and travel agencies is currently experiencing a small boom.

In the digital age of the multichannel approach, however, the synergetic use of online and offline channels is also a critical success factor for the distribution policy in luxury tourism. However, online channels play an increasingly important role almost exclusively for communication and information gathering, but not for sales. While digitalization is increasing rapidly in the budget and first-class segments, it will only continue to develop to a very limited extent in the top-end segment. Tourism in general and the hotel industry in particular are service industries based on the provision of services by personnel, by people. An ITB survey conducted in 2015, for example, confirmed that individualized and personalized service is the most important luxury criterion (also important: time; cf. Ruetz and Aeberhard 2018). While the expansion and optimization of electronic communication and distribution tools are important, the main focus in luxury tourism is on the human being. As a personality, he or she should be perceived as sincere, worthy, respectful, and authentic, and as a service provider (not as a servant) he or she should deliver the most coveted product ever: to conjure up a smile from one person to another and thus give lasting "reward experiences."

4   Storytelling in a Digital Age

Juliet Kinsman

Mastering the Art of "Tell, Don't Sell"

Juliet Kinsman, hotel expert and journalist has fine-tuned the art of creating a compelling tone of voice for a travel brand. As founding editor of Mr & Mrs Smith, she knows better than anyone that it is imperative for hotels that are marketing higher-end products to the consumer, to not only develop the correct communication strategy to influence purchasing decisions and win loyalty but also to speak to the right customer for your brand. Her latest venture Bouteco is a unique independent authority on boutique eco-hotels and sustainable luxury-travel experiences. A social enterprise dedicated to making hospitality a force for good and creating inspiring content with integrity she works with hotels to weave meaningful sustainability stories into their philosophy and in doing so establish a halo effect.

Content marketing is everywhere we turn. Digital communications are being blasted at us from all directions from a panoply of platforms. So, it is more important than ever for a luxury hotel brand to cultivate a voice that the new generation of luxury travelers actually wants to hear.

It is crucial for travel brands to shape a compelling personality and ensure that as part of this they give themselves an engaging tone of voice. It is important to accurately and concisely articulate what a hotel's purpose, vision, and ethos is, and when it comes to the website, social media, e-communications and beyond, any luxury brand will do well to sound more human, more authoritative and distinct from its competitor hotels.

Language Is Everything

A vision and a clear and robust message that communicates a hotel business' core values to exactly the right audience is what hotel marketing is all about. And so, it is vital to communicate what you are and what you are about in the right way. This means ensuring that the text you use, and the way you use those words consistently across all that you do and think about how best to talk to your clients and potential clients to build meaningful relationships with them, to influence their purchasing, and encourage those all-important direct bookings.

Just as we are drawn to charismatic human beings, we are drawn to brands and businesses that feel as though they have been lovingly created by a charismatic human being. Visionaries who run their hotel with passion, and who are personally telling people about their project through the language they use in their marketing. Luxury travelers are desensitized to faceless corporate chains which only cares about profit. The way readers and followers will listen more acutely and trust in what you say, is if you sound genuine. One whiff of corporate messaging and today's ever-skeptical consumers will switch off.

So how to achieve this? Understanding whether you are saying this right is easy, actually. Just be sure to look at all your communications through the eyes of your dream customer. How would they feel when they land on your website? Are the key messages conveyed simply and elegantly, or have you fallen into the trap of exhaustively sharing every detail rather than considering whether anyone would actually stop to take all of that in? Do the images feel natural and beautiful and stir an emotion that makes you actually want to be there? Or do they look like something an estate agent would take to exaggerate size or features? Again, be genuine—your audience will trust and like you better. Now cast your eye over your social media posts—do your musings sound as though they have a smiling, warm human host behind them? Someone who could be a friend? Or are they nagging you to buy something? Again, it is simple, ask: would you want whatever your hotel's account is showing or saying in your timeline or feed?

Entertain and Inform

As a journalist and travel editor, my rule of thumb with all that I write has always been to share insider tips that are original, helpful, and written in a way that is fun to read. Be sure to use a varied vocabulary which piques curiosity and offers a more colorful flavor than standard could-be-computer-generated brochure-speak. Imagine you are at a dinner party, and you are the mood-lifting raconteur who is telling enlightening stories and amusing anecdotes, which not only entertain but they also inform.

Social Media: It Is All About the Storytelling

What distinguishes quality marketing materials is when it reads as though written by an editorial publication. So, think of your different content channels and social media platforms as your "magazines." This means that you need to play editor-in-chief as though you are in a planning meeting and ponder what the "sections" in your magazine would be and what the editorial themes, interests, and relevant topics you want to talk about are, and whether these are relevant and enriching for your target customer. Perhaps your publication would have a food and drink section? Interior design and architecture? Culture and the arts? Once you have decided on which "sections" you would have, make like an expert in those areas and demonstrate insight—even if it is a judicious sharing of others accounts' content. A generosity of spirit when it comes to sharing useful travel tips or lifestyle advice again helps to cultivate interest and trust.

Who Is Your Target Audience?

It is important to decide who you are talking to. Have your dream audience in mind and talk directly to them. That is what every publication does when planning their articles and how they talk to their readers. Always look at your messaging through your customers' eyes, listen to how you sound, through their ears. What benefits do people get from following you on social media or for being in your world? Branding is a lot about marketing your product, of course, but there is a knack to doing this graciously and without coming across as a shameless, self-obsessed, self-promoter. Gratitude and thanks help when sharing praise or accolades that have been directed at you is always a winner.

Do not just be on "send"—be an active listener. There are various types of social media users—lurkers who sit silently on the sidelines (you do not even know they are there, but so many of them are), the "smoasters" (social media boasters) and the indiscriminate broadcasters. Going back to the number-one rule—think of yourself in human form at a social or work event. Display curiosity, ask questions, be interested in others, comment and like and engage—it is good etiquette if nothing else.

Ask yourself: how do we come across? If your dream customer or travel agent looked at your social media profiles, what would they think? Be sure to uplift and inspire. We live in a world surrounded by a blizzard of negative messages and depressing data, why not be the glimmer of inspiration that is a positive influence on others and the world we live in? That does not mean sanctimonious, philosophical motivational quotes or kitten GIFs but it could be a link to a genuinely thought-provoking article. Have meaning and purpose behind what you do.

Become an Influencer in Your Own Right

It is an obvious marketing strategy to harness the power of influence. A simple yet effective way for a brand to reach a particular audience or demographic is through the endorsement of bloggers, influencers, and KOLs (key opinion leaders). The digital revolution amplified this type of influence through the new-media superstars. Research says that 50% of all purchasing decisions are driven by influencer recommendations in some form or other. When the conversation first started around influencers, it was the megastars that grabbed everyone's attention—the YouTube sensations with millions of followers and the Front Row fashion bloggers who struck it lucky. These influencers still exist, but it is become the more low-key specialists and micro-influencer (someone who has < 100,000 followers or less) that offer something the YouTube superstars cannot—genuine engagement.

Positioning Your Brand Through Your Personality

How to make sure that your hotel stands shoulders above its competitive set? Look at how to position your product and the brand in the market in a way that makes it distinct from all its competitors—and impart how to include more detail and make copywriting less generic so as to convert your messaging into more sales but also to ensure you do not upsell or oversell to customers, managing expectations elegantly and in the right way.

Blogger Venkatesh Rao started an interesting discussion around what he terms "premium mediocre." Those establishments where everything seems great on the surface—premium, even—but there is no real story or soul. Since it is getting harder to distinguish between what is meaningful and what is this average unimaginative premium, it is important that you position yourself as authentic and original.

Luxury, as a label, has been diluted and redefined; today's esthetes appreciate originality without gimmicks, craftsmanship, and a consideration of detail, rather than five-star frills or flashy, fancy facilities. The acid test of pleasing-to-the-eye places need not be that they make you cry, but it is a reminder that the most beautiful hotels do not just beg to be seen or touched, they should have a knack of making their residents feel something.

Using Sustainability to Boost Your Sex Appeal

Luxe living and practicing eco values should be at odds—one signals indulgence, the other abstinence, but some of the most finger-on-the-pulse hotels are proving it is glamorous to be greener. The traditional sense of what comprises luxury travel is shifting from opulence, indulgence, and superficial gloss, toward genuine compassion, empathy, and an appreciation for quality and an understanding of what cost that all comes at. And that is what sustainability is all about.

Sustainability is increasingly front of mind for the luxury traveler—not that this catch-all term necessarily seduces them. Ponder how to make your environmental initiatives and a social conscience part of your brand story. More than ever before, hotels need to show they have a big heart to inspire guests to book and to deliver a meaningful guest experience while doing their bit for the world.

More and more hotel-bookers, especially younger customers, actively look for brands that make a positive difference—today's purchasing decisions are proven to be influenced by how sustainable a brand is. It is essential to identify sustainability and community stories around what you are doing as a hotel, create more stories, and share them loudly in the best possible way for your target customer: without a whiff of greenwashing. If someone is choosing between two brands that provide similar service offerings, but one is involved in projects to make a difference to its community/the environment, chances are they are going to choose that one.

Do Good and Drive Direct Bookings

Seeking brands with purpose is a priority of the upcoming generations. Under-30s are re-evaluating their consumer choices and actions. The hotels they spend time in is a badge of their ethics. Though their budgets do not always match their eco-luxe aspirations, these are the luxury hotel bookers of tomorrow and those engaging most meaningfully with digital content and sharing their stories, peer to peer. At the other end of the scale, older, wealthier hotel bookers are increasingly interested in responsible luxury. This group does not want to scrimp on comfort or style, and they are more than happy to pay their way toward a more sustainable holiday. Nature, wildlife, and eco-credentials are top concerns, whereas younger people are more likely to engage with social and community issues. Whatever sparks an interest or motivates people to care about the brands that are kinder, most people turn to a hotel's website to assess its sustainability integrity. This tells us that in order to communicate their philosophy and initiatives effectively, hotels that care should have a dedicated page online. By sharing compelling content detailing articulately exactly what they are doing and why you are ingratiating yourselves with your customer—and chances are they will want to book directly with you as a result rather than go through a third party. You could even follow 11 Howard's lead in New York and offer to make a donation to charity, if people book direct—in their case, The Global Poverty Project. It is win-win then for everyone.

5 Corporate Management in the Luxury Tourism Segment: Example of Top-End Hotels

The Answer Is Yes. What Is the Question?

Marc Aeberhard

Basic Attitude and Differentiation

As already stated in various places, the luxury hotel industry today distinguishes between high-end and top-end establishments. In the following considerations, the (necessary) leadership qualities in the top-end hotel industry are addressed exclusively. In contrast to high-end hotels, the following key points are defined for the top-end hotel industry:

- **The size:** Top-end means small hotels, so that an optimum of privacy on the one hand, personally available space on the other hand, and a maximum of individualized and personalized service can be offered. Therefore, top-end hotels are generally smaller than 30 units (and the number is declining).
- **The room price:** in the industry, the mark of 1500 euros is considered the Average Daily Rate (ADR) or the magic number, and when reached, a hotel rises into the league of super-exclusive companies. It is important to note clearly that this is not about a rate of individual room categories or limited time periods (such as suites or special events), but about the effective ADR of the entire hotel averaged over the year.

In addition, the top-end hotel industry is focusing above average on the definition of new and hidden luxury. This is referred to as an immaterial understanding of luxury. The main dimensions are:

– Space
– Time
– Personal and individualized service
– Safety and security
– Exclusivity
– Health

It is important to understand here that, consequently, a democratization of material luxury is assumed. This means that the existence of material resources and infrastructure (even at the highest level) is assumed. The provision of impeccable hotel infrastructure, world-class inventory, and impeccable interior is taken for granted when addressing the top-end clientele. Or to put it another way: Neither a golden bathtub, nor a crystal chandelier, nor hand-forged silver cutlery is enough to win a guest. It must be assumed that the guest is living in an environment that is equal or even (far) superior to the hotel. The differentiations are therefore made either through destinatory criteria, taste, and/or the fulfillment of intangibles.

Leadership

What is meant by leadership, how leadership should be handled or who uses which management instruments in which circumstance, innumerable publications have been published on this subject, and a differentiated view is deliberately omitted at this point. Instead, the following definition is used as a basis for the following considerations:

Leadership means: The will to align a group of entrusted people with an objective and to achieve this together, taking into account personnel, material and monetary resources, an infrastructural, cultural, social, geographical, and climatic framework with the given temporal and spatial limitations.

To Like Human Beings

In the hotel industry, the most important factor in production is human beings. And while efficiency-enhancing measures and production processes are being sought in many industry branches and the introduction of robot technology is being celebrated, the top-end hotel industry can only use this to a very limited extent for back-of-house technology, which means that the hotel's main services take place and will always take place in direct human-to-human interaction regardless of all developments.

Although countless hotels outdo each other in terms of infrastructure design with even more, even more expensive, even more opulent, it is often forgotten that by reducing a hotel to its hardware in the best case a pretty museum is created. The essence that ultimately turns a building into a hotel is the people and the services they create.

The particular challenge now lies in dealing with the human factor. By definition, humans are fallible, sometimes in a good mood, sometimes in a bad mood, sometimes concentrated, sometimes mentally absent, sometimes clumsy, sometimes precise, to one person someone seems likeable, someone else unsympathetic, sometimes anxious, sometimes courageous, sometimes motivated, sometimes reserved. The human psyche and state of mind is as colorful and varied as a bouquet of flowers, and the hotelier has no alternative but to not only fulfill, but to overfulfil, a pronounced or unspoken expectation of his guests with the help of human beings. This basic attitude of giving one's best at all times and at the same time putting one's own needs aside for the benefit of one's counterpart's needs and thus rendering services in the true sense of the word is the essence of good hotelistic quality. The prerequisite for this is an overall positive feeling, the strong desire and drive to provide services with pleasure, solely inspired by the thought of provoking a momentary feeling of joy and happiness in the service recipient. In short: only those who genuinely like human beings can be good hoteliers.

It Is the Right Person That Makes the Difference

This already is the first decisive management criterion: the selection of the right employee. A proper, well-founded education should be assumed, whereby a corresponding qualification—with certain exceptions (for example in the kitchen or in the technical department)—does not even have to be mandatory. Much more important is the character of a person. Does he bring the necessary passion, the feu sacré? Can he find pragmatic solutions under complicated circumstances (in the interpersonal sphere), is he able to conjure up a cheerful smile in all situations, does he recognize the attention to detail, is he willing to make the extra effort to provide an extraordinary service instead of an ordinary one? Here all training courses and all Standard Operating Procedures (SOP) are of no help; if a person does not want to go beyond, cannot go beyond, or does not pick up on any clues, he or she is the wrong employee. Neither references from 5-star hotels nor expensive professional training can help. The focus is therefore not on static facts, but on the dynamic will and a huge heart full of empathy and the courage to think, recruit, and act outside the box. Experience shows that a vast amount of apparently useless knowledge is washed into the company, which can, however, prove to be a blessing in the most unusual situations. Examples of this: the contacts to the local airport of a Guest Relation Officer, who previously worked as an Air Traffic Controller, were always worth their weight in gold when special check-ins, late arrivals, or priority statuses were required for executive aviation handlings. The training of a butler as a car mechanic also proved to be extremely helpful when the limousine with VIP guests from Hollywood suddenly broke down on a trip to a nature reserve.

In short: In addition to direct professional skills, there should also be room for indirect professional skills in the company, but this also implies that the employee receives the trust of the hotel management and within his Scopes of Work (SOW) has the freedom to fulfill his role. If the right person has been selected for the team, then constricting SOPs and often unrealistic policies & procedures are not necessary, because the employee has a natural need to perform above average.

Praise and Reprimand

The combination of the success factors described so far:

– Right selection
– Think outside the box
– Empowerment
– Love for people
– Above-average service commitment

Only works if two additional factors are used, namely:

– Excellent communication
– Motivational instruments

For empowerment to function properly, a culture of trust must be established within the company. Good, smooth, and open communication is a must. Fast, clearly defined chains of command, uncomplicated communication channels, and flat hierarchies are particularly important here. In addition, it is also part of the repertoire of a successfully managed top-end hotel that the employees' motivation is particularly nurtured. The focus here is not on monetary and material incentives (proper remuneration should actually be assumed), but on praise and reprimand. If the employee is praised for his commitment, for his decisions, if he enjoys the trust and knows that he is not unduly reprimanded even in the case of a mistake made, he delivers top performance and feels valued and confirmed in his importance. This self-confidence is the basis of a healthy "hotel soul."

Being a Role Model

Organizations that take all of this into consideration experience a strong momentum at their base, which in turn relieves the hotel management in an invaluable way, creating space and time for the management for strategic and superordinate management tasks. In order for an organization to function perfectly, it must be possible to assume that the hotel management never loses contact with the base at any time. A descent into the ivory tower would be the beginning of the rapid end and would be met by the base with brutal disobedience. The hotel management thus legitimizes its function by setting a good example at all levels. This is not about fraternization with rank-and-file staff, but about superior professional competence, and especially about humanity and cordiality in the company. Care, gentleness, and sincere appreciation of both people and all things are perceived and acknowledged by employees and guests alike. Seen in this light, a hotelier in a top-end hotel is always on a virtual silver platter. It is expected of him to not only live or represent the hotel soul but to shape it as well.

- It requires an unconditional attitude to quality, attention to detail, and perfection. Standards set by the hotel management are implemented, implemented by the Heads of Department (HOD), lived by the rank-and-file staff and reviewed, controlled and, if necessary, corrected by all levels.
- The guest perspective must never be left out of sight. It is an open secret that after a period of about 1.5 to 2 years in a hotel a certain degree of operational blindness sets in. Not least for this reason it is of existential importance that the hotel management consciously and mentally steps back from daily business, takes on the role of the guest, and examines the business with a critical eye. This is not a matter of highlighting the inadequacies or neglect of employees, but of ensuring that guests can experience an impeccable hotel experience at all times. At the same time, however, it is also a suitable tool to check your own services time and again for changes, innovations, adjustments, and trends.
- It therefore also requires the will to inform oneself again and again where, how, and why other hotels are developing, what the current trends are, and how society, culture, and other sociodemographically relevant factors are developing or changing.
- "Management by Feldherrenhügel" (commanders hill): It was no coincidence that generals were standing on a small hill to have an overview of what was happening in the field and to command troops. Transferred to the present day, this means for the successful hotelier in the top-end sector to always take a step to evaluate operations. Put simply, the question is: Where does the boss stand? And what does a good boss have to do? Here it is:

  - Good planning: Although the hotel's day-to-day dealings with guests in an open system (additional influences by guests, suppliers, authorities, stakeholders and shareholders, technical, climatic, geographical, cultural, social, political, and infrastructural framework conditions) are essentially characterized by uncertainties and constantly changing situations, the elaboration of structures, checklists, ideological memoires, to-do lists, and reserved decisions is nevertheless a favorable management tool for the hotel director. They represent practicable instruments in an everyday life that can hardly be planned, in order not to lose the red thread in the strategic and operative orientation and further development of the hotel.
  - Preparation: In an environment which is characterized by constantly changing situations, preparation at all levels (mise-en-place in the kitchen, prepared registration forms, documents of all kinds, check-in and check-out amenities, weather forecast and travel preparations, reserve of resources and spare parts, etc.) is an extremely effective way to gain time in critical moments, not to be embarrassed by the guest and to always create the possibility of further actions and alternatives.
  - Contingency plans and back-up procedures: But the room for maneuver created by preparation is only useful as long as it succeeds in creating

meaningful alternatives and, above all, implementing them. The highest operational goal of a hotelier in the top-end sector is the unconditional fulfillment of guest wishes (provided that they are within the legal, moral, and nonhazardous framework) and for this purpose a rapid and targeted creation of alternatives is important. A headless hotelier will not only lose the respect of the guests but will also lose his credibility with the staff. Especially in case of emergencies (fire, tsunami, pirate threat, terrorist attack, civil war, theft, mutiny, brawl, strike, death, storm damage, avalanches, injuries, wildlife threats, accidents, total loss of connections, complete failure of building services, vermin infestation, mafia attacks, corruption cases) a decisive, target-oriented, cleanly communicated leadership from the hotel management upward, downward, inward and outward is crucial. It makes the difference when panic is about to break out, it decides about peace and order and finally, it shows the applied coping strategies led the hotel out of chaos and how the reputation of the hotel stands in the market as a result.

- In exceptional situations, giving the guest, but also the employee, the feeling that everything is in order, that worries are unnecessary and that security and discretion are guaranteed at all times are crucial in such moments.
- Network: In order to cope with the tasks already listed, the hotelier should build, expand, and above all actively maintain a broad, solid network of contacts and competencies in advance and in cooperation with his team of employees. Despite all the zeal—and even with the best preparation—reality always holds surprises in store and demands that the impossible be made possible. Here, a well-positioned network often proves to be a savior in an emergency. Be it to borrow or replace a special spare part or rare liquors that are in particularly high demand or help in case of overbookings. A professional network often turns into a strong network of relationships, including interpersonal relationships, and not infrequently into friendship.

• Guidance with open eyes, ears, and all senses: In the understanding that true luxury today is defined, perceived, and lived immaterially, it is quite logical that the hotelier sets a good example and exemplifies the (newly) created understanding of luxury. This essentially includes the creation of an aura in the hotel, which often cannot be described in words, but which is felt by the guest. This includes the creation of soothing tranquility and grounding. The olfactory senses must also be stimulated. A unique smell remains in the guest's memory a long, long time after arrival, literally as a sweet memory and is therefore perhaps one of the most powerful marketing tools there is. The typically overused senses and organs—brain, taste, and eyes—can be expanded to include the dimensions of heart, stomach, ears, sense of smell and touch. This explicit expansion of consciousness creates a holistic experience in the true sense, which has a much more lasting effect than just the rational depiction of structured organizations. Here lies a second, very essential root, which is the foundation of the "hotel soul."

• Follow-up: According to the principle of "living and learning in a growing organization," after every activity and every operative step a moment of reflection and conscious examination of the situation is required. Here it makes sense to learn from the past with regard to the improvement of the offer, the hardware and above all the service, to record knowledge, to distill it, and to introduce it as an improvement into the future production of services. A documentation reduced to the essentials, but clearly structured by means of checklists and feedback forms, are useful tools that flow into everyday hotel life via the HODs as operative inputs, but also for the creation of personalized, individualized guest profiles.

Open-Mindedness

The basic attitude is: "Always expect the unexpected." Everyday hotel life in the top-end area can be described as particularly unpredictable. However, this circumstance cannot be overcome with paper and endless checklists, but to a great extent through personal commitment, coupled with a great deal of common sense, skill, intuition, experience, trust in the hotel team and the unconditional commitment to the basic attitude "The answer is yes. What is the question?" This principle describes in a few words everything a good service provider has to have:

– A positive attitude.
– Unconditional nature.
– The confidence to be able to solve challenges.
– The awareness that there is an answer for every task.
– The knowledge that this promise includes a service chain that requires dependency on downstream service providers.
– Investing in team spirit, creating, shaping, and leading a team and achieving the goals set together with it.
– Experience and communicate a sense of achievement: individual performance is highlighted, but it should also encourage the team to achieve even more.
– To the outside, to the guest, it is important to signal a spontaneous openness that makes it clear unmistakably and naturally that service and the provision of services are understood as a privilege.
– An honest positive attitude and a sincere willingness to provide services promote guest satisfaction above average.
– Competing service providers: if employees are in mutual exchange and the provision of proactive and individualized services is praised with recognition and satisfaction by guests, this encourages employees to outdo each other with even better services and even more attentive service; true to the motto: Who is the best?

One Can Expect Royal Behavior from the King

But here comes another important management function of the top-end hotelier: the unconditional openness and clear commitment to service makes both the individual and entire organizations susceptible to abuse. Here the hotelier has the extremely challenging task of setting boundaries gently and imperceptibly for the guest, but nevertheless clearly. Yes, the principle applies: "The guest is king." But this must be added: "... as long as he behaves royally." A hotel is a highly specialized, emotional service center where services are provided and sold. But to provide services should not be confused with being a guest's servant. Unfortunately, some guests are not aware of this delicate difference. Employees hardly know how to protect themselves from such attacks or how to differentiate themselves from them, so abuse in today's hotel environment is a danger that should not be underestimated. This abuse ranges from theft and loutish behavior to physical and psychological attacks on employees and blackmail on opinion portals. Here the limits are often not only reached, but crossed and so the hotelier also has the role of mediator, judge, advocate, and opinion leader. And it is not wrong for a hotelier to say no, when a guest is not aware of his royal responsibility, so he can promise all other guests wholeheartedly, "The answer is yes. What is the question?".

6 The Pioneering Force of Luxury Travel

Adapting a New Concept of Consumption at the Pursuit of Happiness

Magda Antonioli Corigliano and Sara Bricchi

It is unquestionable that lately the travel market is experiencing a polarization: when the traditional midscale segment suffers, low-cost and luxury tourism at the opposite ends increase their market shares. With more than 54 million international trips[1] in 2016 (IPK International 2017), luxury continues to grow at double pace compared to the average in the tourism industry, now accounting for almost 7% of the total international travel and 12.5% of the travel expenses.

Its growth is not fueled by an increase in per capita expenditure as much as by an increase in the number of travelers: new tourists from the emerging markets, as well as consumers from mature markets turning to luxury travel for the first time, plus new generations, millennials, and generation Z, with luxury buying power and clearer personal priorities and tastes compared to their parents.

According to estimates, globally, there are more than 18 million high-net-worth-individuals (HNWIs) (Capgemini 2018), with an increase of 1.6 million just in the last year, and there are almost 2700 billionaires (Hurun 2018), not counting an increasingly large amount of people possessing less than $1 million in financial assets, but occasionally buying luxury goods and services. In fact, the division

---

[1]IPK considers luxury trips to be those with a minimum expense of 750 euros or more per night for travels for up to 3 nights and of 500 euros or more for travels for more than 3 nights.

between premium and luxury market segments becomes more and more blurred and the diffusion of affordable collections by international luxury brands in the early 2000s surely contributed to this (Martini Media 2015). Nowadays, customers often buy from multiple market levels, looking for low-cost options for basic goods with no or low emotional investment and seeking greater personal value when purchasing products (among which *in primis* recreational experiences), which make them feel better, both intrinsically and in terms of their positioning within their social circles (Miller 2008).

However, the importance of this market segment is not limited to its growth, which has been uninterrupted even during the crisis, with a +4.5 % CAGR in the period 2011/2015 (Amadeus AIT Group 2016), nor to its turnover, amounting, according to Bain & Company, to almost 300 billion euros in 2017 just for hotels, cruises and F&B, with expenses for luxury experiences expected to increase even more, reaching 67.5% of total luxury expenses in 2024 (Boston Consulting Group and Fondazione Altagamma 2018).[2] Its relevance is based mainly on the potential of luxury tourism to anticipated trends and consumer behaviors, which, in the near future, will affect the whole industry.

If it is true that trickle-up and trickle-across phenomena do exist (Veblen 1899), the trickle-down effect is still relevant in the luxury industry. All fashions pass through three stages and tourism is no exception to this, with luxury being the segment where innovation mostly occurs: firstly, a small number of people, usually wealthier, become interested in something new which allows them to stand out while at the same time expressing their personality; secondly, other consumers, usually from the upscale market, follow, desiring to emulate fashion leaders; finally, the trend is introduced to the mass market and, over time, it usually declines (Michman and Mazze 2006).

Therefore, it is extremely important to understand luxury travelers' motivations and their decision-making process as this will not only increase the chance to meet their expectations by offering them a more tailored product/service and delivering a higher added value, but also by being able to scout the drivers, which will characterize the entire travel market in time to develop a proper offer and, with it, some sort of competitive advantage, thus capitalizing on the investment.

As a matter of fact, with a little imagination, a tourist product originally conceived for affluent customers can be redesigned and, limited solely to the essential components, turned into an accessible experience, while conversely, something created for less sophisticated clients, through extreme customization, can be appreciated also by the very high-end of the market (Antonioli Corigliano and Bricchi 2017a).

If in the last few years personalization and experience have been the main drivers with reference to luxury tourism (IE Premium and Prestige Business Observatory

---

[2]According to the European Cultural and Creative Industries Alliance (2015), 70% of the total luxury goods and services turnover, equal to almost 5% of EU's GDP, can be attributed to European firms/companies. The relevance of the industry for Europe finds further evidence in the fact that it represents 17% of the Union's exports. On this matter, it should not be forgotten that tourism is effectively considered to be part of the Cultural and Creative Industries.

2015), now, mass-market brands are also catching up on these aspects. On one side, large brands have the scale to validate the technology expense required to deliver, to a good extent, real-time personalization. For example, one may think of the wearables and the IoT solutions adopted by the major cruise carriers to customize the experience on-board, to locate other members of one's own party, to book extensions and excursions. On the other side, however, the internet and social media have accelerated the diffusion of destination-specific experiences throughout all market segments: a cooking class in a stranger's home on the other side of the world or a tour at the local market can be booked directly from the comfort of your own sofa. And it is not only due to the popularity of Airbnb Experience that hundreds, if not thousands, start-ups have been created in the last few years which promote local experiences offered by residents, usually amateurs, all around the world (Antonioli Corigliano and Bricchi 2018). Experiential travel has become a commodity and, by now, luxury travelers take these aspects for granted (View from ILTM and Skift 2017).

In the same way, wellness, once appreciated mainly by affluent consumers, is now a 3,250 billion euros business per year, with tourism constituting 60% of the entire industry turnover (Global Wellness Institute 2017). This is being experienced by all market segments alike, with the high-end now looking for a more complete wellbeing, a *fil-rouge,* to link all the different experiences of the journey and the pleasures one can derive from travel, with specific wellness treatments not just for the body, but also for the mind and soul.

Luxury can be defined as the frontier which sets the rules for the entire market, with innovators wishing to differentiate themselves, once a fashion has been adopted by everyone.

The days when luxury travelers visited the most exotic destinations and just lay on a beach for days to relax or to spend their time in the most expensive restaurants are long gone. The concept of luxury evolved over time and today, even if for some consumers, especially in the emerging markets, it is still linked to material goods and status/ostentation, for the majority of customers it is now linked to self-esteem and passions (Mastercard 2014), with a clear benefit for the tourism industry. Culture surely plays a role in defining the relationship between the individual and luxury (Moscow 2017), with western societies emphasizing the private dimension of the self and, related to this, the drivers of personal preferences like passions and emotions. At the opposite end, eastern cultures tend to highlight the social role of the individual and, as a consequence, variables like prestige, acceptance, and belonging (Ahuvia and Wong 1998). However, a concept of consumption conceived as "the pursuit of happiness" is rapidly spreading among all the cultures and, according to a survey run by Marriot International and Skift in 2017, even if with considerable differences among diverse countries, more than 50% of the respondents would choose a luxury travel experience over a luxury good (Marriot International Luxury Brands and Skift 2017).

In an era of constant disruption, a time of fear, uncertainty, and anxiety, the definition of luxury, which still retains its aura of quality, elegance, exclusivity, and anticipation of needs, now means different things to different consumers (Ipsos

2017a, b). The time of typified customers is gone. Actually, many affluent travelers purchase products and services depending on their current mood, turning the concept of luxury into something even more complex, flexible/in motion, related to psychographic variables and emotions, which can easily adapt to different contexts, cultures, and generations.

The subjectivity inherent in this definition creates new challenges for the tourist operators, associated both with the fulfillment of these evolving needs and aspiration and, even more importantly, with market segmentation and determination of catchment areas/potential demand. In this context, traditional marketing is losing relevance: it is the actual consumption behavior and not the geographical or sociodemographical variables that define the luxury traveler (Collison Group 2012). And brands, in tourism as well as in all other economic industries, need to interpret customers' aspirations, while staying true to their own DNA, delivering engagement and inspiration throughout all the touchpoints.

If luxury buying decisions are more and more linked to intangible variables connected to self-image, personal values, and philosophical ideas, there is an impact both on how and why HNWIs want to travel and, as a consequence, on which destinations they are going to visit (Airbus Corporate Jets 2014). As a matter of fact, there are less and less reasons in modern society to leave home, with modern-day entertainment options, comfort, and access to information. With many destinations struggling to differentiate, luxury is less and less represented by specific sites or regions in themselves and more and more by what can be experienced in those territories. According to a survey held by Skift in 2017, more than 80% of affluent travelers have a significant emotional connection to the places they visited during their holidays (Skift 2017).

As already said, even memorable, tailored experiences are not enough: for this market segment, personal fulfillment and the self-improvement one gains from travel are what truly matters now. In other words, affluent travelers ask the luxury operators to deliver the appropriate experience for that particular person in that particular moment.

This shift from chasing esteem—that is, gaining recognition and approval from a group with a shared status (Bain & Company and Fondazione Altagamma 2017)—to looking for self-actualization—standing out as individuals through the self-expression in the pursuit of the desired-self—toward the top of Maslow's pyramid (Maslow 1943), naturally led to an evolution from experiential tourism to the so-called transformative travel (Singapore Tourist Board and Skift 2018). Trips are seen as an occasion to learn something new, to enrich one's cultural background and, as the word suggests, the concept recalls the personal and internal journey that occurs to the individual during the physical travel experience. It may be great or basic, depending on the association a traveler wants to create within the context but will bring changes in his/her life.

In this context, luxury experiences are intended as gaining access to something of high quality, something authentic and unique, in short, "a something-money-can't-buy experience": being able to cook with a Michelin star chef and learn a few of his/her secrets, to see a designer at work in his/her atelier and discuss with him/her the last fashion trends, to assist when supercar components are tested in the wind tunnel and have a conversation with the engineers about the aerodynamic, or to meet a football player just before a match and review strategy and tactics with him. The list of these experiences is as endless as the travelers' and tourist operators' creativity.[3] Many high-end consumers are wealthy enough to buy everything they want: value is attributed to things, goods, and experiences, which usually are not for sale. Travelers look for emotions, connection, and insights and they possess a stronger sense of curiosity, wanting to be part of the production process. It is not a coincidence that important brands in the hospitality industry, like Verdura Golf and Spa Resort by Rocco Forte—just to give examples—organize exclusive excursions to nearby villages where their guests can taste local dishes and experience how food is being traditionally prepared (and consumed!) by local women.

Speaking of gastronomy, food and wine still remain among the favorite areas of interest of affluent tourists, as they allow the visitors to get directly and more easily in touch with the local community and culture (Antonioli Corigliano and Bricchi 2018). Cuisine is not only a relevant element in establishing customer satisfaction for every kind of travel but more frequently is also a driver in the choice of the destination and/or the main reason for the travel. And seemingly endless is the number of experiences a traveler can have, as consumer or as cook, in this field. In the same way, shopping is more and more conceived by luxury tourists as a means to come into direct contact with local traditions, expressed through arts and crafts (Antonioli Corigliano et al. 2014): as a matter of fact, other than luxury international brands (whose turnover still derives from international travelers by more than 50%—(Bain & Company and Fondazione Altagamma 2017), affluent tourists look for handmade local products, those being not only wine and food, but also clothing from emerging designers, or artwork from young artisans (Antonioli Corigliano and Bricchi 2017a).

However, cultural tourism also gains more and more relevance in the high-end of the market. The boundaries of cultural tourism have progressively expanded in the last few years and now encompass not only heritage but also contemporary and performing arts, including encounters with local artists and the cultural community. Moreover, arts are in fact contributing toward the characterization and distinguishing of the destinations in a mix between historical landmarks, urban renewal projects, talented creatives (Antonioli Corigliano and Mottironi 2016). From being an elitist experience, (contemporary) art is becoming increasingly more popular among the affluent consumers: the more people are exposed to it, the more they are intrigued and interested in learning about it. It is not by chance that luxury brands, especially from the fashion industry, try to stress even more the intrinsic link between luxury, on the one side, and art and culture on the other: the opening of corporate museums,

---

[3]It has to be remembered that these kind of luxury experiences are valued and applied also in top management training and executive courses from business schools all around the world: in-company visits, especially in industries which are different from those that the participants are a part of, allow them to face real backroom situations and stimulate "out of the box"- and lateral thinking.

like in the case of Ferragamo or Fendi, where fashion is the starting point to celebrate art and creativity, or the creation of cultural foundations, as in the case of Prada and Luis Vuitton, surely go into this direction. As Holt already stated in 1998, affluent customers see in "cultural capital" a way to express their personality and esthetic sensitivity; they do not despise comfort and functionalities, but having already satisfied their material needs, they tend to affirm their character through the taste and the enjoyment of cultural consumptions, which by definition, tend to be of more immediate gratification and amusement with recurrent consumption (Holt 1998). In this respect, travel is also a way to enrich one's own cultural background, especially in an area like Europe, where culture and beauty are fundamental values of the collective identity, as well as a characterizing feature, not only of the luxury industry but more in general of the entire cultural style and way of living.

In a sense, it is the whole trip which is more and more conceived as an experience: from the "digital experience" in the booking phase (the internet is the first source of information for luxury travelers—Resonance 2016), to the transportation and the stay (at a hotel or a private property, but—in any case—immersed in the local culture), and finally to the post-trip phase with the sharing, both off- and online, of the experiences had.

It is important to remember that emotional attachment leads to higher consumer engagement even when the guests return home. In fact, nowadays, the service encounter stage is no longer the final stage in the travelers' consumption process; instead, it is the focal point for/of his/her own as well as other people's future decisions. Again, the value of luxury tourists is also apparent in the role they can have for both structures and destination as brand ambassadors, being able, with their choice of either re-purchasing or changing, to influence, also with the favor of ICTs, others' decisions.

Fundamental is the role played by travel intermediaries: if 70% of luxury travelers book an independent tour/stay (Visa 2015), the majority of them also opts for a travel consultant, at home and/or once in the destination, to translate needs and dreams into tailor-made experiences and to approach local luxury brands/operators.

This new perception of luxury pushes tourist operators to rethink their role in the industry: to offer an added value product to modern travelers (Visa calculated an average expense per person per trip of 3250 Euro among HNWIs), is firstly necessary to understand which are their *desiderata* in the specific situation and how to fit in their travel plans, acquiring the flexibility necessary to make adjustment to travel plans and itineraries at every moment and in real time. The more luxury brands are able to align their products with the shifting moods of their customers, the more opportunities they have to increase revenues and consumers' loyalty in the luxury marketplace.

Data surely changed the operations in the travel industry as everything can be measured, scored, rated, and ranked and, with digital natives gaining purchasing power, technology becomes a prerogative also for luxury brands. However, even if AI can take advantage of digital footprints and, thanks also to big data, can help to understand travelers preferences, improving customization or service offered, if social media make it easier to spread luxury buzz among potentially interested consumers, thus enlarging the catchment area, or if VR, anticipating the experience the traveler is going to have, contributes to reduce the information asymmetry inherent in the tourism industry, ICTs are mainly taken for granted by luxury tourists, without adding value to a product/service.

It should not be forgotten that tourism is a labor-intensive industry, in the sense that personnel is of great importance in both, creating added value and shaping the offer (Fourseasons 2012). Especially in the luxury segment, it is the human touch that makes the difference. It is therefore important that the front-end workforce, on top of the hard competences, also possesses soft skills, among which, *in primis*, is the ability to read between lines, anticipating customers' reactions, and figuring out *desiderata* and expectations (Antonioli Corigliano and Bricchi 2017b). Professionalism of human resources is the starting point to reach quality standards, which are always increasing. It is therefore important, especially in the luxury market, to pay attention and invest in people, co-workers and employees, equally or more than in other types of assets, with specific actions taken in terms of talent scouting, training, employer branding, and retention.

Literature

Ahuvia AC, Wong NY (1998) Personal taste and family face: luxury consumption in confucian and western societies. Psychol Market 15:423–441

Airbnb (2018) Airbnb announces strategy to enable magical journeys for all. Airbnb, 22.2.2018. https://press.airbnb.com/de/airbnb-verkuendet-strategie-um-magische-reisen-fuer-alle-zu-ermoeglichen/. Accessed 9 Oct 2018

Airbus Corporate Jets (2014) Billionaire study. https://www.ausbt.com.au/files/Airbus%20Billion aire%20Study%20Report.pdf. Accessed 31 July 2018

Amadeus AIT Group (2016) Shaping the future of luxury travel. Future traveler tribes 2030. https://amadeus.com/documents/en/travel-industry/report/shaping-the-future-of-luxury-travel-future-traveller-tribes-2030.pdf. Accessed 31 July 2018

Antonioli Corigliano M, Bricchi S (2017a) Luxury tourism: current status and potential. In: Becheri E, Micera R, Morvillo A (eds) Report on Italian tourism, XXI edn 2016/2017. Rogiosi Editore, Naples, pp 649–664

Antonioli Corigliano M, Bricchi S (2017b) Catering, luxury and territory: drivers of the italian way of living, HOST - Fiera Milano. http://host.fieramilano.it/sites/default/files/Ristorazione_lusso_territorio%20%283%29.pdf. Accessed 31 July 2018

Antonioli Corigliano M, Bricchi S (2018) Are social eating events a tool to experience the authentic food and wine culture of a place? In: Bellini N, Clergeau C, Etcheverria O (eds) Gastronomy and local development: the quality of products, places and experiences. Routledge, London, pp 245–261

Antonioli Corigliano M, Mottironi C (2016) Tourism. Egea, Milan

Antonioli Corigliano M, Mottironi C, Baggio R (2014) Shopping as an urban tourism determinant: technological tools for its promotion. In: Garibaldi R (ed) Il Turismo Culturale Europeo. Città Ri-Visitate. Franco Angeli, Milano, pp 208–217

Bain & Company and Fondazione Altagamma (2017) Altagamma 2017 worldwide luxury market monitor. https://altagamma.it/media/source/Altagamma%20Bain%20WW%20Markets%20Monitor%202017.pdf. Accessed 31 July 2018

BCG (2018) Global wealth 2018: seizing the analytics advantage. https://www.bcg.com/de-de/publications/2018/global-wealth-seizing-analytics-advantage.aspx. Zugegriffen 6 Nov 2018

Boston Consulting Group and Fondazione Altagamma (2018) True-luxury global consumer insight. https://altagamma.it/media/source/True%20Luxury%20Global%20Consumer%20Insight%202018%20-%20Sintesi%20in%20Italiano_Vsent.pdf. Accessed 31 July 2018

Capgemini (2018) World Wealth Report. https://www.capgemini.com/it-it/wp-content/uploads/sites/13/2018/06/Capgemini-World-Wealth-Report-10.pdf. Accessed 31 July 2018

Collison Group (2012) Motivating the affluent middle. Understanding the behavior and priorities for mass affluent consumers. http://www.travelbizmonitor.com/images/Collinson_Group.pdf. Accessed 31 July 2018

European Cultural and Creative Industries Alliance (2015) European citizens' perception of the high-end cultural and creative industries. http://www.meisterkreis-deutschland.com/sites/default/files/tns_sofres.pdf. Accessed 31 July 2018

Festa J (2015) A look how luxury hotels use partnerships to elevate their brand. USA Today, 24.10.2015. https://eu.usatoday.com/story/travel/roadwarriorvoices/2015/10/24/luxury-hotel-partnerships/83073372. Zugegriffen 6 Nov 2018

Fitzmaurice R (2017) The most exclusive private members' clubs in Europe, ranked by price. Business Insider UK, 20.10.2017. https://www.businessinsider.de/europes-most-exclusive-private-members-clubs-2017-10?op=1. Zugegriffen 6 Nov 2018

Fourseasons (2012) The luxury consumer in the new digital world. https://www.fourseasons.com/content/dam/fourseasons/web/pdfs/landing_page_pdfs/2012_TRD_Report_final.pdf. Accessed 31 July 2018

GfK (2016) Purchasing power in Europe: positive developments in Central and Eastern European countries. GfK press release, 8.11.2016. https://www.gfk.com/de/insights/press-release/kk-europa/. Accessed 6 Nov 2018

Global Wellness Institute (2017) Global wellness economy monitor. https://static1.squarespace.com/static/54306a8ee4b07ea66ea32cc0/t/58862a472994ca37b8416c61/1485187660666/GWI_WellnessEconomyMonitor2017_FINALweb.pdf. Accessed 31 July 2018

Holt DB (1998) Does cultural capital structure American consumption? J Consum Res 25(1):1–25

Hurun (2018) Hurun global rich list 2018. http://www.hurun.net/EN/Article/Details?num=2B1B8F33F9C0. Accessed 31 July 2018

IE Premium and Prestige Business Observatory (2015) IE luxury barometer 2015. https://observatoriodelmercadopremium.ie.edu/wp-content/uploads/sites/59/2013/11/Barometro-2015-ENG-_2-14.pdf. Accessed 31 July 2018

IPK International (2017) World travel monitor. http://www.ipkinternational.com/en/world-travel-monitor. Accessed 5 Nov 2018

Ipsos (2017a) World luxury tracking: do you speak luxury? Consumers new luxury culture. Ipsos, 30.10.2017. https://www.ipsos.com/en/world-luxury-tracking-do-you-speak-luxury-consumers-new-luxury-culture. Zugegriffen 6 Nov 2018

Ipsos (2017b) 2017 affluent outlook. https://www.aaaa.org/wp-content/uploads/2016/01/Ipsos-Affluent-Outlook-2017.pdf. Accessed 31 July 2018

Kapferer JN, Bastien V (2012) The luxury strategy. Kogan Page, London

Kotler P, Keller K (2016) Marketing management, 15th edn. Pearson, Hallbergmoos

Langer DA, Heil OP (2015) Luxury essentials. University of Mainz, Mainz

Marriott International Luxury Brands & Skift (2017) Skift trends report: the luxury evolution. https://skift.com/2017/12/15/skift-trends-report-the-luxury-evolution/. Accessed 31 July 2018

Martini Media (2015) The Martini report. The affluent audience online. Vol. 2: Luxury goods. http://martini.media/wp-content/uploads/2015/03/MartiniReport_Vol2_LuxuryGoods.pdf. Accessed 31 July 2018

Maslow AH (1943) A theory of human motivation. Psychol Rev 50:370–396

Mastercard (2014) Mastercard affluent report. https://newsroom.mastercard.com/asia-pacific/files/2014/05/MasterCard-Affluent-Report-2014.pdf. Accessed 31 July 2018

Michman RD, Mazze EM (2006) The affluent consumer: marketing and selling the luxury lifestyle. Praeger, Westport

Miller S (2008) The middle of nowhere. Brandweek 49(1):18–20

Mirror Online (2017). How to find the accommodation of your dreams. Spiegel Online, 20.1.2017. www.spiegel.de/reise/aktuell/schicke-hotels-so-finden-sie-ihre-traumunterkunft-a-1131222. html. Accessed 6 Nov 2018

Moscow F (2017) Strategies in luxury markets, marketing, digitization, sustainability. Egea, Milan

Resonance (2016) Future of luxury travel report. http://resonanceco.com/reports/2016-future-lux ury-travel-report/. Accessed 31 July 2018

Roland Berger Master Circle (2005) Luxury in China. Roland Berger

Roll M (2018) Hermès—the strategy insights behind the iconic luxury brand. Martin Roll Business & Brand Leadership, Mai, p 2018

Ruetz A, Aeberhard M (2018) Between Bling-Bling and Bamboo Grove. The international luxury hotel industry: perception, development, perspectives. Results of an explorative study at the ITB Berlin. In: Ehlen T, Scherhag K (eds) Current challenges in the hotel industry. Erich Schmidt, Berlin, pp 133–148

Schroder J (2018) Transformation successful? Prada reports strong profit growth. Fashionunited, 8.8.2018. https://fashionunited.ch/nachrichten/business/transformation-geglueckt-prada-meldet-kraeftiges-gewinnwachstum/2018080615371 .Accessed 6 Nov 2018

Shankman S (2018) The future of luxury sits at the intersection of travel and fashion. Skift, 13.3.2018. https://skift.com/2018/03/13/the-future-of-luxury-sits-at-the-intersection-of-travel-and-fashion/. Zugegriffen 6 Nov 2018

Singapore Tourist Board and Skift (2018). The rise of transformative travel. https://skift.com/2018/03/30/new-skift-report-the-rise-of-transformative-travel/. Accessed 31 July 2018

Skift (2017) The megatrends defining travel in 2017. https://skift.com/2017/01/12/the-megatrends-defining-travel-in-2017/. Accessed 31 July 2018

Veblen T (1899) The theory of the leisure class. Macmillan, New York

View from ILTM and Skift (2017) Building brand love and loyalty in luxury hospitality. https://view.iltm.com/wp-content/uploads/2017/05/ILTM-Apr20-final.pdf. Accessed 31 July 2018

Visa (2015) Visa global travel intentions study 2015. https://pata.org/store/wp-content/uploads/2015/09/Visa-GTI-2015-for_PATA.pdf. Accessed 31 July 2018

Wittig MC, Summer Skirt F, Hatchet P, Albers M (2014) Rethinking luxury. LID, London

Marc Aeberhard founded Luxury Hotel & Spa Management Ltd. in Zurich in 2004 and has been acting as Managing Director ever since. The company has access to a global network of travel trade partners, lifestyle and travel media and (U)HNWI and works closely with public relations and sales & marketing agencies in Frankfurt, Munich, Paris, Dubai, Milan, New York, Hong Kong, and London. Furthermore, the native Swiss is a member of the consulting networks of the Gerson Lehmann Group, USA and Hotellerie Suisse, Bern. He also takes on an active role in the "Luxury" task force of the management of ITB Berlin, Germany. As author and co-author of various specialist publications, his name can be found regularly. He also holds guest lectures in Berlin, Istanbul, Lausanne, Lucerne, Munich, Singapore, Stuttgart, Thun, Vienna, Worms, Zurich, etc. The graduate hotelier graduated with distinction from the Ecole Hôtelière de Lausanne (EHL) and previously completed his studies in business administration as lic.rer.pol. (MBA) at the University of Bern. The luxury hotelier has more than 20 years of experience in the fields of hotel opening, management and renovation/refurbishment in Abu Dhabi, Germany, France, Maldives, Morocco, Seychelles, Sri Lanka, Switzerland, Thailand, Ukraine, and Cyprus of small hotels in high and top end. Many of the hotels have been awarded international prizes. All projects are based on the definition of New Luxury and work according to the principles of the Triple Bottomline.

Prof. Dr. Magda Antonioli Corigliano is a professor of Tourism Economics, Director of the ACME (M.SC.course) and Senior Professor at the Management School at Bocconi University-Milan. She is a member of various European and national scientific committees on Tourism, member of AIEST, ATLAS, and IFITT. Magda Antonioli Corigliano was Special Adviser for the EU Commissioner for tourism (2010–2014), President of the Taskforce on Tourism of the Alpine Convention (2013–2014), member of the Technical Committee for Tourism at the Italian Ministry for Culture and Tourism (2014–2018), and of the Advisory Board of BIT (Milan Fair). She is the author of various publications on Tourism, Industrial and Environmental Economics, and Policy.

Sara Bricchi is the academic tutor at the Master in Economics and Tourism Management. After a double degree in management for cultural industries and creative business processes, since 2009 she collaborates with Bocconi University and SDA Bocconi School of Management. She does research in the tourism field, mainly on destination marketing and management, digital strategies for the industry, trends in offer/demand, and policies. Within the university's courses in tourism, she gives speeches and lectures on the same topics.

Juliet Kinsman is an editor and journalist. Meeting philanthropic hoteliers and working with award-winning luxury-travel brands inspired her to set up Bouteco (www.bouteco.co). As well as sharing enlightening stories of sustainability with the world, she works with design-led hotels to help them position themselves more strongly thanks to their sustainable spirit—especially when it comes to their communications. Founding editor of Mr & Mrs Smith and an author of Louis Vuitton City Guides, Juliet Kinsman has spent much of the last two decades celebrating the world's most special places to stay in the likes of Condé Nast Traveller through to newspapers and supplements globally, and she appreciates how more than ever there is a need for independent, trustworthy tips when it comes to recommending quality hotels.

Prof. Dr. Keiko Kirihara is responsible for International Marketing and International/Intercultural Management in the Department of Tourism/Transportation at the University of Applied Sciences Worms. She is also committed to an intensive exchange with the industry to promote women's careers within the framework of the FocusFrauen initiative that was founded by her. Prior to her work in research and teaching, Keiko Kirihara had a long career in the consumer goods industry, most recently as Director International Market Research for Schwarzkopf Professional at Henkel KGaA, where she was responsible for international market research.

Luxury and the Tourism Offer

Marc Aeberhard, Roland Conrady, Stephan Grandy, Thomas P. Illes, Adam Parken, Norbert Pokorny, Ralf Vogler, Maria Wenske, and Jens Wohltorf

M. Aeberhard (✉)
Luxury Hotel & Spa Management Ltd., Zürich, Switzerland

R. Conrady (✉)
University of Applied Sciences Worms, Worms, Germany
e-mail: conrady@hs-worms.de

S. Grandy (✉)
Premium Products & Lufthansa Private Jet, Frankfurt, Germany
e-mail: stephan.grandy@dlh.de

T. P. Illes (✉)
Zürich, Switzerland
e-mail: illes@marinepress-fotomarine.com

A. Parken · J. Wohltorf
Blacklane GmbH, Berlin, Germany
e-mail: press@blacklane.com; press@blacklane.com

N. Pokorny (✉)
Art of Travel GmbH, Munich, Germany
e-mail: N.Pokorny@artoftravel.de

R. Vogler (✉)
Hochschule Heilbronn, Heilbronn, Germany
e-mail: ralf.vogler@hs-heilbronn.de

M. Wenske
Düsseldorf, Germany

1   Current Differentiation Criteria in the Luxury Hotel Industry[1]

Marc Aeberhard

The World Was Shaken ...

Hardly any other event in recent history has moved mankind as much as the terrorist attack on the Twin Towers in New York in September 2001 and the financial crisis triggered by the collapse of the investment bank Lehmann Bros. in 2007. Both events hit societies in the heart and shook fundamental foundations that had become self-evident in recent decades, especially in the Western world (the USA and Europe): Security and the belief in eternal growth. Both assumptions were abruptly shattered by the two events and have highlighted the vulnerability of highly developed post-industrialized societies. The effects on the economy, politics, and society are known and will not be discussed further here. What is exciting, however, is what these turning-points meant for the hotel industry.

September 11, 2001: Existential Fear Reawakened

The attack on New York fired a fundamental fear of being injured, killed and, consciously or unconsciously, being the target of an attack. In short: the existential fear of being at risk of life and limb has led to a rapid and lasting change in thinking in the choice of hotels. Not long after the attacks, large US hotel companies and chains around the world were shunned, because the attacks were politically motivated and everything US suddenly seemed at risk. Furthermore, large hotels were suddenly regarded as unsafe because there was a (justified or exaggerated) fear that a hotel with five, six, or more entrances could not be properly controlled, that there was no suitable or sufficient overview of incoming and outgoing people. As a result,

---

[1]The word hotel derives from the Latin "hospes," which means hospitality. Especially in the Middle Ages, thousands of believers, who often made dangerous, daring and long pilgrimages, depended on an infrastructure that offered them protection from cold, wetness, wind, weather, and hunger. This infrastructure was (and still is) provided and maintained by the monasteries along the pilgrim routes. What once offered protection from the elements and consolation for the psyche of man has now been preserved in its basic idea, even though the person seeking protection today seeks far less physical shelter than psychological ones. The idea, the conviction to live hospitality is lived regardless of the needs even one and a half millennia after the foundation of monasteries. The related words like hospital, hospice, hotel, hostel, hospitality etc. still show the deep and strong anchoring of the term in our everyday life.

small, fine hotels—especially in cities—offering a particularly personal service unexpectedly gained in popularity. They were literally overrun by business people. The small discreet city hotels in London with their butlers, who had been mocked as completely outdated until then, became the epitome of reserved, charming but authoritarian service quality overnight, whose watchful eyes not even a fly would slip by unnoticed. The birth of the City Retreats had come—small, fine boutique houses, which perhaps in the area of the general infrastructure could not keep up with large houses (for example with regard to sports facilities, banquet rooms, reception halls or the like), but which scored points in terms of security, cordiality and personalized and individualized service and often outstripped the large hotels by miles. The special strength of the luxurious small(est) hotels is being able to arrange and create this feeling of security, warmth, coziness, intimacy, and safety. The worn-out slogan "home away from home" suddenly became revived.

Financial Crisis: Loss of Confidence in Brands

Hardly recovered from the events and consequences of September 11, 2001, another event tore the world from its pedestal in 2008: with the bankruptcy of the world's fourth-largest investment bank, the myth of unsinkable blue-chip companies collapsed. A global financial crisis not only destroyed billions in portfolios and assets, but also destroyed faith in and trust in major brands. Who would have expected before the Lehmann Bros. collapse that companies like UBS or the Deutsche Bank could start to stagger? That large airlines were grounded[2] or entire states went bankrupt? The debt coefficient, whether of individuals, organizations, companies, or states, is constantly rising and immeasurable. The real economy has been overtaken by a virtual, decoupled economy. However, the events in 2007 and thereafter put a damper—albeit only temporarily—on the situation, forced many to their knees and led to substantial write-downs/corrections. Many things have been reconsidered: investment strategies have been re-evaluated, companies have been re-evaluated, in some cases, banks have been forced to new capital structures. Disillusionment is now followed in many areas by a critical questioning: Man has learned to critically confront beautiful appearances, great promises, and strong names and to check every promise for its truthfulness. The slogan "walk the talk" becomes the new credo here.

This has fundamental effects on the hotel industry as well: big names, regardless of origin, content, or orientation, are no longer enough on their own. The time of Unique Advertisement Propositions (UAP) (significant communication difference) was and is over. Such congenial advertising slogans like that of Hilton from the

---

[2]Grounding: An airline's flight license is revoked and its aircraft may no longer take off but remain on the ground. Reasons for grounding can be inadequate technical maintenance of individual aircraft, which are consequently declared unfit for flight, but it can also be economic reasons, such as the threat of insolvency at Swissair or Skyworks.

1990s "take me to the Hilton" have suddenly become waste paper, because Hilton has become completely interchangeable with Intercontinental, Hyatt, Sheraton, Shangri-La, Swissôtel, Sofitel, Taj, Mövenpick, etc. The same applies to premium hotel chains such as Ritz Carlton, Four Seasons, St. Regis, Kempinski, or Mandarin Oriental: no clear Unique Selling Proposition (USP—significant product difference) is discernible at any of them; at best, there are nuances that permit marginal differentiation.

In summary, this means for the hotel industry that from experiences of the two events in the 00-ies an epochal shift in the perception of hotel products began to take place. While throughout the period before 2001 and 2008 a hotel was judged on its material merits, its architectural design, its opulent interior, its extensive infrastructure, and its global communication concept, for more than 10 years now completely different dimensions have slipped into the forefront. In particular, but not conclusively, in addition to (exchangeable) convenience, security/protection, personalized service, authenticity (uniqueness), security, etc. must now also be included in the assessment or in the creation of the offer. It becomes clear very quickly that the usual evaluation grids like star classification, Richies Reports, etc. could no longer deliver sufficient results or guidelines. Luxury hotels of the new kind are defined and built completely different today.

The Pioneers

A visionary pioneer who correctly interpreted the signs almost 40 years ago and has been consistently implementing them in his product range was/is Adrian Zecha with the Aman Resorts. When asked what the three most important success factors for a hotel are, Conrad Hilton answered decades ago: "Location, location, location." This is still true today, but the definition of a good location has suddenly changed completely. Whereas in the past hotels were located close to railway stations, airports, or important trading houses, today remote mountain valleys, lonely islands or small, isolated oases can make for great locations. Aman Resorts have succeeded in capturing, shaping, and living a new dimension of luxury by building decentralized hotel complexes in remote destinations: Privacy. In combination with breathtakingly beautiful surroundings, the hotel group has succeeded in establishing a new understanding of the luxury hotel business, a term that naturally presupposes material infrastructure and equipment, but builds on it and goes far beyond. Discretion, high room prices, and a smaller number of units lead to a shortage of supply and exclusivity that many consider to be highly desirable. As a result, a select circle of guests was formed, who were soon referred to in the hotel scene as the Aman-Junkies.

The model has been and still is copied in many places. Among the pioneers are Sonu Shivdasani and Eva Malström, who with the creation of the Soneva Resorts at the end of the 1990s on the Maldives began to coin the concept of Barefoot Luxury: a very special mixture of modern Robinson feeling and luxury hotel business. A world in which the guest is immersed in a mixture of luxury tree hut, eco-gourmet experience, and sustainable fun club. The Oberoi Group launched its Vila line and created unique small hotels in Rajasthan from old Maharajah palaces with almost unparalleled service quality. With his Virgin Collection, Richard Branson has also begun to develop a keen sense of the new dimensions of luxury. The authenticity of the destination, the harmonious integration into social environments, and the careful handling of architecture play an important role in this. Every house, every unit thus becomes an ergonomically integrated part in a larger whole. It goes without saying that such hotels are small and often only have a few units.

These pioneers are now being followed by more and more interesting concepts that incorporate the premises of the new understanding of luxury into their strategic mission statements and implement them. Often it is only the small hotel groups such as Wilderness Safari or Flame of Africa who set up and operate exclusive lodges in the African wilderness, and increasingly individual hotels all over the world, who have consistently established a genuine USP and adhere to it. Some good examples are Frégate Island Private in the Seychelles, Laucala Resort in Fiji, Hacienda Katanchel in Yucatan, Huka Lodge in New Zealand, etc.

The New Understanding of Luxury

All these companies have some characteristics in common: their commitment to the Triple Bottomline. This stands for:

- Economic profitability
- Social responsibility
- Ecological sustainability

The basic understanding here is based on the deep conviction that a successful hotel can only function in harmony with the surrounding system. A healthy business management basis is of course indispensable, because, without sufficient economic performance, the long-term preservation of a company simply cannot be justified. The question, however, is rather how the surplus is used. And this is how the hotels and small hotel groups that have already been presented differ noticeably from most mainstream hotel organizations. The inclusion of the social environment is not a well-packaged PR story, but everyday life: be it the recruitment and training of employees for the business from the surrounding neighborhoods, the consideration of local economic power (agricultural products, craftsmen, building materials, art, and culture, etc.) or charitable commitment in the form of support for community projects such as the creation of school infrastructure, development of medical facilities or contributions to care for orphans and the elderly. Also, as far as the commitment to ecology is concerned, much more demanding criteria apply to this "new" form of luxury hotel. They have understood that an intact natural or cultural environment is the basis for operational success. Often this is even the USP of a house, and thus the preservation, protection or if necessary even the restoration of an ecological system stands in the center of the hotel's business.

In addition to this strategic triple bottom line approach, many such hotels differ from the traditional "old school" luxury hotels in a number of operational features. Here we are talking about New Luxury; this is the explicit addition of immaterial dimensions to the materially shaped understanding of luxury. The existence of a first-class infrastructure is a prerequisite for this. Much more important are now in the expanded definition of luxury:

- Space and time
- Personalized and individualized service
- Exclusiveness
- Safety and security
- Health

Distinctive Features: High-End Versus Top-End Hotel Business

From these remarks, it is now clear that the luxury hotel industry must be split up.

On the one hand, there is the typical and well-known high-end hotel business, which is characterized by:

- Well-known brand names and strong brand communication.
- Interchangeability.
- Unclear USPs, often price factors and location factors decide. However, the location factor does not refer to the hotel as an end in itself (the hotel does not become a destination per se), but to the location of the hotel in relation to the final destination, for example, a customer address, a beach, etc. (so the hotel is a means to an end).
- Large hotels (up to several hundred rooms).
- High, but affordable price segment (rarely an ADR of over 500 euros).
- Global distribution and high local presence.
- Well-structured operative units which, however, only allow a limited degree of personalization and individualization.
- Segregated offers: Subdivision into room categories, executive floor, suites partly with adapted service culture.
- Spatial limitation: a high number of people per unit area reduces individual privacy.

This is now contrasted with the top-end hotel industry, which is distinguished by its outstanding features:

- Exclusivity; often only very few units, rarely more than 30–50 units.
- The hotel per se becomes a destination.

- Maximum privacy protection, retreat, and hideaway concepts guarantee paparazzi-free zones, data protection, and physical protection are guaranteed. The idea of retreat is increasingly being adopted in cities as well. The City-Hideaway becomes an oasis in the urban bustle, often only with small inconspicuous, hardly recognizable entrances.
- The discretion and incognito of the guests are given the highest possible attention; trophy walls with signed photos of stars and starlets are an absolute taboo and testify to bad taste and social mediocrity.
- Decentralization of the units creates a lot of space for privacy.
- Consistent profiling creates tailor-made service concepts for each individual guest 24/7.
- The guiding principle "anything, anytime, anywhere" applies.
- The ADR is generally over 1500 euros (the extent to which additional services such as F&B (Food and Beverage) services etc. are included is of secondary importance).
- Accessibility can sometimes be difficult, which can be seen as the special attraction of a house, or as an arduous arrival/departure. The answer to this is often complex transport concepts which, however, sometimes contradict ecological compatibility.

Travel Motive: Accelerate Versus Decelerate

The travel motives are as different as the thoughts of the travelers. The only important thing for the design of luxury hotels is that pure travel profiles no longer exist; today several motifs often combine to form a motif cluster. For example, a business trip to Milan can be accompanied by a visit to the Scala Opera, a visit to a museum, or an urgent conference call with business partners in New York during a beach holiday in Bali. In its infrastructure design, the hotel must provide for multitasking and polyvalence seamlessly and in real time (albeit discreetly and in the back of the house). Each hotel thus gets a business hotel and a leisure hotel component. But much more important than the distinction between city and holiday hotels is the distinction between accelerated and decelerated hotels. Accelerated hotels are characterized by the omnipresence of technology and digitalization. This begins at check-in with the unsolicited WLan code and ends with a true keyboard of plugs, sockets, and wireless access points in every corner of the house. Here the guest is offered the perfect high-tech environment, optimal connections, and trouble-free digital interaction. Such hotels, however, do not manage to let the guest arrive in the true sense of the word. Be it that he does not want to arrive at all or has already left before he arrived.

Hotels that focus on deceleration are quite different. In these hotels too, time plays a central role in the self-conception and perception of the house, but not as a constant of acceleration, but exactly the opposite. These hotels see themselves as antipol to hectic and promise the guest an increased perception of the present and the self through a deliberately chosen slowing down of all processes. The risk in deceleration hotels lies in stressing out the guest by underchallenging him. The sudden loss of non-stop information, the conscious reduction of sensory overload provokes in some a void that a hotel must know how to fill. This process, which is controlled and consciously lived, for example by omitting music and TV screens in public (and private) spaces, the reduction to essential equipment parts, the symbiotic fusion of nature and hotel, leads to a reawakening of the senses, a strengthened self-perception, a measurable pulse slowdown, and a reduction of stress hormones and thus a resurgence of moments of happiness and contemplative complacency. This psychological effect is often so intense that small qualitative inadequacies or infrastructural deficiencies are sometimes overlooked. The focus is completely shifted from material obsolescence to immaterial apotheosis.

Lifestyle: Stiff Versus Légère

A third dimension, how the new luxury hotel industry enters the market today, is the celebrated lifestyle. This may go hand in hand with a generation change, but it may also serve as a testimony to current events. So, it is not surprising that hotel managers in Zurich's grand hotels wearing ties and collars look very stiff and outdated, while fashionably dressed hipster hotels in New York portray a completely different image of hotel and hotel lifestyle.

While only a few years ago differences in status, rank, and social affiliation were documented by the suit (white collar vs. blue collar), the breakup of the old-established dress codes resulted in considerably smaller differences looking at it from a purely superficial view. Like the general director as well as the assistant in a conservative corporate environment such as a finance firm or an insurance business or old-fashioned hotel chains, they both wear suits, almost like uniforms; but in a rather casual environment, the differences between the top and the bottom have also diminished at least on the outside. The difference lies in the detail. It requires a trained eye and stylistically experience to distinguish quality differences in small details. Today, class and style are no longer decided on ties or corduroy trousers, but on details such as the quality, tradition, and origin of the buttons of the naval blazer, the buttonhole on the jacket sleeve, the position of the initials in tailor-made shirts, manufacturers and materials of accessories such as watch brands and writing instruments and the quality of shoes. These nuances are often only discreetly detectable, but they quickly decide between wannabe and genuine, so the somewhat oversized gold watch in the wrong social context can quickly be interpreted as a sign of boasting.

In the hotel industry, this social shift means that the choice of fittings, furniture, and equipment (FF&E, equipment) has to be coordinated in every detail with the statement of the hotel. In particular:

- Toiletries
- Bed and table linen
- Terry towels
- Dishes, glasses, and cutlery
- Upholstery fabrics, bedding
- The quality of the writing paper

a hotel must be able to make a clear statement about its style.

But the subtleties that a hotel must satisfy today also extend far into the food and beverage (F&B) sector. The whiskey selection is not measured by the big names, but by the first-class quality of the small distilleries combined with the knowledge of the Chef de Bar. It is not the amount of imaginative Martini cocktails that makes a good bar, but the quality of the tonic varieties, which are perfectly matched to the hand-picked gin collection. The fact that a good cook can prepare fine pieces of Wagyu and Simmentaler cattle flawlessly may be assumed, he will get measured by the refinement with which he implements "Snout to Tail," how he manages to stage textures, colors, shapes, tastes, and temperatures on a plate. Large fuss with lots of silver, bar pianists, lavish chandeliers, and many expensive labels distorts the view of the essential. Lifestyle today does not see itself as a broad mix of things, but as an elaborate in-depth focusing.

Understanding, Not Copying

It is well known that product life cycles have become very short. Pop-up concepts are spreading and the customer is in the comfortable position to choose from an almost infinite range of products. This puts the lazy hotel business under massive pressure and demands everything from it in terms of a current, hip offer. It forces the hoteliers to keep their eyes and ears open at all times and to be able to react quickly to any trends and adapt the offer accordingly. The keyword here, however, is "react." To react to a trend means to be one step behind already. Business conduct thus becomes an endless lag, a constant, never-fulfillable need to keep up. The energy required is enormous and quickly leads to irreparable wear and tear on people, materials, and images. Management by shooting star is a short bright glow in the sky, a short glow followed by a long sink beyond recognition.

However, things are different when a hotel manages to proactively design the scene and act as the actual trendsetter. These companies are rapidly becoming opinion leaders. These hotels do not copy, they are being copied. Market and image leadership can have various reasons; this can be history and reputation of a hotel, success or design, service or traditions. Examples include The Peninsula in Hong Kong, The Raffles in Singapore, the Orania in Berlin, The Delano in Miami, the Royal Mansour in Marrakech, etc.

Summary

In summary, the paradigm shift in the luxury hotel industry manifests itself multidimensionally:

- A clear differentiation between high-end and top-end (luxury and beyond)
- A shift from a material to an immaterial understanding of luxury (anything that money cannot buy)
- Segmentation into accelerating and decelerating hotels
- A subtle characterization of lifestyle, measured by the small but subtle differences between the ordinary and the extraordinary is the little extra.
- Does a hotel copy or is it being copied?
- And last but not least in the quality of a hotel, to be loud or quiet. In the (newly understood) luxury hotel business, discretion is a cardinal virtue. But there are many hotels that boast loudly and with a lot of PR noise about who stayed in their house, when and for how long, who keep golden guest books, in-house newspapers or social media walls and illustrate more or less important and well-known celebrity stories. Such behavior is typically found in high-end hotels. In the (true) top-end establishments, on the other hand, it is a matter of honor to protect the identity of their guests, to guarantee discretion and paparazzi-free zones, and to maintain secrecy about the clientele, their identity, and customs. Today this is an important booking criterion for many UHNWI.

2 Interview: Aviation—Beyond First Class

The Example of Lufthansa Private Jet

Stephan Grandy and Roland Conrady

The following reflects Stephan Grandy's answers to questions asked by Roland Conrady.

Question 1: How Is the Aviation Market Segmented? Where Does Middle-Class End, Where Does Premium Begin and Where Does Luxury Begin?

The classic segmentation according to cabin classes such as Economy, Business, and First Class still exists, although it no longer represents a clear classification, but rather gives an indication of the group in which the seat should be classified. On the one hand, new categories were introduced, such as the Premium Eco at Lufthansa or the Business Plus at American airlines; on the other hand, these designations say little about the price paid by the customer. Within Lufthansa, there are many booking classes and fares that include segmentation by service. These services, such as reserving a seat, advance booking period, extra piece of luggage, and much more, have been integrated into the booking classes and have thus created different valences, ranging from a simple ticket for just the flight to all kinds of amenities such as ordering special meals.

So where does middle class end and where does premium start and change into luxury? There are certainly different definitions for this, which often have their justification. Lufthansa's "New Premium" concept is less about labeling a certain booking class or seat category as "Premium" but more about offering the customer the best service at all touchpoints along the "Customer Journey." "Brilliant Basics" and in some situations to create Magic Moments beyond the expectations of the customer. And this regardless of the actual seat or booking class. Five aspects are of particular importance here: Personalization, Simplicity, Human Touch, Consistency, and Pleasure. The focus is put on these factors and improvements are sought.

Question 2: How Do You Define Luxury in the Aviation Segment?

Lufthansa Private Jet (LPJ) is addressing this New Premium concept in their classic Premium concept. Lufthansa Private Jet today already stands for precisely this premium understanding, for fulfilling the needs of wealthy private individuals, the so-called High Net Worth Individuals (HNWI):

- **Personalization:** The customer decides when, where, with whom, and with which aircraft category he wants to fly.
- **Simplicity:** One call or one e-mail; two options Limousine Service and/or Flight Attendant, three aircraft categories—these are the only decisions our customers still have to make. LPJ does everything else.
- **Human Touch:** Personal customer contact by our LPJ service team, personal greeting on board and on flight support by the pilots.
- **Consistency:** 24/7 availability, guaranteed availability, consistent pricing, and International Commercial Aviation Industry Safety and Security guidelines.
- **Pleasure:** From home to the nearby airfield with the limousine, from the General Aviation Terminal of a hub with the limousine directly to the wide-bodied aircraft for intercontinental flights. On arrival at the destination, the chauffeur is waiting at the aircraft ready to take the guest directly to the hotel or his destination.

How do we define luxury in the aviation segment? Today's luxury is less and less about high-priced champagne parties, in diamond necklaces, or in general in showing off wealth. Luxury refers on the one hand to meeting customers' high expectations of a product (e.g., selected materials, fine workmanship) and on the other hand to enjoyment and the attitude to life itself. The aim is to have positive experiences and to optimally fill the available time with this attitude to life. Luxury and time are strongly linked here. To use free, available time optimally for the things that are important, have content, and engage the senses.

With regard to the aviation industry, queues at gates, taxi stands, public transport stops, etc., bureaucracy, high expenditure for booking and processing belong on the negative list. Everything should be fast, easy, and immediately consumable.

From our point of view, the following basic needs of our customers are met by LPJ in an outstanding way:

- **Time** (availability/speed): The guest is usually under time pressure, decides at short notice and his valuable time is optimally used (no queues, little bureaucracy, little effort in the organization for the customer). Lufthansa Private Jet guarantees the availability of the aircraft if booked up to 10 h before departure.
- **Flexibility** with regard to travel planning or travel components. Rebookings can also be made at very short notice.
- **Service/Convenience** ("ease"): The handling, execution, and organization of the flight are adapted to the customer's needs. Lufthansa Private Jet guarantees that their customers reach their destinations. The Lufthansa Private Jet Service Team will coordinate for the customer in case of storms, bad weather, or similar. The team is available 24/7.
- **Cost control** in relation to the total price of the trip. There is a fixed price guarantee and no upfront investment, as is usual with hourly ticket offers or fractional ownership offers. There is no specification of the services rendered and the guest is always highly flexible.

The described "Lufthansa Private Jet" business model (Private Aviation) differs from the classic commercial passenger air traffic (Commercial Aviation) as shown in Table 1.

Table 1 Comparison of commercial and private aviation (LPJ)

	Commercial aviation	Private aviation (LPJ)
Destinations	Average number of destinations (commercial airports) according to airlines' flight schedules	Very high number of destinations (commercial airports and small/mini airports) independent of airline schedules
Flight times	Departure and arrival times according to airline schedules	Departure and arrival times according to passenger requirements
Travel time	Higher travel time	Shorter travel time due to faster flight (higher climb performance), faster ground processes, and direct flights
Aircraft types	Narrow-body and widebody aircraft	LPJ: Small Size Jets (6 seats), Mid-Size Jets (8 seats), Large Size Jets (10 seats)
Safety and security	IOSA certification	IOSA and LH-CARA certification
Airport terminals	Passenger terminals with standard passenger processes (security checks, gates, etc.), often with queues	General Aviation terminals, without queues
Price	Price per seat	Price per flight (regardless of the number of seats used)

A Short Story to Illustrate the Special Character of a Trip with Lufthansa Private Jet
Due to a challenge in the field of air traffic control (ATC) in France, creativity and cooperation were particularly important in this case:

Lufthansa Private Jet customers flew from Barcelona to the USA via Frankfurt. In order to make the connection to the long-haul flight in time, a small safety cushion was installed. On the morning of the flight, the French government together with Air Traffic Control had imposed a ban on overflights in Europe. This meant for LPJ that despite a time buffer the long-distance destination could no longer be reached, because the flight time got extended by 1 h.

The service team immediately started to work out alternatives. There were flights in the afternoon, but they were all fully booked. Six first-class passengers were not willing to fly in a different class, they had to get home because of important business.

It was decided to divert to Munich (MUC). The connecting flight was reached in time and space was still available for all guests. With the help of the colleagues in MUC, all tickets were reissued and the check-in including bag tags was carried out. The close cooperation between the LPJ team, LH MUC Station, and VIP Services MUC worked optimally. Within a very short time, everyone was ready to receive the guests. Also, the Ground OPS crew did a great job.

The guests were thrilled and 1 year later they flew with LPJ again, this time without any incidents.

Question 3: Which Providers with Which Offers Can Be Found in the Luxury Aviation Segment?

The market for business aviation providers within Europe is highly diversified. There is a large number of small and medium-sized enterprises offering capacity from 2 to 10 aircraft, some in the order of 20–30 aircraft, and very few large operators, such as NetJets, with up to 100 aircraft in their fleet in Europe. In recent years there has also been an increase in the merger of smaller providers into conglomerates, such as Gama Aviation, which emerged from several European companies.

In addition, there are regional operators that limit themselves to a limited number of bases and offer flights exclusively or at least for the most part from their base. Finally, there is a large number of small suppliers with one to two aircraft in their portfolio offering pure flight services. Flight, booking, service, billing are all from one source, without any great administrative effort.

From the user's point of view, which is structured according to available options and frequency of use, there is a different approach. Firstly, there are the *owners of* an aircraft, who often complete up to 500 flights a year. Second, customers who do not fly quite as much often use the *fractional ownership* model established by NetJets—a model that allows customers to get high availability of their "own" aircraft at significantly lower costs. Here the customer buys a share of an aircraft, together with other users. This model only makes sense if the provider owns several aircraft of the same model and can share the aircraft among the users without overlap. If the customer flies significantly less, a third model is available, the so-called *hourly ticket model*. Here the customer buys 25, 50, or more flight hours at a certain price in advance and the provider charges via a card. The fourth model is *LPJ's business model*.

Question 4: What Does the Lufthansa Private Jet Business Model Look Like?

While you are waiting in line at Düsseldorf airport for your flight to Mallorca at 5 am, squeezed into an economy seat with a cereal bar and bad coffee to get to your holiday destination or hurry to the next business appointment, who has not dreamed of this: being picked up from home in a limousine, no waiting at check-in, no queuing at the counter, cared for by smiling staff, and then hovering above the clouds in a luxurious leather armchair en route to the destination. For some, this is the normal way to fly: for Lufthansa Private Jet customers.

The launch of the premium product Lufthansa Private Jet was the answer to the growing need of a small but constantly growing target group: wealthy private individuals (HNWI), whose striving for time flexibility, individuality, and first-class service is served by Europe's largest airline with this offer for over 10 years now. The Lufthansa Private Jet segment rounds off Lufthansa's range of individually customized flight plans with private aircraft in the upper product segment.

Right from the start, Lufthansa focused the product on the special needs of this clientele. On the one hand, Lufthansa Private Jet provides a limousine service from home or from the office directly to the aircraft, on the other hand, individual requirements can be expressed regarding food, flight attendants or other wishes. The Lufthansa service team is available 24 h a day, 7 days a week to meet all requirements.

When purchasing a "Lufthansa Private Jet" flight, the passenger does not purchase a charter contract as is the case with competitors, but rather a Lufthansa First Class ticket, which gives him or her all the benefits that correspond to a regular Full Fare First Class ticket, such as access to the First Class Terminal in Frankfurt and to all Lufthansa First Class lounges worldwide. In addition, the booking passenger receives 10,000 award miles, status miles, and HON Circle miles for each "Lufthansa Private Jet" flight. Frequent Travelers, Senators, and HON Circle Members earn an additional 25% Executive Bonus on their mileage credit.

The aim is to offer "Lufthansa Private Jet" customers maximum flexibility by making the desired aircraft type available at short notice up to 10 h before departure in Europe and up to 12 h before departure in North America. Free cancellation is possible up to 48 h before departure. In principle, all destinations can be flown to, except if the approach requirements do not allow this. The coverage includes more than 1000 destinations in Europe and over 6000 within North America.

The range of aircraft includes three aircraft categories: Small Size, Mid Size, and Large Size Aircraft:

1. Small Size—maximum six seats, such as Embraer Phenom 300
2. Mid Size—maximum eight seats: for example, Cessna Citation Excel/XLS, Hawker 750[3]
3. Large Size—maximum ten seats: for example, Dassault Falcon 2000/2000EX

---

[3]In planning: Cessna Citation Latitude.

The costs vary depending on the choice of aircraft type and route. A flight from Munich to Milan in a small-size aircraft costs around 7900 euros.

Lufthansa has been working with NetJets as a strategic partner since the product launch in 2005. NetJets is one of the world's largest providers of private jets.

Question 5: What Are the Main Competitors?

In principle, the LPJ business model is based on the sale of single flights, unlike NetJets, which focus mainly on the sale of fractional ownership or hourly tickets in Europe. There is therefore no competition here between Lufthansa Private Jet and NetJets. The situation is different for regionally oriented companies that operate out of their bases, such as Air Hamburg from Hamburg, DC Aviation from Stuttgart, or Windrose from Berlin.

On the other hand, there are a large number of brokers who buy free capacity on the market and offer their flights via platforms or service centers. Brokers often use capacities offered by the airlines as ferry flights (empty flights). They arise when the aircraft returns to the base or is positioned by the customer. Avinote or Fly Victor are particularly relevant here.

Question 6: What Is the Customer Structure of Lufthansa Private Jet and Its Competitors (Including Regional Distribution, Business/Private Travelers, etc.)?

Lufthansa Private Jet's customer structure is very diverse. On the one hand, there are of course some celebrities, sportsmen, or actors, but by far the largest part of our customers are business people. And here especially businessmen from small and medium-sized enterprises. But the "typical" "Lufthansa Private Jet" customer cannot always be assigned to exactly one segment such as business travel or private travel. Today the transitions are fluid. For example, there is the managing director of a medium-sized company who flies to Milan for a business appointment, goes to the opera with his wife in the evening, and flies back to Munich the next day with his wife.

Our core target group for both First Class and Lufthansa Private Jet are guests with available assets of more than 5 million euros. These individuals, also known as High Net Worth Individuals (HNWI), are the growth drivers in the luxury business segment. There are over 14 million HNWIs worldwide and this group is growing every year—although to varying degrees in different regions. While growth in the USA and Germany is between 3 and 8% per year, growth is significantly higher in China (20%) and the Middle East (33%).

Question 7: What Demands Do Customers Have for the Product?

As already described, our customers generally have high, sometimes extremely high expectations. The quality of each product must meet the highest demands for all senses, haptic, optical, smell, or taste. The aim is to savor the experience and experience an optimal balance between stimulation and relaxation. Having time to enjoy things, but of course, also time for business is a bottleneck and thus becomes the most important good. We give the passengers back the time they would normally spend in queue after queue. Our customers expect the aircraft to be positioned and ready on time, even if the schedule has changed at short notice. They expect that "one" knows them and responds to their needs, that it is easy for them to travel and that they are able to enjoy it.

- **Time:** The bottleneck is essentially "free" time that one creates for oneself when things are outsourced as a service. This free time is the most valuable commodity for customers. It is used to do exactly what customers really want to do. This is achieved through a product range that ensures that all destinations can be reached at short notice without queues, without administrative effort, and without time-consuming organization.
- **Flexibility:** Flexibility is highly valued in order to react quickly and unbureaucratically to changing customer requirements. This includes short-term rebooking options, date changes, and cancellation options.
- **Service/Convenience**: Certain travel processes should simply just "work." The customer does not want to have to think about what to do next to get from A to B. He wants to reach his destination "easily," have it comfortable, and spend the travel time as pleasant as possible. Product offers that guarantee the arrival at the target location and a perfect organization contribute to this.
- **Transparency**: Although HNWIs are willing to spend money on tailor-made experiences, they appreciate cost transparency and a high degree of control. No pre-investments and fixed services as with hourly tickets or purchased shares, instead fixed price guarantees and insight into pricing.

Question 8: How Will Demand Develop in the Coming Years?

The demand for flight services in 2018 shows a clear upward trend within Europe. Overall, business air traffic in the last 2 years recorded an increase to 5.2% in March 2018 compared with the previous year. Growth in 2017 was driven to a large extent by increased charter activity to Southern Europe.[4]

---

[4]Cf. data of the WingX database in www.wingx-advance.com.

Question 9: How Are Customer Expectations Likely to Change Over the Next Few Years?

The most important developments are as follows:

- Growing cost awareness.
- Flexible, more individualized, exclusive offers ("money can't buy" offers).
- Increased awareness of sustainability.
- Tendency toward larger aircraft types, more comfort, more exclusive destinations.
- Door-to-door services are more in demand (here the media plays a major role with advertising for exclusive holiday resorts).
- More advantages with the Miles&More program, e.g., more miles earned with HUB transfers, longer flights, etc.
- Companies have discovered the market and will value time savings higher than money savings in the future—top managers can visit several, more distant branches within a very short time and be back home on the same day. At the same time, they have the flexibility to adapt their flight schedule at short notice to their business needs and to change it if necessary. The plane becomes a flying office.

Question 10: How Do First-Class Offers of Renowned Airlines Differ from Private Jet Offers?

Lufthansa is the only airline that strongly integrated a business aviation product into its first/business and economy class product portfolio. Other airlines also cooperate with private jet operators, such as Air France/KLM with WIJet, or have even strongly integrated it in the name of the parent company, such as Delta Private Jet, but they are still separate charter contracts. With LPJ the customer essentially gets a First-Class ticket, meaning that the ticket is treated like a normal First Class ticket, with the possibility to visit the First-Class lounges of Lufthansa and also the partner airlines. Furthermore, in both examples, the flights and the services offered are limited to the respective hubs of the airlines. WIJet is offered almost exclusively from Paris and serves only as a feeder to Air France's intercontinental services. The same applies to Delta Private Jet. In contrast, LPJ offers its feeders and defeeders to all hubs such as Frankfurt, Munich, or Zurich and also Vienna. This feeder/defeeder share is 35% for LHG. 65% are therefore point-to-point connections within Europe and North America.

Question 11: What Are the Cost Differences Between First Class and Private Jet Products?

The example of a trip from Munich/Germany (MUC) to Karlovy Vary/Czech Republic (KLV) illustrates price and travel time differences between a train trip, a commercial aviation flight, and a private jet flight (Table 2).

Conclusion Assuming that six people travel together, an LPJ journey costs 1375 euros per person. In comparison, a commercial aviation trip costs 860 euros per person, which is marginally less. On the other hand, the travel time for an LPJ journey is considerably shorter at 1.5 vs. 5 h. The short travel time allows arrival, business appointment, and return on the same day, which is not possible with a commercial aviation trip to remote locations. A business aviation trip often involves significantly lower opportunity costs because the management team can work longer and valuable management time is not "wasted" on travel. Taking opportunity costs into account, business aviation travel is often more cost-effective than commercial aviation travel.

Question 12: How Do You See the Future Development of the General Economic and Social Conditions for the Luxury Aviation Segment?

Due to the further growth in passenger numbers in air traffic, the chaos at airports will continue to rise and thus intensify the need for alternatives to business aviation. The number of HNWIs worldwide will increase significantly, with the strongest growth expected in Asia and the Middle East. However, the market potential in

Table 2 Comparison of itineraries, travel times, and travel prices

	Itinerary	Travel time	Price
Train journey (1st class)	Munich—Karlovy Vary (1st class, 2 changes)	5 h	140 €/person
Commercial aviation flight (Business Class)	Drive to MUC (plus parking fee if applicable)	1 h	30 €/trip
	Security Controls/Boarding	1 h	–
	Flight MUC—PRG	1 h	700 €/person
	Limousine PRG—KLV	2 h	450 €/trip
Private Aviation (First Class, Small Size Jet)	Dive to MUC (LPJ Limousine Service)	1 h	–
	Flight MUC—KLV	36 min	8250 € (1 traveler) 4125 € (2 travelers) 2062 € (3 passengers) 1375 € (4 travelers)

Europe and North America has not yet been exhausted either. Further growth in the luxury aviation segment is therefore to be expected. Dampening effects could arise as a result of increased sustainability and environmental awareness. Another major challenge could be the low barriers to market entry for new providers in the luxury aviation segment.

3 Cruises: Beyond Upper Premium

A Crystal Chandelier Does Not Equal Luxury

Thomas P. Illes

Cruises: When Does a Ship Belong to the Luxury Class?

Cruises are polarizing. On the one hand, the deep-sea cruise industry continues to grow very dynamically. On the other hand, there is an increasingly critical attitude toward holidays at sea. In media reporting, cruises are regularly viewed with little differentiation or are one-sidedly lumped together and associated either with mass entertainment or—the extreme on the other end of the spectrum—with elitist decadent luxury, this is due to the often very incomplete knowledge of the overall picture of the market on the part of most editors.

The overused term "luxury" is often wrongfully used to describe cruises in an almost inflationary way. However, some shipping companies and tour operators themselves also play their part in this, contributing a great deal to the uncertainty in the categorization of ships with completely exaggerated advertising and marketing messages.

Only just under 10% of ocean-going ships belong to the luxury category. Nevertheless, the myth of the "luxury liner"—which has become somewhat dusty in the meantime—persists, probably in the glamor of the ocean liners of bygone days. And even so for the group of ships in the standard volume segment, the floating megaresorts representing modern mass tourism, with a passenger capacity of meanwhile far more than 6000 guests, utilizing economies of scale optimally, which operate the usual seven-night race routes in the Mediterranean between the Canary Islands or, the cruise destination of the first hour, the Caribbean, at partly very favorable entry prices.

On the other hand, nobody would give every hotel the label "luxury hotel." And it also makes sense that airlines, for example, have significant differences in quality and prices between First, Business, and Economy. So how is luxury defined on the high seas?

Below is a summary of the most important quality features.

Ship Design

On luxury ships, solid materials, understated, unpretentious elegance and high-quality design elements such as tastefully selected art or real plants create an atmosphere of—genuine—class and authenticity. Less is often more here. There are no shrill color orgies, cheap imitations, and overly intrusive wow effects, nor are there any hip facilities that are commonly found on mainstream ships, such as water slides, karting tracks, climbing walls, surf/parachute simulators, ice bars, and the like with the goal to provide adrenaline kicks. The focus is on the sea voyage experience with exclusive quality in terms of destinations, nature, culture, education, wellness, gastronomy, and service. There is disproportionately much space per passenger (keyword PSR—see info box) available—including beautiful outside decks, mostly with real teak covering. Because space costs money—even on ships. Newer luxury ships have a quiet, low-vibration propulsion system and good sound insulation. Apart from a few exceptions (e.g., Hapag-Lloyd Cruises' ultra-modern "Europa 2"), ships in the luxury segment so far have tended to be more conservative. However, new projects from existing and new shipping companies promise a change of attitude in this respect.

Passenger Capacity

Oversized ships with a large number of passengers and luxury do not get along well. The upper limit for offering an individual, personal travel experience with a luxury claim is currently just under 1000 passengers. Most luxury ships are in the 200–700 passenger segment, some even lower. However, not all smaller vessels belong to the luxury category.

Cabins

Outdoor or balcony cabins and suites are now industry standard. Ships that still have interior cabins cannot be assigned to the luxury class.

Lifestyle

Analog to noble boutique hotels on shore, the atmosphere on board is mainly quiet, relaxed, and familiar. This also means: as few announcements as possible, no loud background music. Although it is sometimes possible to dance and "rock" on luxury ships, there is no demand for permanent animation and mass entertainment, which often occur on mainstream ships. Moreover, in the course of a changing concept of luxury, it is no longer primarily flaunting one's wealth by e.g. eating caviar and drinking champagne that define a luxury product at sea. Today's upscale clientele—

at least those from the traditional markets of the past—are increasingly looking for authentic, inspiring holiday experiences combined with positive emotionality beyond purely material values and all too intrusive opulence, even at sea. Above all, this also includes a lot of flexible freedom for personal development and shaping one's life—something that is not always easy or even impossible to achieve on large mainstream ships. The situation is somewhat different for ships designed for the growing Chinese (source) market (with its very prestige-oriented clientele), where luxury class ships are not yet in use. The latter is interesting because China has many hotels of the luxury category ashore, but the—still very young—cruise market in the Middle Kingdom or Asia in general is not yet considered mature enough to justify the profitable use of genuine luxury ships. Nevertheless, even in these up-and-coming, new markets, the label "luxury" is often used for advertising, which can lead to disappointed reactions and problems among many demanding and quality-conscious customers (and has repeatedly done so).

Family Friendliness

With the exception of the "Europa 2" and the ship-in-ship concepts of MSC and Norwegian, luxury ships still do not tend to be family-friendly and offer little excitement on board for children and teenagers. In the course of a changing guest demography and because family travel is becoming more and more important in the luxury segment, shipping companies are starting to adjust their offers accordingly.

Service

Luxury ships are characterized by a better trained and more attentive crew as well as an increased number of crew members per guest (keyword PCR—see infobox). The emphasis is put on hospitality that is lived as naturally as possible: personal, flexible, individual, unagitated, professional service with plenty of time for guests and interaction at eye level. Not an anonymous, submissive, gimmickry service drill. While on mainstream ships there is often a lot of noise in the main restaurants and due to a tight schedule a hasty standard service has to be provided, here too the credo of peace and relaxation applies—stress should be an absolute foreign word on luxury ships. However, within the framework of cost optimization programs of some shipping companies with continuously increasing demands on the service crew while the number of crews remains the same, this is no longer consistently guaranteed even in the luxury segment.

PSR and PCR: Space and Crew Size

Two incorruptible parameters that indicate whether a ship is a luxury product or not are the Passenger Space Ratio (PSR) and the Passenger Crew Ratio (PCR).

The PSR provides information about the space available to an individual passenger on board in purely mathematical terms. For this purpose, the gross tonnage (GT) of the ship is divided by the maximum number of guests. While large ships in the volume segment have an average value of around 30, the current leader "Europa 2" has almost three times that value, i.e., 83. Exceptions are special ships such as icebreakers or cargo ships with passenger transport, which carry only a few passengers in relation to tonnage/cargo and therefore have a very high PSR but are not luxury ships. However, the bar could soon become much higher: The luxury expedition yacht "Crystal Endeavor," which is currently under construction in Germany for Crystal Cruises, is to lead this ranking in 2020 with a PSR of 100.

The second indicator, the Passenger Crew Ratio (PCR), measures the relationship between passengers and crew members and thus allows conclusions to be drawn about the service quality offered on board. On the "Europa 2", this value is 1.4, which means that almost every passenger has one crew member. For mainstream vessels in the volume segment, this figure is around 3–4, meaning that a crew member has up to four times more passengers to handle. Or to put it another way: With the same number of passengers, there are four times fewer crew members on board. This is not compatible with real luxury. "Crystal Endeavor" also promises a new service milestone here with a PCR of almost 1:1 (the maximum of 200 guests will have approximately the same number of crew members at their disposal).

Gastronomy

A diverse kitchen that does not allow repetitions in the menu even on long journeys around the world is a must in the luxury segment. If possible, the dishes are prepared freshly and made to order and presented in an appealing way. Fresh, high-quality, often local ingredients are used, which the chef sometimes buys together with the guests at local markets as part of a special cruise experience. On mainstream ships, where for cost reasons much is prefabricated industrially, this would not be feasible with the mass of passengers simply because of hygiene and logistics—on ships there are much more stringent regulations and rules than on land. There is almost without exception only one dinner session, the restaurants have long and flexible opening hours. You are free to choose where, when, what and with whom you would like to eat—free 24-h room service in your own cabin/suite is also available. The restaurants have many tables for two (which again costs space) and a large selection of drinks with an exquisite wine selection. Details such as freshly squeezed orange juice are a matter of course.

Sports and Wellness

As a special feature, some ships offer an on-board marina with free water sports facilities at the stern.

Entertainment

A high-quality, sophisticated entertainment program coupled with numerous opportunities for further education and learning in popular leisure areas are the standard. Usually, there is no permanent animation, which often occurs in connection with mass tourism. Further quality features are the use of renowned editors, prominent guest artists who are well established (and accordingly more expensive) on land, as well as a range of free films or DVDs and a varied and interactive TV and music program in the cabins and suites.

Routing

More exotic smaller ports are regularly called at and varying routes are put together off the usual standard routes of large mainstream ships—for example, world voyages or expeditions. Longer periods at port, sometimes overnight, are also very popular. In addition to a more authentic destination experience, the extension of the port lay times can avoid or at least partially mitigate overtourism, which is often associated with the large volume ships and, due to the more moderate size of the ships, offer a better port location close to or directly in the city center, which is not available to the large ships. Land excursion programs are more exclusive and individual and can take place in small groups.

Dress Code

Contrary to a popular opinion that luxury ship passengers are overdressed, the trend in the luxury segment is now moving in the opposite direction. Of course, swimsuits do not belong in (indoor) restaurants even on luxury ships, but usually suits and ties can be left at home. A luxury clientele that is constantly modernizing itself but is often still faced with dress codes in hectic work environments prefers now a "smart casual" or "elegant casual" dress code that is more suited to their holiday needs. In contrast, on many mainstream ships "costume balls" and galas in the tradition of the glamorous ocean liners of the past are sometimes still celebrated in the course of a theatrically staged simulation of emulated luxury and as an integral part of the travel experience expected by these guests. Exceptions such as Hapag-Lloyd Cruises' classic-conservative luxury ship "Europa" (not to be confused with the "Europa 2" of the same shipping company, which is committed to contemporary "casual luxury" and targets a younger clientele) confirm the rule.

Ancillary Costs

Most luxury ships include a variety of drinks, specialty restaurants, tips, and extras such as water bottles for excursions, shuttle buses to the city center, daily newspapers, etc. for a relaxed travel experience, some even include shore excursions, Internet, and the like.

Special Features

On luxury ships a lot of attention is paid to details: fine tableware, high-quality, comfortable deck chairs, decent coat hangers, discreet but effective reading lights, real towels in the public bathrooms, exquisite care products in the bathrooms, etc. Privileges such as a publicly accessible command bridge are sometimes also part of an individual travel experience. Aggressive, on-board sales promotions, as can often be observed on the large and sometimes very hectic mainstream ships and where parts of the public spaces are regularly temporarily converted into additional, cramped sales stands, have no place on board of luxury ships. In the event of unforeseen events, such as the weather-related cancellation of a port call, the company—unlike a large volume ship—actively searches for alternatives or compensation.

Examples of Shipping Companies in the Luxury Category

Examples of shipping companies in the luxury category are: Hapag-Lloyd, Crystal, Silversea, Seabourn, Regent Seven Seas, SeaDream, Sea Cloud, Ritz-Carlton Yacht Collection, Scenic.

The Niche: Upper Premium

Shipping companies in this high-quality niche between premium and luxury—such as Oceania, Azamara, Ponant, Windstar, Paul Gauguin, or Viking Ocean—are characterized on the one hand by a very high product quality; in some areas, the highest mark is definitely reached. Nevertheless, a classification into the luxury category fails in the totality. The gastronomic offerings and service quality of Azamara, Ponant, and Windstar, for example, are not always at the level of established luxury lines. Oceania is able to score points in this domain. On the other hand, their older ships have well-equipped standard cabins that are too compact for luxury and relatively little outside deck space. This also applies to the Azamara ship trio, which is the exact same so-called "R-class" of the former Renaissance shipping company. Furthermore, decisive criteria such as the passenger/space ratio and/or the passenger/crew ratio (PSR or PCR) as well as other areas of ship hardware and, in some cases, the range of services included do not always meet the current standards of luxury ships.

Ship-in-Ship Concepts

Products such as the MSC Yacht Club or The Haven by Norwegian promise the service of a luxury product within a separate closed area of a large mainstream ship.

Despite advertising statements to the contrary, they are only able to compete to a limited extent with established luxury providers. This alone is due to the often very uniform standard routes that are typical for mega-ships and that are shared with all the other guests—here, the ship is seen as the destination. The mega-ships have a diverse infrastructure, more energetic on-board areas with an extended range of entertainment, a pronounced family friendliness, a generally younger and more heterogeneous audience and, due to the size of the ship, even more impressive vantage points with panoramic views at lofty heights. As an alternative—but with an older audience than on the cheaper ships—the suite categories on large ships with more varied worldwide routes, such as the "Grill Class" on the Cunard ships, are worth checking out. Newly created suite categories with special guest privileges are also trendy among other shipping companies in the volume segment.

Boom Segment Expeditions

Another niche within luxury tourism at sea is the recently growing segment of high-quality expeditions to remote areas, often only accessible by ship, such as Antarctica. Such voyages are sometimes also carried out by regular cruise ships. However, due to the higher environmental, logistical, and safety requirements that apply in these sensitive areas, conventional cruise ships are less suitable for providing an authentic and environmentally friendly travel experience than expedition cruise ships specially designed for polar regions with, for example, ice-strengthened hulls.

For a long time, the worldwide expedition ship fleet was regarded as outdated and, in terms of on-board comfort, less and less sufficient to meet today's requirements of an ever more demanding luxury clientele.

On the one hand, the operators of such ships shied away from the high bed costs for new ships, which were generally very high due to operational or statutory restrictions on passenger capacity to a maximum of 200–250 guests. Real expedition tourism is incompatible with mass tourism.

On the other hand, the building slots of the few established shipyards capable of building cruise ships (of which there are only a handful worldwide in Italy, Germany, Finland, and France, cruise shipbuilding is still firmly in European hands) are fully booked for years with mainly larger ships. The interest of these shipyards in accepting orders for small ships that take more time to construct and thus block their shipbuilding sites was correspondingly low.

But now there has been a paradigm shift. Two developments, in particular, are responsible for the fact that the expedition cruise segment has for some time been awakening from its Sleeping Beauty slumber and is experiencing a real boost in development thanks to numerous rebuilds and new buildings as well as several newly announced projects:

On the one hand, demographic change coupled with changing values is contributing to the fact that more and more younger and high-income sections of the population are seeking the privilege of traveling to remote areas as part of an eventful, educational, and emotional broadening of their horizons, without wanting to do without classic luxury features such as comfortable and spacious balcony suites (which have been scarce on expedition ships up to now), extensive top-class gastronomy and first-class service—and are also willing to pay the corresponding price for them.

At the same time, a number of shipyards that had previously concentrated on the construction of highly complex offshore special ships for oil production and wind power generation were faced with dwindling orders and felt compelled to rethink their business models and reorient themselves. This was a real stroke of luck for the expedition cruise industry, as it suddenly opened up a selection of highly competent shipyards that were desperately looking for new orders and offered their expertise proactively in the form of innovative design studies and project drafts tailored to the special needs of expedition ship operators at very interesting prices. These yards have a great deal of experience in building ships which must be able to ensure a reliable and safe, yet as comfortable as possible, operation, even under difficult and sometimes harsh conditions, as they regularly occur in the wintry North Sea (where many oil and gas platforms are located) and/or in polar regions. These are characteristics that are also in demand on expedition cruise ships with their sometimes very unusual routings.

Not every expedition ship is a luxury ship—the luxury criteria described above also apply to expedition ships. In line with customer needs, the market for luxury expedition ships and yachts is also growing steadily and has produced a whole new species of high-tech passenger ships with the highest hotel comfort and features such as on-board Zodiacs or even helicopters and/or submarines.

Bottom Line

Opinions about what is meant by luxury may well differ. An official star categorization, as it is common in the hotel industry, does not exist in the cruise industry.

Regardless of all advertising promises, there are still clearly defined parameters that provide little room for discussion and decide whether a ship belongs to the luxury category or not. A crystal chandelier alone does not equal luxury. It is a complete package within the industry standards established in deep-sea tourism, which leads to the well-deserved label luxury at sea, the combination of all relevant ingredients with naturally high weighting in the areas of service and gastronomy.

In the context of different customer needs, however, the ideal ship looks a little different for everyone—the ideal of all-around perfection does not stand up even to ships. Luxury cruise operators must be able to name at least one positive point for each negative point, for the ship to belong to the luxury category. If two or three of these criteria are missing, the vessel fails to reach the highest quality class.

As a positive example, the two cruise yachts "SeaDream I" and "SeaDream II" are mentioned here: The hardware is, despite good maintenance and care, already relatively old. Also, there are no balcony cabins and the pool is very small. The unique selling points of the ships are a marina with free water sports facilities, an excellent passenger/crew ratio of 1.2:1, excellent, award-winning gastronomy with outstanding service and extensive all-inclusive services and with a maximum capacity of only 112 guests, they offer a great deal of privacy; and thus precisely those features that many cruise enthusiasts would prefer to see over the deterrent ambience of modern, anonymous but often considerably cheaper mass ships.

If there is uncertainty about the categorization of a particular ship, it is advisable to compare the daily rates. Although there are now also massive discounts available in the luxury segment, depending on capacity utilization, the daily rates are a good indicator. If one night per person on a ship costs the same as a whole week elsewhere, this is usually no coincidence in a very competitive market environment and speaks for a real luxury-oriented equivalent.

The large price differences are not only explained by a much more expensive operating cost structure compared to large mega-ships. Above all, it is the disproportionately high construction costs that impact the smaller ships. For example, construction costs per bed on most mainstream ships for 3000–6000 guests vary between $150,000 and $250,000, while on a small expedition ship for 200 or fewer passengers they can be over $1 million. Consequently, an economically justifiable return on investment can only be achieved by correspondingly higher daily rates.

4 Road Traffic: The Limousine Service for the Twenty-First Century

Jens Wohltorf and Adam Parken

The chauffeur and his car have always been part of luxurious passenger transport on the road. A ride in a limousine embodies the grown tradition of exceptional customer service in a high-class vehicle. At the same time, new technologies have dramatically improved the efficiency, affordability, and accessibility of this long-established service.

A ride in the luxury class gives passengers the feeling that they do not have to worry about anything anymore and provides moments of relaxation while traveling. A chauffeur who can anticipate wishes and needs and interpret body language correctly, a defensive driving style, and a perfectly maintained vehicle that guarantees a smooth and comfortable ride are decisive for this. In addition, excellent knowledge of the area in order to avoid traffic jams on the one hand and to be able to help passengers at their destination on the other is a must. And last but not least, a team behind the scenes that makes on-site service possible.

Rides with taxis or ride-hailing services are in the truest sense of the word commodities. The passengers expect a basic service and basic vehicles, whereby the quality can vary from city to city and from driver to driver. Rides with limousine services, on the other hand, remain something special. Customers also typically show a greater brand awareness and place more value on service and quality than on a low price.

So, what makes modern limousine services a real luxury experience?

Travelers Prefer Chauffeurs for Longer Distances

Most travelers have several options to get from A to B. If an individual journey is to be independent of fixed timetables, the majority opt for taxis or ride-hailing services for shorter inner-city journeys—for example, if the route spontaneously leads from bar to bar. Typically, the distance covered by such journeys is less than ten kilometers.

If, on the other hand, the journey is so long that the much higher quality service and comfort of a limousine service comes into play, travelers will gladly choose this option. Airport transfers, for example, are a popular option. Both business and leisure travelers choose this option to enjoy the assurance that the entire trip has been arranged door to door.

Pickups at the airport include a personal greeting by the chauffeur—either in the baggage claim area or shortly afterward. The chauffeur also observes long before the scheduled arrival time whether a flight is delayed, arrives early or on time and ensures that he is there at the right time. He helps with the luggage and ensures that the travelers find the necessary peace after a strenuous flight. For families with children, the chauffeur has the appropriate child seats ready.

Bookings by the hour are also common. These offer the necessary flexibility—for example, for managers or celebrities who attend several appointments one after the other or also for a group of friends who travel through a city.

In addition, transfers between the metropolises are a growing segment for limousine services. A chauffeur-driven journey can be a time- and cost-saving alternative to train journeys or short-haul flights—especially if the passengers are traveling in a group. Typical routes include Munich–Zurich and New York–Philadelphia.

Event organizers also use chauffeur services in a variety of ways. They book vehicles and drivers by the hour in order to be able to offer their guests a convenient transfer between different event locations. They ensure a smooth arrival and departure by providing chauffeurs for airport transfers. And occasionally they use larger buses for the public, supplemented by limousines for VIP guests.

No matter whether in the context of events or elsewhere, VIP guests are often those who demand individual, tailor-made services. These may include, for example, a particular make or model of vehicle, the desire to have one and the same driver for several days, a particular interior of the vehicles, or special procedures for greeting guests. Frequently, they also request an airport concierge service before or after the trip, which may include private terminal areas, faster check-in, and other airport support.

The Requirements

Almost everywhere in the world, professional drivers need several certificates of qualification to be allowed to be active as an officially licensed chauffeur. This includes passenger transport tickets for drivers and/or vehicles issued by local, regional, or even national authorities. The prerequisite for holding these certificates permanently is an absolutely clean record—both professionally and personally. Usually, chauffeurs also undergo a police check for previous convictions and unannounced drug tests.

Chauffeur services require commercial passenger insurance for all drivers and vehicles. The vehicles must be registered and equipped for commercial use. In addition, drivers and/or vehicles may require special permission to use certain waiting areas at airports, railway stations, ferry ports, convention centers, stadiums, or other locations.

The abundance of these necessary formalities creates a basic trust in limousine services. Travelers can be sure that the driver and vehicle are officially authorized, inspected, and employed by a local company.

Leading limousine services ensure the greatest possible care and reliability at all times—often referred to as "duty of care" in the industry. Put simply, these are guidelines that ensure the safe transport of passengers. The US National Limousine Association defines "duty of care" as follows (cf. Ride Responsibly n. d.):

Drivers should be properly inspected, licensed, and trained. That includes:

1. Verification of criminal record and safety inspection by a certified body.
2. Drug testing prior to hiring and unannounced testing if required by the U.S. Department of Transportation.
3. Driver training program that includes customer service, safety, and defensive driving.
4. Certified medical examination at local, state, or national level, depending on regulatory requirements.
5. Control, training and, where necessary, disciplinary action.
6. Vehicles should be properly licensed, safe, and insured for commercial use. That includes:
7. Commercial liability insurance, according to the requirements of local, state, or national authorities.
8. Proper registration papers.
9. Technical inspections, in accordance with the legal requirements at local, state, or national level.

The Service Mentality

All these components, however, only provide a legally prescribed minimum that every supplier must meet. Only quality and uncompromising service orientation make a limousine service a special offer and experience.

Frequently, the traveler's gaze first falls on the vehicle. Certain brands such as Mercedes-Benz, Audi, BMW, or Cadillac have always stood for luxury. Their brand emblems and external appearances signal to the passengers that the limousine, business van, or SUV in the respective vehicle class belongs to the top of the range.

The vehicle must be absolutely flawless inside and out. Heating or air conditioning must be switched on to anticipate the needs of passengers. In the vehicle, the sound system should be switched off or calm music should be played at low volume. The front passenger seat should be fully forward to ensure maximum legroom for the passenger.

There should be bottled mineral water available for the travelers. Ideally, passengers will also find chargers, WLAN, sweets, snacks, and refreshing wipes. In other words, they should take a seat and immediately see that everything they could wish for is already there.

The chauffeur greets the guest by name and waits to see if he initiates a handshake. He then receives the luggage and opens the rear door so that the passenger can board the car. Only then does the chauffeur close the door and stow the luggage. As he moves around the vehicle, he ensures that the passenger can see him at all times. As soon as he has taken a seat behind the steering wheel, he inquires whether the temperature and music are pleasant and asks the guest to confirm his destination. If necessary, he also confirms a specific place to get dropped off, for example, a terminal at the airport or an entrance at the exhibition grounds.

In the meantime, the chauffeur should have studied the body language of his guest sufficiently. From this, he can then conclude whether the guest wants to talk and whether it should be rather small talk or a deeper conversation.

Discretion forms the basis of all interactions with the passenger. Chauffeurs should never ask for personal information or disclose information from previous guests.

It is part of the chauffeur's job to serve actors, musicians, board members, diplomats, politicians, or prominent athletes. These customers pay just as much for the protection of their privacy and for discretion. They must be able to make business phone calls while driving or simply be able to relax.

As a result, all this contributes to the passenger feeling perfectly cared for as soon as he or she sits down in the vehicle. Any distraction disappears as soon as the door closes. Travelers can concentrate on the purpose of their journey. You can enjoy the silence or the undisturbed exchange with fellow travelers. They have these possibilities because the chauffeur permanently anticipates their needs.

More than Just Standard

Passengers often have additional wishes or opportunities arise to impress them. A typical example is recommendations. Passengers could, for example, ask for a restaurant for a business lunch, the latest bar, a family park, or many other things. Or the travelers are interested in the history of the city or in the sights along the route.

In addition, chauffeurs can help travelers if the airline has lost their luggage by knowing where to find the appropriate counters at the airport. For international guests, they can also help as a translator in this case.

Some cities and routes also offer great potential for extraordinary moments. A chauffeur in New York could, for example, choose a route for a guest who is visiting the city for the first time that sets the skyline of the city particularly impressively in scene—while Frank Sinatra's "New York, New York" is playing in the vehicle.

In the Engine Room

Limousine services must deliver the luxurious customer experience not only in the vehicle and on the road, but also online and on the phone.

For example, the booking process should be state of the art. Limousine services are one of the last sectors of the travel industry to be modernized. Some of the largest and oldest chauffeur companies still have complex, outdated websites.

On such pages often many clicks are necessary and several form fields must be filled out, until the customer gets the final fare displayed—if this is even possible online at all. Often the traditional limousine operators cannot display final prices because fees, taxes, tips, and other extras cannot be taken into account.

However, travelers should be able to see the final fare, including all charges, within a few seconds and book the trip within a minute—whether on the website or through an app. Providers should also be able to confirm the booking immediately.

In addition, limousine services should have cooperations with partners in the travel industry. This makes it possible for travelers to book chauffeur services together with flights, trains, hotels, or other travel services and thus close the last gap in their door to door journey.

The terms and conditions should be clear and concise. These include the rules for waiting times for different types of pick-ups and the conditions for cancellations and changes. And last but not least, these conditions should be customer-friendly.

Travelers should be able to expect that bookings, cancellations, or changes are possible up to 1 h prior to departure. Chauffeurs should ensure 1 h of free waiting time for pick-ups at airports or train stations and 15 min for pick-ups at other locations.

In addition, it should be possible for travelers to contact the chauffeur. For example, they should receive their contact details by text message, e-mail, or app notification before the journey begins. This can become important, for example, if the baggage arrives late. Many traditional providers only provide the number of the head office, which usually leads to frustrations with customers and is not very efficient.

To ensure all this, suppliers should ideally offer 24-h customer service in different languages. Just like the chauffeurs, the customer advisors should be able to anticipate customer wishes, always proceed discreetly and make the passengers feel like they are in the best of hands.

The Overdue Modernization

The term "chauffeur" goes back to the French word for "heater"—someone who warmed up the vehicle engine before the trip. Later the responsibilities developed further and was finally limited to driving.

In some countries—such as the USA, Great Britain, Germany, and France—some companies have achieved nationwide distribution. These large, traditional operators carry out most of their journeys in two different ways.

Firstly, they own or lease vehicles driven by their own employees. Sometimes the vehicles also belong to franchisees, who in turn hire drivers on behalf of the limousine service. This variant is particularly common in large cities in the provider's home market. The fleets can then consist of hundreds of limousines, vans, and SUVs. Secondly, they pass on trips to subcontractors.

Traditional limousine services increase their profits by trying to increase vehicle utilization and minimize driver waiting times. However, it is not uncommon that the utilization rate is only 20% due to downtime and lack of efficiency.

In this respect, many providers usually face two major challenges: On the one hand, they have high operating costs due to their vehicles and employees. On the other hand, they must realize large profit margins to compensate for low capacity utilization and long downtimes. Both of these factors mean that they usually require their customers to make cancellations several hours in advance—sometimes even 24% before the start of the journey.

In the era of the modern chauffeur service, however, a different model is possible: Here a globally active company offers local trips at competitive, attractive prices. Such a model provides international travelers with access to the vast and high-quality range of local, smaller limousine services available anywhere in the world. These smaller suppliers have a local market share of approximately 70%. They join the global network in order to benefit from the global reach and state-of-the-art scheduling technologies and thus significantly increase their capacity utilization.

This significant improvement in capacity utilization will enable global network providers to offer significantly lower prices. They offer limousine trips at fixed prices to travelers and then offer these confirmed trips to the respective local partners. These in turn select the trips they want to operate.

Such a technology brings together the demand of travelers with the locally available chauffeurs and vehicles. Local suppliers will be able to make much better use of their existing capacities and expand their business. In global terms, this means that millions of trips that would have been empty become paid trips.

The modernization of the sector also creates a direct link between flights and limousine trips, which is increasingly in demand by travelers. This link between the road and the aircraft door is usually called airport concierge service in the industry. This service includes the personal greeting of the traveler upon arrival at the departure airport and at the aircraft door at the destination airport, fast track for security and entry, assistance with connecting flights as well as assistance with luggage.

Already today, some chauffeur services with their own staff are offering such services at the airport. Similarly, there are airport operators or airlines that either offer such services themselves or subcontract them. Even modern network providers call such services their own. In any case, it is a matter of seamlessly closing the travel chain for the traveler at the key point of the airport so that he or she saves time and arrives stress-free. Business travelers, celebrities, families, and senior citizens are among the most frequent users of these services.

The Next 50 Years

Looking to the future, it is foreseeable that technology and automation will be the key drivers for the further development of the user experience with limousine services. According to analyst IHS Markit, more than 33 million autonomous vehicles will be sold worldwide in 2040. Credit Suisse predicts that by 2040, 14% of all vehicles produced will be self-driving.

Even if the predictions differ to some extent, the development is clearly marked out and the advantages are clear: fewer deaths and injuries in road traffic, fewer vehicles overall, and improved mobility of people at the same time are extremely tempting.

In the future, chauffeurs will be relieved of certain tasks such as freeway driving and parking. However, they will continue to take over most of the driving. But the further technological development progresses, the less they will drive themselves. The added value of their personal service, however, will increase even more as a result.

When the chauffeurs give up their sovereignty over the steering wheel and pedals, their profession will change. The self-propelled vehicle will allow chauffeurs to focus their time and attention on additional services. They will still welcome the passengers, help them with their luggage, and provide them with their knowledge of everything local.

In the era of autonomous driving, however, they can contribute more to the travel planning of their passengers than simply making recommendations. They can make reservations at restaurants and book other activities for their guests during the trip and thus assist in the optimal arrangement of their stay. Especially if the travelers do not speak the local language, this is a valuable service. In addition, the personal network of the chauffeur can help to provide access to highly sought-after tourist attractions.

In this sense, chauffeurs can also become cultural mediators for their international guests. They can use their knowledge of local customs and practices to make life easier for travelers and avoid cultural pitfalls.

They can also offer guided city tours with their vehicle. They can concentrate fully on their passengers without having to pay attention to dense road traffic. The chauffeurs can then explain places of interest and answer all questions without being distracted by driving.

If you think one step further, chauffeurs could also offer tours, some of which take place in the vehicle and some outside it. The tour group could then leave the vehicle at different points of interest. Together with the chauffeur, sights could be visited, gastronomic offers could be taken advantage of or architectural highlights could be visited. In the meantime, the vehicle could park on its own, do a few laps around the block, or pick up the group at another point—all without the chauffeur having to worry about it.

Further additional services could become possible in the area of security and personal protection. These are widespread today, especially among celebrities and wealthy private individuals. In the future, however, many travelers may want the feeling of security that appropriately trained personnel offers. In this scenario, it would be the chauffeur's job to ensure a safe way through crowded locations and keep people from approaching the vehicle. They could also ensure that relatives are informed about the safe arrival at their destination.

Chauffeurs could also increasingly take on interpreting tasks. This could be requested by different types of travelers—especially executives, senior citizens, and celebrities—who would like to be accompanied by a local, multilingual expert.

Limousine services have developed an outstanding reputation over time, based primarily on unbeatable service in high-class vehicles. Irrespective of all the technological advances described, these pillars will continue to be crucial.

Travelers will continue to expect premium vehicles with luxurious and comfortable amenities. At the same time, they will demand personal service from highly trained professionals. Just as the profession of chauffeur changed over a century ago as engines evolved, so it will change again in the future when vehicle technology takes the next steps forward. The inner drive of every good chauffeur—to anticipate the wishes of the passengers, to shine with local knowledge, and to express the solidarity with one's own city—will enable a multitude of outstanding, individual services even in the era of autonomous driving.

5 Rail Transport: Luxury Passenger Trains

Ralf Vogler and Maria Wenske

Traveling by rail can be viewed from different angles. Thus, the primary focus of rail tourism is on the aspect of transporting people to their actual destination (Cf. Blancheton and Marchi 2013). In contrast to other means of transport in tourism, rail tourism combines transport with an experience component. The journey itself thus becomes the destination. Special train forms have developed to increase the experience components. In panoramic, adventure, and luxury trains, the primary tourist motif is even more clearly in the foreground (Cf. Freyer 2015, p. 236). Conceptually, this is already evident in the fact that luxury passenger trains are also referred to as rail cruises (Cf. Freyer 2015, p. 232; Schultz 2014, p. 88).

Luxury in the Context of Luxury Train Travel

In 1913 Werner Sombart was one of the first to deal with the concept of luxury in his work on modern capitalism (*Studien zur Entwicklungsgeschichte des modernen Kapitalismus*). Thus, he defined luxury as any "expenditure that goes beyond what is necessary" (Sombart 1913, p. 71). In order to understand the concept of luxury, it is therefore indispensable to define what is necessary. According to Sombart, this can be done either subjectively or objectively. Accordingly, luxury can have both a qualitative and a quantitative character. Qualitative luxury is equated with the "use of better goods," whereas quantitative luxury is identified as a "waste of goods" (cf. Sombart 1913, p. 71 f.). Sombart derives the concept of luxury goods from qualitative luxury. A luxury good represents "a refined good," whereby this refinement stands for everything that is "redundant for a makeshift fulfilment of purpose" (Sombart 1913, p. 72). Even in the twenty-first century, these observations of 1913 by Sombart are still very popular and are the starting point for numerous supplementary explanatory approaches. All have in common the view that luxury is both dependent on place and time and also depends on the personal point of view of a person (Cf. Dederl et al. 2017, p. 159; Hennigs and Wiedmann 2017, p. 165; Prüne 2013, p. 164). Thus, depending on the person, the same good can be functional or luxury (cf. Hennigs and Wiedmann 2017, p. 165).

Despite the subjective and epoch-dependent interpretation of the concept of luxury, luxury goods are generally characterized by six factors. These include outstanding product quality, a high price, exclusivity or uniqueness, an esthetic form or color, a long history, and a lack of necessity (Cf. Hennigs and Wiedmann 2017, p. 165; Prüne 2013, p. 168). In addition to classic personal luxury goods such as clothing, leather goods, jewelry, cars, and electronics, more and more services such as exclusive travel, hotels, and restaurants are becoming luxury goods (Cf. Hennigs and Wiedmann 2017, p. 165). Today, the experience of luxury rather than the possession of luxury goods is perceived as a true luxury (cf. Dederl et al. 2017, p. 148 ff.). This development is primarily due to demographic factors and the generation of *baby boomers* who are now retiring, as well as the *generation Y* or the *millennials* who are in their twenties. Both generations are typical consumers of experience luxury. For them, pleasure is in the foreground, not the sole possession. (Cf. Abtan et al. 2014) As can be seen in Table 3, the Boston Consulting Group GmbH (BCG) differentiates between classic personal luxury goods and the newer luxury experience. This is characterized by the fact that it conveys unique and individual experiences (see Dederl et al. 2017, p. 153 f.).

Table 3 Luxury experiences versus personal luxury goods

Experience luxury	Personal luxury goods
Experience	Owning
Together with others	For yourself
Not always visible	Usually visible to all
Immediate pleasure	For medium and long-term use

Source: Based on BCG (2016), p. 9

Tourism Classification of Luxury Trains and Characteristics

Luxury trains are initially assigned to Special Interest Tourism (SIT). This special form of niche tourism is regarded as an antipole to mass tourism (cf. Novelli and Robinson 2011, p. 6) and is oriented toward very special motifs of the travelers (cf. Douglas et al. 2001, p. 3; Novelli and Robinson 2011, p. 13). According to Blancheton and Marchi (2013, p. 31), SIT also focuses on sustainable tourism practices.

As a result of the high interest of passengers in railways and the low speed of luxury trains we also speak of Slow Rail Tourism. This slow railway tourism is based on comfortable train journeys, the purpose of which is the discovery of landscapes, the search for new experiences, or the enjoyment of the journey itself (Cf. Blancheton and Marchi 2013, p. 32).

A journey with a luxury train can last from 1 day up to several weeks. The routes often run through exceptional remote landscapes, such as mountains, deserts, or along coasts, which are cut off from general transport infrastructure (Cf. The Train Chartering Company Ltd. 2018a). The unique nature and design of the trains and the route is characteristic of each case. Despite their uniqueness, it is striking to note that luxury travel trains combine the combination of accommodation, restaurants, and other service elements, some of which require a high level of personnel (The Train Chartering Company Ltd. 2018a) with high-priced marketing. A trip on a luxury train can cost up to 1500 euros per night per person (cf. Blancheton and Marchi 2013, p. 36) and it is therefore not surprising that the target groups of such trips are mostly people with a very high income and level of education. Special occasions such as a honeymoon, an anniversary, or the passion for trains are often mentioned as motifs for journeys on a luxury train.

The World of Luxury Trains: Product Overview

The Luxury Train Club and the Society of International Rail Travelers (IRT) list a total of 37 trains worldwide as luxury passenger trains (see The Train Chartering Company Ltd. 2018a; IRT n.d. c). Since the respective list also includes explicit panorama trains such as the Bernina or Glacier Express (cf. RHB 2014), this figure seems too high.

Based on a study of the authors from 2018 and the aforementioned criteria, a total of 17 luxury trains can be identified worldwide. Most of them can be found in Europe, as can the most famous luxury train in the world—the *Venice Simplon-Orient-Express* (VSOE). Other trains are the Spanish *Al Andalus* and the *Transcantábrico Clásico* as well as the *Transcantábrico Gran Lujo*. The *Belmond Grand Hibernian* runs in Ireland, the *Belmond Royal Scotsman* in Scotland and the last part of the VSOE's route in Great Britain is taken over by the *Belmond British*

Pullman. Finally, the *Golden Eagle Danube Express* and the *Golden Eagle are* also classified as European luxury passenger trains. In addition to the *Eastern and Oriental Express*, there are three other luxury trains in Asia. These include the Japanese *Seven Stars* in Kyushu and the Indian luxury trains *Deccan Odyssey* and *Maharajas Express*. The *Royal Canadian Pacific* operates in North America, but will only be available to private charterers in 2018. The *Belmond Andean Explorer* as well as the *Belmond Hiram Bingham* are the South American luxury trains and both run in Peru. Finally, two luxury trains can also be found on the African continent. These include the *Blue Train* on the one hand and the *Rovos Rail* on the other. What all trains have in common is that they represent a luxury product. This applies in particular to the nonessential, the high price, and the exclusivity.

In Table 4 the 17 luxury trains mentioned above are listed together with the average prices per person and night.

Although the VSOE is regarded as the most famous and best-known luxury train in the world and can look back on a long literary history, it should not be described in more detail here. More appropriately in the context of luxury travel, the authors see the *Rovos Rail* in South Africa as an example of a nostalgic-historical and authentic luxury train of and the *Seven Stars* in Kyushu, Japan as a symbol of a modern interpretation of luxury. Therefore, these two luxury trains are presented in more detail below.

Table 4 Luxury trains with average prices per person per night

Name	Region	Capacity	Price per person per night
Al Andalus	Europe	74	From 600 EUR
Belmond Andean Explorer	South America	48	From 650 EUR
Belmond British Pullman	Europe	20–26	From 600 EUR
Belmond Grand Hibernian	Europe	40	From 1250 EUR
Belmond Hiram Bingham	South America	84	From 450 EUR
Belmond Royal Scotsman	Europe	36	From 1400 EUR
Blue Train	Africa	54 or 82	From 750 EUR
Deccan Odyssey	Asia	80	From 550 EUR
Eastern and Oriental Express	Asia	82	From 1050 EUR
El Transcantábrico Clásico El Transcantábrico Gran Lujo	Europe	52 28	From 500 EUR
Golden Eagle Danube Express	Europe	72	From 750 EUR
Golden Eagle	Europe	132	From 550 EUR
Maharajas' Express	Asia	88	From 900 EUR
Rovos Rail (Pride of Africa)	Africa	72	From 700 EUR
Royal Canadian Pacific	North America		Charter only
Seven Stars in Kyushu	Asia	28 (30)	From 1650 EUR
Venice Simplon-Orient-Express	Europe	177	From 950 EUR

Source: based on data from Belmond Ltd 2018a–e; Golden Eagle Luxury Trains Ltd. 2018a, b; Indian Holiday Pvt. Ltd. 2018, n.d.; IRCTC 2017; IRT n.d. a–c; Kyushu Railway Company 2018b; Lernidee Erlebnisreisen GmbH 2018; Renfe n.d. b, c; The Blue Train 2018; The Train Chartering Company Ltd. 2018c

Rovos Rail

Rohan Vos founded the company named after him, *Rovos Rail Tours (Pty.) Ltd.* in 1989 which today is one of the most luxurious railway companies in the world (See Rovos Rail 2018; Ebert 2016, p. 114; Viedebantt 2015, p. 131; Howard 2014, p. 174). Since the 1980s, Rohan Vos has been collecting and restoring historic railway wagons, the oldest of which date back to 1911. Today the Vos family owns one of the largest private collections in the world with more than 75 wagons. In addition to five restored steam locomotives, the company also owns three guest houses, a private railway station in Pretoria, which is also the company headquarters, and the Victoria Hotel opposite the railway station (Cf. Rovos Rail 2018; Müller-Urban and Urban 2017, p. 128; Ebert 2016, p. 116; Viedebantt 2015, p. 132; Howard 2014, p. 174). The trains are powered by diesel or electric locomotives. Due to the lack of coal and water along the lines, the elaborately refurbished steam locomotives are now only used in the area of the Pretoria railway station (Cf. Rovos Rail 2018; Viedebantt 2015, p. 132). In addition to the *Rovos Rail*, there is also the event train *Rovos Rail Tours (Pty.) Ltd.* which can accommodate up to 250 guests and is only suitable for day trips. The *Rovos Rail*, also known as "The Pride of Africa" (IRT n.d. a; Farren 2006, p. 160), can accommodate up to 72 passengers (see Rovos Rail 2018; Ebert 2016, p. 120; Howard 2014, p. 174). It consists of the locomotive, a generator car and a staff car, various sleeping cars, a non-smoking lounge car, two restaurant cars, a kitchen car, and a smoking lounge, followed by a car with an open viewing platform (Cf. Rovos Rail 2018; Müller-Urban and Urban 2017, p. 128; Viedebantt 2015, p. 131). Characteristics of the *Rovos Rail is* the authentic and nostalgic design of the train, which underlines the long history of the company and takes passengers back in time. The restaurant car and the lounge car can be seen in Fig. 1.

Travelers have a choice of three suites, all equipped with a safe, minibar, heated towel rail, dressing gowns, slippers, goggles, earplugs, laundry bag, and shoe bag. Furthermore, in the bathrooms with original fittings, you will find all kinds of hygiene articles and a hairdryer. A 24-h room service is also available for travelers. The *Royal Suite* is the largest suite at 16 square meters and has a Victorian bath and separate shower, a living area, and two single beds or a double bed (see Fig. 2).

Fig. 1 Rovos Rail, restaurant car and lounge car (source: Rovos Rail 2018)

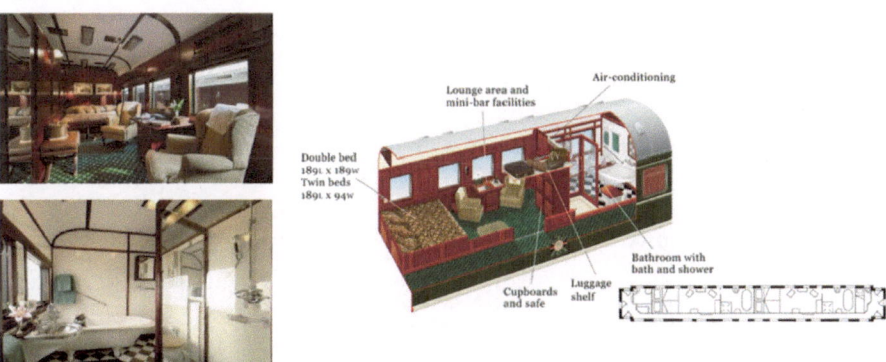

Fig. 2 Royal Suite (Quelle: The Train Chartering Company Ltd. 2018d)

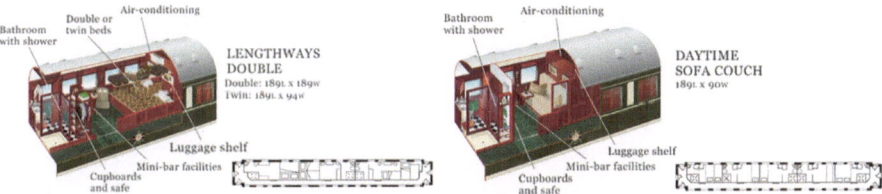

Fig. 3 Deluxe Suite and Pullman Suite (source: The Train Chartering Company Ltd. 2018d)

The *Deluxe Suite* is 10 square meters, also equipped with two single beds or a double bed, a living area, and a bathroom with shower. The smallest suite with 7 square meters is the *Pullman Suite, which has* a bathroom with shower. The comfortable daytime sofa transforms into a double bed or two single beds in the evening (see Fig. 3). The *Rovos Rail* offers its guests a free laundry and ironing service.

In the two lounge cars, you can find board games and books as well as a small gift shop. During the trips to Dar es Salaam, these are also used for lectures. In the two restaurant cars of the *Rovos Rail,* travelers can enjoy the finest food. All traditional dishes are freshly prepared on board. Lunch and dinner are announced ceremonially with a gong (Cf. Rovos Rail 2018). A dinner at *Rovos Rail* consists of five courses (cf. Ebert 2016, p. 121). Formal clothing is therefore also expected, which should at least consist of a jacket and tie for men and an evening dress for women. During the day, travelers are advised to wear rather chic but casual clothing. Comfortable footwear and sunscreen are a must, especially during excursions outside the train. When booking trips with the *Rovos Rail,* please note that they are not recommended for children under 13 years of age. There are no radios or televisions on board of the train and the use of mobile phones and laptops is limited to the private suite (Cf. Rovos Rail 2018). This is how the nostalgia of the journey should be preserved (Cf. Viedebantt 2015, p. 131).

In 2018 the *Rovos Rail* offered seven routes in addition to special travel packages and the possibility of renting the train privately. The journeys last from 48 h up to

Fig. 4 Dar es Salaam Route of Rovos Rail (2018) (source: Rovos Rail 2018)

15 days. The longest and probably most famous tour is the one from Cape Town to Dar es Salaam or vice versa, as you can see in the picture in Fig. 4 (Cf. Rovos Rail 2018). During this trip, travelers will spend eleven nights aboard the *Rovos Rail*, two nights at the five-star Dew Game Lodge in Madikwe Game Reserve including four game drives and all meals, and one night at Victoria Falls Hotel (Cf. IRT n.d. d). Also included is a visit to the authentic Victorian village of Matjiesfontein, a city tour, and visit to the Kimberley Diamond Mine Museum and the world's largest man-made excavation site, the Big Hole. Other highlights include a visit to Victoria Falls, a bush walk at Chisimba Falls in Kasama, a game drive in Selous Reserve, and a Zambezi River Sunset Cruise (Cf. Rovos Rail 2018; IRT not published). This tour

is exemplary for the experience character and exclusiveness of the train, which radiates a long history through its use of original wagons and appeals in particular to a clientele that attaches importance to "classical" luxury elements.

Seven Stars

The *Seven Stars* in Kyushu is Japan's first luxury sleeping car train. Koji Karaike, chairman of the Japanese Railway in Kyushu, had the idea for this train 20 years ago. With the help of designer Eiji Mitooka and local craftsmen, the *Seven Stars* became a reality in 2013 (Kyushu Railway Company 2018a). The design of the train consists of a combination of modern Japanese and Western elements, a lot of attention to detail, and elaborate woodwork. The *Seven Stars* consists of a total of seven wagons and offers space for a maximum of 30 passengers. In addition to the five sleeping cars, there is also a restaurant car and a lounge car (see Fig. 5).

There are 12 *suites* and *two deluxe* suites on board. One suite is barrier-free and therefore bookable for people with physical disabilities. All suites have two single beds that can be converted into seating during the day, air conditioning, W-LAN, a free minibar, and private bathroom with shower and Arita porcelain sink. In addition to towels, bath slippers, and a bathrobe, the bathroom also contains various personal care and hygiene articles (see Kyushu Railway Company 2018a; The Train Chartering Company Ltd. n.d. b.). Each of the 12 *suites* is 10 square meters in size and has been designed with different types of wood, designs, and colors. The *Deluxe*

Fig. 5 Lounge car (source: Kyushu Railway Company 2018a)

Fig. 6 Deluxe Suite (source: Kyushu Railway Company 2018a)

Suite A is 21 square meters in size, has a huge panorama window, and can accommodate up to three people. It is also equipped with a projector for watching a selection of DVDs. The *Deluxe Suite B* with 17 square meters is slightly smaller than the *Deluxe Suite A*, but also suitable for up to three travelers (Cf. Kyushu Railway Company 2018a; IRT n.d. b). Both Deluxe Suites differ in their design and the wood species used (Fig. 6).

The lounge car of the *Seven Star* is equipped with a bar, sofas, and rotating chairs as well as a large bay window at the end. During the day, the lounge car with its modern Japanese atmosphere serves as a rest area, while in the evening live music and other entertainment take place. A typical Kyushu cuisine with seasonal delicacies awaits the traveler in the restaurant car (Cf. Kyushu Railway Company 2018a). The dress code of the *Seven Star* prohibits sportswear, torn jeans, T-shirts, and shorts on board. What is desired is a smart casual look that consists of shirts, jackets, suits for men and dresses, blouses and skirts for women. For dinner, it is recommended for men to wear tuxedos. Sneakers may only be worn during excursions (Cf. Kyushu Railway Company 2018a).

The special thing about the *Seven Stars* is that it is not easy to book a trip. Due to the great demand, travelers have to fill out a registration form for when they want to make which trip. If there are more registrations than there is space on the train, the seats will be drawn according to a lottery procedure. With this measure, the supplier underlines the exclusivity and uniqueness of this luxury train. The *Seven Stars* regularly offers two different routes as well as three premium tours. The final itineraries will be communicated to travelers only 2 weeks before departure. On the respective routes, there is always at least one night in a traditional Japanese guesthouse, called Ryokan. The prices, therefore, depend on the choice of the corresponding Ryokan. The start of all routes is Hakata. There, travelers can enjoy a welcome drink and sweet treats in the private Kinsei Lounge before boarding the *Seven Stars* (see Kyushu Railway Company 2018a). The longest regular route is the *4 day, 3 night trip* through the regions of Fukuoka, Oita, Miyazaki, Kagoshima, and Kumamoto (Fig. 7).

Highlight is the overnight stay in the most traditional Ryokans Kyushus in Yufuin Onsen. Other program items vary depending on the time of travel and may include a geisha evening, a visit to a ceramics workshop, a sake tasting, or a tea ceremony. Thus, the train also meets the demands of luxury travelers in terms of history and tradition.

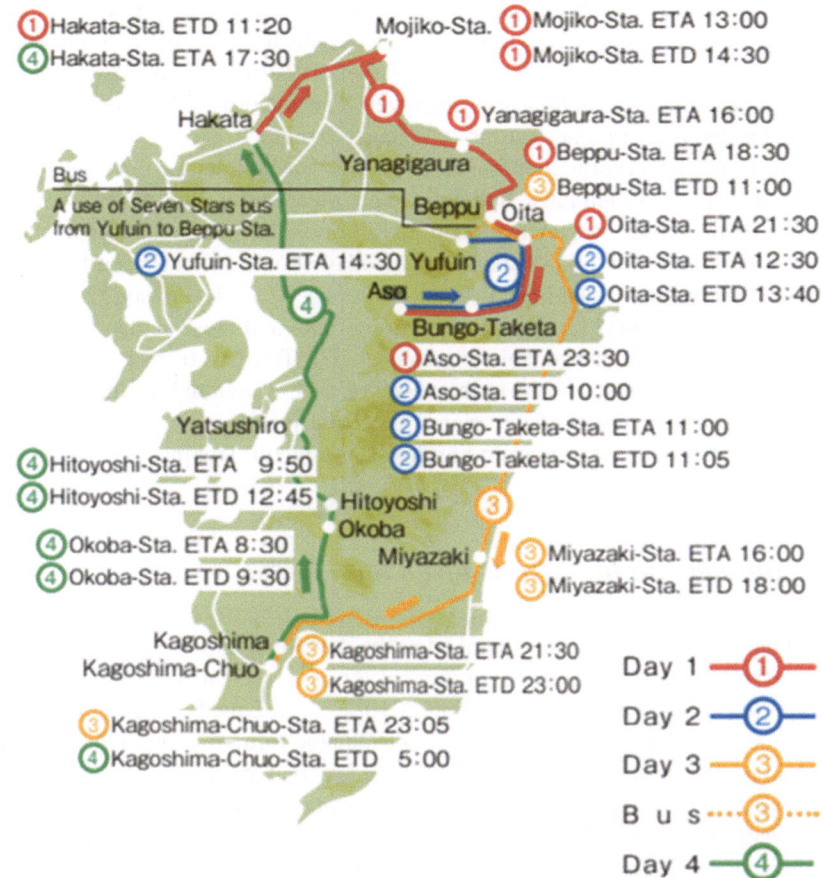

① Hakata-Sta. ETD 11:20
④ Hakata-Sta. ETA 17:30

Mojiko-Sta. ① Mojiko-Sta. ETA 13:00
            ① Mojiko-Sta. ETD 14:30

Hakata

① Yanagigaura-Sta. ETA 16:00

Yanagigaura   ① Beppu-Sta. ETA 18:30
                           ③ Beppu-Sta. ETD 11:00
Bus   Beppu Oita
A use of Seven Stars bus   ① Oita-Sta. ETA 21:30
from Yufuin to Beppu Sta.

② Yufuin-Sta. ETA 14:30 Yufuin   ② Oita-Sta. ETA 12:30
                       Aso   ② Oita-Sta. ETD 13:40
                                Bungo-Taketa
                           ① Aso-Sta. ETA 23:30
                           ② Aso-Sta. ETD 10:00
Yatsushiro   ② Bungo-Taketa-Sta. ETA 11:00
④ Hitoyoshi-Sta. ETA 9:50 ② Bungo-Taketa-Sta. ETD 11:05
④ Hitoyoshi-Sta. ETD 12:45 Hitoyoshi
                           Okoba
④ Okoba-Sta. ETA 8:30   Miyazaki   ③ Miyazaki-Sta. ETA 16:00
④ Okoba-Sta. ETD 9:30   ③ Miyazaki-Sta. ETD 18:00

Kagoshima   ③ Kagoshima-Sta. ETA 21:30   Day 1 —①—
Kagoshima-Chuo ③ Kagoshima-Sta. ETD 23:00
                                          Day 2 —②—
③ Kagoshima-Chuo-Sta. ETA 23:05
④ Kagoshima-Chuo-Sta. ETD 5:00   Day 3 —③—

                                          B u s ····③····

                                          Day 4 —④—

Fig. 7 4 days, 3 nights of Seven Stars 2018 (source: Kyushu Railway Company 2018a)

Satisfaction of Needs and Motives by Luxury Trains

Like luxury goods in general, the significance of luxury train travel from the perspective of tourist consumers can be characterized by six factors: outstanding quality, a high price, exclusivity or uniqueness, esthetics, a long history and it needs to be non-essential (Cf. Hennigs and Wiedmann 2017, p. 165; Prüne 2013, p. 168). In the luxury passenger train segment, being non-essential and a high price (cf. Table 4) are satisfactory. Esthetics and a long history are covered by the need for nostalgia (cf. Prideaux 1999). Most of the trains listed in Table 4 are either historical or based on historical trains.

The satisfaction of exclusivity and quality motives is favored by the increasing importance of individual luxury experiences (cf. Dederl et al. 2017, p. 148 ff.), which each consumer defines on the basis of their demands as qualitative (cf. Dykins

2016, p. 4). This development is primarily attributable to demographic factors. Especially baby boomers and millennials are typical consumers of the luxury experience offered by luxury trains (Cf. Abtan et al. 2014). This is also reflected in the consumer structure of luxury train travel. The majority of customers can be attributed to the baby boomers group (Cf. Bauer 2012).

Apart from the criteria mentioned above, a development from conspicuous consumption to conscientious, authentic, and meaningful luxury travel can be observed in the luxury travel market (Cf. Euromonitor International Ltd. 2017). Luxury travel, and thus luxury train travel, today should primarily be ECO. This refers to "Exclusivity, Customization and Options" (Cf. Resonance Consultancy Ltd. 2016, p. 3). Luxury travelers are becoming more and more demanding and are looking for unique and sustainable experiences (Cf. Tourism Insider 2018). Offers should be exclusive and tailored to the needs of each individual. In the new era of luxury travel, it is a matter of experiencing unique journeys, to which only a few have access, by means of corresponding luxury trains (cf. Dykins 2016, p. 15).

Rail Tourism and the Effects of Luxury Trains

Traditionally, railways are closely linked to the development of tourism in a region. Rail transport laid the foundations for modern tourism (see van Truong and Shimizu 2017, p. 3097; Dickinson and Lumsdon 2010, p. 110; Duval and Lohmann 2011, p. 17). Today, *rail tourism* is considered a sustainable form of tourism that is limited to certain regions and countries of the world (Cf. Duval and Lohmann 2011, p. 17). Dickinson and Lumsdon (2010, p. 105 ff.) stress the environmental advantages of rail tourism over cars and planes and the crucial role it plays in destination development in general. This view is supplemented by Villanueva Cuevas (2012, p. 10), who sees further advantages in comfort and safety for passengers and a competitive advantage for the railways, especially for distances between 200 and 500 km. Tillman (2002, p. 40), on the other hand, points out that historic railways contribute to preserving cultural heritage and stimulating tourism in a region in the economic sense. In addition to these positive aspects of rail tourism, Su and Wall (2009, p. 650) expressed concerns about tourism development, regional economic development, and the preservation of culture in less touristy regions. In these destinations, there is often a gap between the tourist demand and the supply, so that even the comparatively small number of tourists can lead to an overtaxing of the region. In summary, however, railway tourism is generally considered to have rather positive effects based on the available literature. *Luxury trains* are often regarded as unique prestige experiences where the train is the actual destination of the journey (Cf. Hall 1999, p. 182). They are a special form of niche tourism within rail tourism and little is known about the concrete effects of luxury trains. It can probably be said that the Compagnie Internationale des Wagons-Lits (CIWL), whose successor companies still exist today (cf. Sachslehner 2001, p. 169; Farren 2006, p. 7; Schultz 2014, p. 100), as a historical provider of luxury passenger trains, contributed significantly—with the founding of Europe's first hotel chain in 1894, the Compagnie Internationale des Grands Hotels (CIGH)—to the development of tourism in certain regions. At that time, hotels were built or bought that were located at terminal stations of CIWL passenger trains (cf. Mühl and Klein 2006, p. 293). Among other things, this led to the creation of jobs and thus contributed to the economic development of the various regions (Cf. Villanueva Cuevas 2012, p. 10).

Among others, Magadán Díaz and Rivas García (2012) examined the cultural effects. They studied the *Transcantábrico* in Spain and found that the *Transcantábrico* helped raise awareness of the luxury rail passengers for the cultural heritage of the regions Galicia, Asturias, Cantabria, the Basque Country as well as Castile and León.

In Galicia, it was also found that, from an economic point of view, tourism transport products such as the *Transcantábrico* contributed to rising demand and even to increasing and structuring supply (Cf. Magadán Díaz and Rivas García 2012, p. 21). This form of tourism in Galicia also preserves the region's existing resources and respects its sociocultural identity. It is also stressed that the *Transcantábrico does* not create social inequalities, since it is not withheld from the inhabitants of the regions (See Magadán Díaz and Rivas García 2012, p. 22). Tourism representatives within the Basque government have expressed the hope that the *Transcantábrico Gran Lujo* in particular will help the region to increase the number of overnight stays in five-star hotels. Furthermore, the average expenses in restaurants can be increased, as some meals are consumed outside the train (See Magadán Díaz and Rivas García 2012, p. 24). Magadán Díaz and Rivas García (2012, p. 25) conclude their studies by saying that the *Transcantábrico* favors a form of tourism that can contribute to more sustainable growth.

Overall, it can be assumed that luxury passenger trains, as a high-quality niche product, can contribute to avoiding the negative external effects known, for example, from the cruise industry (cf. Carić and Mackelworth 2014) and also bring wealthy tourists to the regions. It should be noted, however, that it is not known to what extent luxury trains actually contribute to *promoting the local economy* and thus to *improving the quality of life* for the local population. This is particularly true since the routes of luxury trains often run through remote regions. In countries like India or South Africa, communities far away from the usual tourism destinations can benefit from the trains as well.

6   Tour Operators and Intermediaries

Norbert Pokorny

Of Happy Rats, Priceless Happiness, and the Power of Knowledge

Emotions make life worth living and are part of our soul.

How exactly these components are connected, what comes first in emotional processes, and what is caused by what, and how, has been the subject of scientific research for over a century.

And—typically human—on the trail of emotions our fears have been studied the most and are also best understood today.

Only 50 years ago, researchers discovered by chance in rats that they could not get enough of the electrical stimulation of a certain area of the brain. In these experiments, the rats were allowed to trigger this particular electrical stimulus for this brain region by pressing a lever themselves—and pressed the lever more and more frequently. Some animals even forgot to eat and drink. And died, obviously addicted to that positive rush feeling that made them forget the world around them.

Emotional states of happiness are—and this has been proven—just as important for a long life as a healthy lifestyle. Happy people are more successful at learning and working, often more creative, more popular, more sociable, mentally healthier, less selfish, and less aggressive.

When Luxury Becomes Boring and Self-Evident

Everyone has certainly experienced for themselves how intense and powerful these special moments of happiness are. But this feeling of happiness, as further studies of the neurobiology of happiness have shown, is not made for "constant operation."

This circumstance affects the so-called "luxury customer" in a very special way. Extremely wealthy people who have lived an abundant life for years, perhaps even generations. These luxury customers were the first to be able to afford new technologies, build large villas in prime locations, and buy second homes in France, Tuscany, and the Caribbean. They parked two or more luxury cars in their fleet and each time a new model came onto the market they ordered another one.

This luxury-oriented clientele used to have more traditional demands on travel. It simply meant residing in the best houses in the world, having the suitcases unpacked by butlers in the largest and most expensive suites, and ordering the most expensive champagne in the house.

But today this lifestyle seems to degenerate more and more among luxury customers to a "continuous operation of the feeling of happiness." What do you want with the umpteenth luxury car, even more expensive shoes or handbags in the long overcrowded closets? What do you gain with another property, which you toast to with another expensive bottle whose sparkling content hardly promises pleasure anymore? And is it really inspiring and fulfilling to spend your most precious time in the traditional ambience of a deeply traditional luxury hotel?

Even though the desire for luxury accommodation has remained while traveling, there has been a significant change in awareness. The definition of luxury is changing. Values that have played a secondary role for luxury customers in the past are becoming more important.

But what values are important to this clientele? In times when the multiplication of money on the financial markets has become so difficult. In times when you realize how much you work or have worked. In which, among other things, one has not been part of the development of one's own children. In times when relationships are at stake because priorities have been in other areas. What values are becoming important today because stress at work tugs at health and time-outs for themselves are falling by the wayside?

Time Has Become the True Luxury. Luxury Is Quality Time

The answer is: true luxury is increasingly manifested in spending scarce time with people who are close to you. Luxury customers are moved by a deep longing to establish deep and intensive connections during their holidays. There is a growing desire for new, shared experiences that can lift a relationship to a new, more intense and unforgettable level.

The first step toward Quality Time is for many luxury customers, who usually still have to work hard to achieve this status, the free and individual planning of their vacation.

However, this change in awareness or the new definition of luxury has not quite arrived yet in the tourism industry. But for the luxury travel sector in particular, this insight is essential, because luxury customers are highly valued here. Because he usually rewards competence and is not very price-sensitive. Because earnings are more lucrative in an industry that lives on a percentage of the travel price. Because trust has a special significance for this type of customer and they are therefore much more likely to become regular customers.

Today it is no longer enough for the luxury customer who wants to make use of the service and know-how of a travel agency to simply be offered the best suite in the best hotel. Because meanwhile the Internet can do that better, faster, and cheaper.

The good news: The new definition of luxury is playing into the cards of tourism!

To enjoy special and unique experiences alone or together with family and close friends as well as to broaden one's own horizon is certainly an essential part of this new luxury thinking.

This means, however, that tourism is more than ever required to take account of the changing wishes and requirements of luxury clientele both in its products and in its communication with these customers.

Understanding Luxury Correctly: Rethinking Tourism

However, it is precisely this change that the vast majority of people in the tourism industry find very difficult.

So, what is the "secret" of winning over luxury customers? And more importantly, to bind them to your company?

The answer is manifold. It contains components from communication, knowledge, analysis, strategy, and the new understanding of luxury.

To understand the general developments in the luxury market is essential in order to raise wishes regarding the product and travel as well as the requirements for the communication of the luxury customer to the necessary level. For this reason, an intensive look beyond the horizon of luxury tourism is essential in order to successfully serve wealthy customers.

By the way, if you look beyond your own nose, you will quickly notice that many other luxury brands from other industries are also concerned with the strongly changing definition of the term luxury. Because the luxury customer does not only make new demands on his luxury product or service in tourism.

In tourism, this means that in future luxury will cover a much larger field in tourism than the mere stringing together of so-called luxury products.

In any case, the "new" luxury trip takes into account the customer's strong desire not only to consume but rather to be part of something. Part of something that allows him to see the world and his life from a new perspective. Maybe by testing new boundaries. Or fulfilling long-awaited dreams. All this with the aim of spending this quality time together with those people to whom he has a special connection that he would like to build up or intensify. To create a new basis of familiarity or cohesion and to draw new motivation and happiness from it.

The clear trend, for example, toward "multi-generation travel" underlines this special wish of the customer for new travel content.

Two New Developments Present Travel Agents with New Challenges

How can tourism, especially travel agencies or tour operators, meet these new requirements? And how are the travel agency's tasks changing in terms of offers and servicing luxury customers?

For many years, even decades, travel agency consulting lived from the knowledge of travel agents about the conditions at the destination and, at best, the hotel product. The travel agent's know-how was the USP of a travel agency, especially when dealing with luxury customers.

In addition, the market for luxury hotel products was manageable. Only a few hotel chains were able to offer such a clientele the desired standards. And the traditional demands of these clients were mostly limited to the size of the suite, the location of the hotel, and the quality of the restaurants.

However, the last decade has brought two brute changes: the Internet and an ever-increasing range of luxury products. Today, the World Wide Web enables customers with purchasing power to get their own overview of hotels and travel destinations. And to form one's own opinion about the quality of the offer on evaluation portals.

In addition, more and more companies and private individuals are investing in luxury hotel products, so that the supply in this area multiplied in quantity and quality—and an end to this development is far from in sight.

This development also presents travel agents with completely new challenges. Especially the most experienced luxury customers today express their travel wishes with a knowledge that often goes beyond that of the travel agent. The customer's disappointment with the advice is therefore almost inevitable.

What is certain is that the USP of a travel agent from past decades no longer has any relevance as a USP today. However, this does not mean that the knowledge of the travel agent is no longer required. On the contrary! Product knowledge is the basic prerequisite for competent advice. The ability to learn about the multitude of luxury products is a difficult undertaking in the existing structure of a travel agency with constantly growing product portfolios.

Expert Knowledge Alone Is No Longer Enough for Travel Agents

An answer to this development can only be consistent specialization. However, this only makes sense if a travel agency decides to consistently build up and further develop the repertoire of luxury travel. And that requires patience.

Anyone who has already dealt with this topic more intensively may have experienced that the coveted luxury customer is difficult to advertise to and even more difficult to win over as a regular loyal customer.

In times in which many travel agencies have to pay close attention to costs and efficient customer processing with low margins and high competitive pressure, patience for such a segment often does not fit into the concept.

Nevertheless: Only the consistent and long-term work of the travel agents with luxury products and at the same time their interest in the development of the luxury sector as a whole can serve as the basis for a successful appearance and implementation.

Creative, "Priceless" Moments for Luxury Customers

And yet in the future, this status alone will no longer be sufficient to operate successfully in the market, especially as an increasing number of competitors are fighting for this coveted customer segment. The mere processing of bookings on the call of the customer is often not enough anymore for many wealthy customers. The challenge is rather to inspire these wealthy customers by creating unique and extraordinary moments for them. Or rather, tailor everything to their individual needs. An extremely important point to stand out from competitors or the Internet.

To create such special moments, however, a bold change in thinking is necessary. Because of the customer's desire to experience exactly such moments, for example, he is more than ever prepared to do without traditional luxury.

Overnight stay in a tent without bathroom and toilet? If this enables a unique experience, it is often not a problem—even for demanding customers. But of course, the prerequisite is that you know your customers to a T. Only then you know how far you can really get him out of his comfort zone.

In order to be able to offer such special experiences, so-called "can't-buy moments," in which the wealthy customer is inspired with new content, one needs a large network, which can likewise only be built up through consistent and long-standing work.

Knowledge Through Trust: Communication on Innovation Course

Communication with customers must also be put to the test. In this field, too, innovative, new approaches are certainly in demand.

For example, is it still sufficient today to confirm to a luxury customer that he has booked a private guide for his safari? And that he only finds this fact on the voucher and in the itinerary?

Would not it be much more important to give the customer direct access to this more than an essential part of the booked trip? By being able to communicate personally with "his" private guide even before the trip, perhaps even before booking the trip?

But is the travel agent willing to give up his status as the sole contact person for the customer? And to see oneself in the future "only" as a communication mediator between the customer and the relevant persons of the upcoming trip (butler, guide, hotel manager, concierge, etc.)? And is he prepared to leave the communication and decision for or against a product or a journey 100% to the customer?

One thing is certain: creating transparency is a highly valued commodity for wealthy customers, that everyone wants to have a share of. Because transparency creates trust. And trust is the basic prerequisite for retaining luxury customers in the long term. The more trusting a cooperation with such a customer is, the more profound the knowledge about him becomes. And it is precisely this knowledge of the customer that will in future be the new USP of a travel agency specializing in luxury travel.

However, the definition of luxury evolves: The well-off client will search more intensively for specialists with special, individual, and inspiring knowledge. For travel agents who understand him and his wishes and think outside the box of the previous definition of luxury. Which shows him new ways to conquer his very own, perhaps unimagined world of new luxury—and to invest money in priceless moments.

Individualizing these parameters and assigning them to each customer is time-consuming because you have to analyze every single one of them. Anyone who takes this into account and has the patience to go down this road will have the door wide open to a new, almost inexhaustible luxury market.

Literature

Abtan O, Achille A, Bellaïche J-M, Kim Y, Lui V, Mall A, Mei-Pochtler A, Willersdorf S (2014) Shock of the new chic: dealing with new complexity in the business of luxury—new customers and the new ways they buy. The Boston Consulting Group (BCG), Boston

Bauer I (2012) Australian senior adventure travellers to Peru: Maximising older tourists' travel health experience. Travel Med Infect Dis (2):59–68

BCG (2016) Selected key 2016 and beyond business trends in the luxury industry. The Boston Consulting Group (BCG), Boston

Belmond Ltd (2018a) Belmond Eastern & Oriental Express. https://www.belmond.com/trains/asia/eastern-and-oriental-express/. Zugegriffen: 21 Feb 2018

Belmond Ltd (2018b) Belmond Andean Explorer. https://www.belmond.com/trains/south-america/peru/belmond-andean-explorer/. Accessed 21 Feb 2018

Belmond Ltd (2018c) Belmond Hiram Bingham. https://www.belmond.com/trains/south-america/peru/belmond-hiram-bingham/. Accessed 21 Feb 2018

Belmond Ltd (2018d) Belmond Grand Hibernian. https://www.belmond.com/trains/europe/ireland/belmond-grand-hibernian/. Accessed 21 Feb 2018

Belmond Ltd (2018e) Belmond Royal Scotsman. https://www.belmond.com/trains/europe/scotland/belmond-royal-scotsman/. Accessed 17 Feb 2018

Blancheton B, Marchi J-J (2013) The three systems of rail tourism—French case. Tourism Manag Perspect 5:31–40

Carić H, Mackelworth P (2014) Cruise tourism environmental impacts—the perspective from the Adriatic Sea. Ocean Coast Manag 102:350–363

Dederl M, Kanitz C, Mei-Pochtler A (2017) From having to being—the future of the global luxury industry. In Thieme WM (ed) Luxusmarkenmanagement—Grundlagen, Strategien und praktische Umsetzung. Springer Gabler, Wiesbaden, pp 147–161

Dickinson J, Lumsdon L (2010) Slow travel and tourism, Tourism, environment and development series. Earthscan, London

Douglas N, Douglas N, Derrett R (2001) Special interest tourism. Wiley, Milton

Duval DT, Lohmann G (2011) Critical aspects of the tourism-transport relationship, Contemporary tourism reviews. Goodfellow, Woodeaton

Dykins R (2016) Shaping the future of luxury travel—future traveller tribes 2030. Amadeus AIT Group SA, Madrid

Ebert H-D (2016) Dream trips by train—The most famous trains, the most beautiful routes. GeraMond, Munich

Euromonitor International Ltd (2017) Global luxury travel trends report. http://www.euromonitor. com/global-luxury-travel-trends-report/report. Zugegriffen: 13 Jan 2018

Farren J (2006) Legendary luxury trains—dream journeys around the world. Knesebeck, Munich

Freyer W (2015) Tourism—introduction to the tourism economy, 11th edn. De Gruyter Oldenbourg, Munich

Golden Eagle Luxury Trains Ltd (2018a) Golden Eagle Danube Express. http://www. goldeneagleluxurytrains.com/trains/golden-eagle-danube-express/. Zugegriffen: 24 Feb 2018

Golden Eagle Luxury Trains Ltd (2018b) Golden Eagle. http://www.goldeneagleluxurytrains.com/ trains/golden-eagle/. Accessed 24 Feb 2018

Big S (2017) Handbook tourism and transport—transport operators strategies and concepts, 2nd edn. UVK, Constance

Hall DR (1999) Conceptualising tourism transport. Inequality and externality issues. J Transport Geogr 7(3):181–188

Hennigs N, Wiedmann K-P (2017) The rising demand for luxury brands: a global phenomenon with local characteristics. In Thieme WM (ed) Luxusmarkenmanagement—Grundlagen, Strategien und praktische Umsetzung. Springer, Gabler, Wiesbaden, pp 163–176

Howard M (2014) Unforgettable journeys—around the world in legendary trains. DuMont, Ostfildern

Indian Holiday Pvt. Ltd. (2018) The Deccan Odyssey | luxury train travel. https://www.deccan-odyssey-india.com/. Zugegriffen 08 Apr 2018

Indian Holiday Pvt. Ltd. (n.d.) Maharaja Express: world's leading luxury train. https://www. maharajas-express-india.com/. Zugegriffen 09 Apr 2018

IRCTC (2017) Maharajas' Express. http://www.the-maharajas.com/. Date Added: 09 Apr 2018

IRT (n.d. a) Rovos Rail Pride of Africa. https://www.irtsociety.com/train/rovos-rail-pride-of-africa/ . Zugegriffen: 6 Apr 2018

IRT (n.d. b) Kyushu seven stars. https://www.irtsociety.com/train/kyushu-seven-stars/ . Accessed 10 Apr 2018

IRT (n.d. c) World's top 25 trains. https://www.irtsociety.com/worlds-top-25-trains/. Accessed 16 Mar 2018

IRT (n.d. d) El Transcantábrico. https://www.irtsociety.com/train/el-transcantabrico/. Accessed 11 Apr 2018

IRT (n.d. e) About us. https://www.irtsociety.com/about-us/ Accessed 16 Mar 2018

Kyushu Railway Company (2018a) Cruise train seven stars in Kyushu. http://www.cruisetrain-sevenstars.com/. Zugegriffen 10 Apr 2018

Kyushu Railway Company (2018b) Cruise train seven stars in Kyushu premium. http://www. cruisetrain-sevenstars.com/premium-journey/. Zugegriffen 03 März 2018

Lernidee Erlebnisreisen GmbH (2018) Al Andalús: Santiago de Compostela—Seville. https:// www.lernidee.de/de/reise.html?r=2005. Accessed 13 Feb 2018

Magadán Díaz M, Rivas García JI (2012) Transcantábrico. The strategic role of itinerant tourism. Septem Ediciones, Oviedo

Mühl A, Klein J (2006) Travel in luxury trains—the international sleeping car company; the big express trains and hotels—history and posters. EK Publishing House, Freiburg

Müller-Urban K, Urban E (2017) The golden age of railways—the epoch of luxury trains 1850 to 1960. Transpress, Stuttgart

Novelli M, Robinson M (2011) Niche tourism—contemporary issues, trends and cases. Routledge, London

Prideaux B (1999) Tracks to tourism: Queensland Rail joins the tourist industry. Int J Tourism Res 1:73–86

Prüne G (2013) Luxury and sustainability—development of strategic recommendations for the marketing of luxury goods. Springer, Wiesbaden

Renfe (n.d. a) Al Andalus. http://www.renfe.com/trenesturisticos/alandalus.html. Accessed 12 Feb 2018

Renfe (n.d. b) El Transcantábrico Clásico. http://www.renfe.com/trenesturisticos/transcantabrico_Clasico.html. Accessed 11 Apr 2018

Renfe (n.d. c) El Transcantábrico Gran Lujo http://www.renfe.com/trenesturisticos/transcantabrico_GranLujo.html. Accessed 11 Apr 2018

Resonance Consultancy Ltd (2016) 2016 future of luxury travel report. Resonance Consultancy Ltd., Vancouver

RHB (2014). Rhaetian Railway panorama trains. https://www.rhb.ch/de/panoramazuege. Date Added 5 Apr 2018

Ride Responsibly (n.d.) Duty of care. http://www.rideresponsibly.org/duty-care/. Zugegriffen 29 Oct 2018

Rovos Rail (2018) Rovos Rail. https://www.rovos.com/. Accessed 6 Apr 2018

Sachslehner J (2001) On rails through the old Austria—Nostalgic reminiscences of the k.k. The world of railways to legendary steam locomotives and the first big lines to sophisticated luxury trains and golden travel romanticism. Pichler, Vienna

Schultz A (2014) Basics of tourism—airlines cruises trains buses and rental cars, 2nd edn. Oldenbourg, Munich

Sombart W (1913) Studies on the development history of modern capitalism. First volume: Luxury and Capitalism. Duncker & Humblot, Munich

Su MM, Wall G (2009) The Qinghai–Tibet railway and Tibetan tourism. Travelers' perspectives. Tourism Manag 30(5):650–657

The Blue Train (2018) The Blue Train history. https://www.bluetrain.co.za/. Stand 05 Apr 2018

The Train Chartering Company Ltd (2018a) About the luxury train club. https://www.luxurytrainclub.com/about/. Zugegriffen 16 Mar 2018

The Train Chartering Company Ltd (2018b) Seven stars In Kyushu. https://www.luxurytrainclub.com/trains/seven-stars-in-kyushu/. Zugegriffen 10 Apr 2018

The Train Chartering Company Ltd (2018c) About train chartering. https://trainchartering.com/about.html. Zugegriffen 30 May 2018

The Train Chartering Company Ltd (2018d) Rovos—pride of Africa. https://www.luxurytrainclub.com/trains/pride-of-africa-rovos-rail/. Zugegriffen 8 Aug 2018

Tillman JA (2002) Sustainability of heritage railways: an economic approach. Jpn Railway Transport Rev 32:38–45

Tourism Insider (2018) Luxury tourism expects strong growth in the coming years. http://tourism-insider.com/2018/01/luxus-tourismus-erwartet-starkes-wachstum-in-den-nachsten-jahren/. Accessed 6 Feb 2018

van Truong N, Shimizu T (2017) The effect of transportation on tourism promotion. Literature review on application of the Computable General Equilibrium (CGE) Model. Transport Res Procedia 25:3096–3115

Viedebantt K (2015) On rails around the world—the 55 most beautiful journeys by rail. Bruckmann, Munich

Villanueva Cuevas A (2012) Railways and sustainable tourism: The need for comprehensive and inclusive public intervention. Septem Ediciones, Oviedo

Marc Aeberhard   founded Luxury Hotel & Spa Management Ltd. in Zurich in 2004 and has been acting as Managing Director ever since. The company has access to a global network of travel trade partners, lifestyle and travel media and (U)HNWI and works closely with public relations and sales & marketing agencies in Frankfurt, Munich, Paris, Dubai, Milan, New York, Hong Kong, and London. Furthermore, the native Swiss is a member of the consulting networks of the Gerson Lehmann Group, USA and Hotellerie Suisse, Bern. He also takes on an active role in the "Luxury" task force of the management of ITB Berlin, Germany. As author and co-author of various specialist publications, his name can be found regularly. He also holds guest lectures in Berlin, Istanbul,

Lausanne, Lucerne, Munich, Singapore, Stuttgart, Thun, Vienna, Worms, Zurich, etc. The graduate hotelier graduated with distinction from the Ecole Hôtelière de Lausanne (EHL) and previously completed his studies in business administration as lic.rer.pol. (MBA) at the University of Bern. The luxury hotelier has more than 20 years of experience in the fields of hotel opening, management, and renovation/refurbishment in Abu Dhabi, Germany, France, Maldives, Morocco, Seychelles, Sri Lanka, Switzerland, Thailand, Ukraine, and Cyprus of small hotels in high and top end. Many of the hotels have been awarded international prizes. All projects are based on the definition of New Luxury and work according to the principles of the Triple Bottomline.

Prof. Dr. Roland Conrady Since 2002, Prof. Dr. Roland Conrady has been a professor at the tourism/transport department of the Worms University of Applied Sciences. His research and teaching focus on aviation, tourism, and digitalization. Since 2004, Roland Conrady has also been the Scientific Director of the world's largest tourism convention, the ITB Berlin Convention. He was president of the German Society for Tourism Science (DGT) e.V. and is a book author (among others Conrady, R./Fichert, F./Sterzenbach, R., Luftverkehr, Munich 2019). Roland Conrady is a member of various advisory boards of companies and politics. Previously, he was head of the study program "Electronic Business" and Professor of General Business Administration at the University of Applied Sciences Heilbronn. After graduating as Dr. rer. pol. from the University of Cologne in 1990, he held various management positions at Deutsche Lufthansa AG until 1998.

Stephan Grandy is Senior Manager of Lufthansa's premium products and Team Leader for Lufthansa Private Jet as part of Lufthansa Group's B2B Sales Global Product and Program Management. He can look back on 25 years of professional experience in the Lufthansa Group, including 8 years in the marketing of premium products and 3 years as project manager. In addition, he worked for 4 years as a key account manager for American Express. He completed his studies in business administration at the University of Cologne.

Thomas P. Illes is a cruise analyst and expert, maritime trade journalist, and university lecturer in deep-sea tourism. He is regarded as one of the most internationally renowned and critical experts on the cruise and passenger ship industry and consults companies within and outside the shipping industry in the areas of strategy, process optimization, HR, branding, design, benchmarking, and corporate communications.

Adam Parken is Director of Communications and Public Relations at Blacklane. He manages the company's external and internal communications worldwide. Before joining Blacklane, he worked for Oracle, the telecommunications company Tekelec, and PR agencies.

Norbert Pokorny is co-founder and Managing Director of the luxury tour operator "Art of Travel." For more than 25 years, he and his team have been designing individual luxury trips that are either as down-to-earth or as extravagant as the wishes of his wealthy customers. Since the foundation of his company, he has closely followed the changing concept and perception of luxury and how the wishes of luxury customers are developing. His curiosity for trends, developments, and visions are always the driving force behind his work.

Prof. Dr. Ralf Vogler is a Professor of Transport Management, Business Administration and Law at the Faculty of International Business at Heilbronn University of Applied Sciences. Since 2016 he has been the head of the "Tourism Management" program. His main focus in tourism research and teaching includes transport and mobility as well as tourism policy.

Maria Wenske (BA) is a graduate of the Tourism Management course at Heilbronn University of Applied Sciences and has been involved in research on various issues of mobility, in particular the effects of luxury trains on tourism destinations. Before starting her studies, the trained tourism assistant worked in various tourism and hospitality companies in Germany and Spain.

Dr. Jens Wohltorf is Founder and Managing Director of Blacklane, the worldwide premium mobility provider. He studied industrial engineering and received his doctorate at the Technical University in Berlin with research stays at the University of California in Berkeley and MIT in Boston. Prior to founding Blacklane, Jens Wohltorf was a Principal with the Boston Consulting Group.

Luxury Relevance of Selected Megatrends in Tourism

Marc Aeberhard, Stefan Gössling, Mario Krause, and Jörg Meurer

1 Sustainability and Luxury

Luxury Tourism and the Environment

Stefan Gössling

This section discusses from a human ecological perspective whether luxury tourism is compatible with the sustainable use of resources and ecosystems. First, luxury is defined: In biophysical terms, luxury is the description of a use of energy, water, and raw materials that goes beyond what is necessary for survival, combined with the generation of waste and emissions. Luxury is therefore a form of waste that questions the stability of ecosystems. Can mankind afford luxury (tourism)?

Luxury, in direct translation from Latin, can have the following meanings according to the dictionary (cf. Pons 2018):

1. Abundant fertility (from plants and from the earth)
2. Splendor, unnecessary effort

M. Aeberhard
Luxury Hotel & Spa Management Ltd., Zürich, Switzerland

S. Gössling (✉)
School of Business and Economics Linnaeus University, Kalmar, Sweden
e-mail: stefan.gossling@lnu.se

M. Krause (✉)
Deutsches Zentrum für Individualisierte Prävention und Leistungsverbesserung, Hannover, Germany
e-mail: mk@the-krause.com

J. Meurer (✉)
KEYLENS Management Consultants, München, Germany
e-mail: joerg.meurer@keylens.com

© Springer Nature Switzerland AG 2020
R. Conrady et al. (eds.), *Luxury Tourism*, Tourism, Hospitality & Event Management, https://doi.org/10.1007/978-3-030-59893-8_7

3. Opulence, debauchery, feasting, immorality

The three meanings reflect the variations of the concept. Throughout almost the entire history of mankind, i.e., over some 3 million years (cf. Steitz 1993), the availability of food has been the most important element of survival, reproduction, and development—for both animals and plants, the efficient use of energy is a basic biological principle (cf. Pyke 1981). While even today the basic needs of large parts of humanity cannot be met (cf. Pinstrup-Andersen 2009), the possibility of "unnecessary effort" only arose with the progress of technological development, which made it possible to accumulate prosperity through (worldwide) trade (cf. Wolf 1982). Only in the past 150 years, land for food production has to a large extent lost its importance for people in industrialized countries. Primary production in agriculture and fisheries today is employing only small parts of the population in industrialized countries. But the availability of food, even out of season, is taken for granted.

Where basic needs are satisfied, it is possible to accumulate resources. Wealthy individuals—kings, nobles, merchants—were able to accumulate wealth thousands of years ago, for example in the form of gold or other precious metals. This wealth often manifested itself in "splendor" and "unnecessary effort," in the sense of the second definition in Pons (2018). In industrial societies, this became possible for the upper class in the 1920s. In the classic "The Great Gatsby," John Fitzgerald (1925) describes the situation in the USA in the 1920s, when the possibilities of accumulating capital made (physical) labor largely superfluous for securing basic needs for the upper class. This could also be interpreted as the point in time of a broad social reorientation toward materialism, which then found a statistical expression at the end of the 1940s with the introduction of the gross national product. As Latouche (1993) describes, it was now possible to compare wealth on the basis of income and thus to divide entire societies into "poor" and "rich."

Opulence, debauchery, feasting, and immorality, the third definition of Pons (2018), uses very judgmental terms. Biologically speaking, "feasting" would be a state in which more energy is consumed than is necessary for an organism to survive. Associated forms of "opulence" probably did not become widespread in large parts of industrial societies until the 1960s. Vacations were certainly not taken for granted, especially in other European countries and large sections of the German population in the 1960s (cf. Hennig 1997; Löfgren 2002). This also makes it clear that the definition of "luxury" on the basis of the third Pons' definition must remain *relative*: What is "opulent" or "excessive" is measured by the actions and possibilities of others, usually more affluent, people. With regard to this luxury would be the understanding of being able (and allowed) to consume more than others.

This raises the biophysical question of luxury as a question of overuse. Theoretically, it is possible for an organism to be wasteful if access to resources is guaranteed. In ecosystems, this will be accompanied by population growth of certain species until the available energy and nutrients are exhausted and the population partially collapses. Unlike animals, however, only humans accumulate energy and matter in the form of possessions; some animal species accumulate food reserves, but only to overcome periods of limited access (for example, in winter). However, human systems, like animal species, become unstable when resources are overexploited: Diamond (2005) describes in the book "Kollaps" the overuse of resources as one of the reasons for the disappearance of cultures.

The availability and distribution of resources are currently at stake in the context of the Sustainable Development Goals of the United Nations (see UN 2015), which aim, among other things, at "zero hunger." A large number of people remain directly dependent on ecosystem services, while key supply strategies such as inshore fishing are threatened with collapse (cf. Jackson et al. 2001). Two problems arise at this interface: The availability of renewable and finite resources—land, water, raw materials—is finite, there is a scientific consensus that ecosystems are overexploited (cf. Wackernagel and Rees 1998). The probability of a systemic collapse thus increases (cf. Rockström et al. 2009) and places high demands on systemic change to restore system stability (cf. Davis et al. 2018).

How finite resources are was determined in the 1970s by the deforestation of tropical rainforests (cf. Skole and Tucker 1993). Since the 1990s, the "ecological footprint" (Wackernagel and Rees 1998) has illustrated the extent of overuse of ecosystems. The 1990s also saw a growing realization that ecosystem stability is of enormous importance for human life and the economy. "Ecosystem Services" established itself as a term used to illustrate the dependence of human activities on ecosystem stability (cf. Daily 1997).

Since about 2010, there have been discussions about the Anthropocene, defined as the Earth Age, in which human activities influence the biophysical processes of the planet globally and sustainably. Steffen et al. (2011) analyzed, for example, how population growth, industrialization, and the intensity of human resource use have contributed with enormous dynamics to changes in the concentration of greenhouse gases in the atmosphere, the overexploitation of global fish stocks, or the restriction of natural processes (e.g., through the construction of dams).

At the same time, Rockström et al. (2009) defined the areas in which human intervention in bio-geophysical processes already has strong negative consequences, including climate change, loss of biodiversity, and intervention in global nitrogen cycles. In other areas, such as phosphorus cycles, ocean acidification, freshwater use, and changes in land use, critical limits will be exceeded in the near future. Ecosystem services are thus increasingly being called into question. In addition, there are the risks of tipping points, i.e., points of biophysical chain reactions that can lead to a new ecosystem equilibrium that is significantly less beneficial for humans (cf. Lenton 2011).

Three final sentences can be drawn at this point:

1. The use of resources and ecosystems is already no longer sustainable, "natural capital" is consumed faster than it can regenerate.
2. A rapidly growing world population is consuming more and more intensively.
3. Biophysically, every form of nonessential consumption is thus "luxury," i.e., a form of waste that negatively influences future living conditions.

Luxury in Tourism

What is luxury in tourism? For many of the three and a half billion poorest people, it would certainly be a luxury to travel for leisure purposes at all. The perspective often presented by the World Tourism Organization that leisure travel is a normality for many people is statistically refutable: Only an estimated 2.5% of humanity fly across their national border within a year (cf. Peeters et al. 2006). Even in industrialized countries, the entire population does not take part in holiday trips: the German travel analysis, for example, puts the proportion of the population who do not go on vacations at around 23% (cf. Schmücker et al. 2015).

Luxury in tourism could be defined from the point of view of the average traveler as everything that an individual cannot afford or can only afford very rarely and thus always represents a comparison to other people with more leisure time or better financial possibilities. From the perspective of a tourist in an industrialized country, luxury tourism will represent an unusual, costly experience. This is also evident in publications geared toward wealthy travelers, such as the Condé Nast Traveller magazine with its focus on exclusive travel and destinations. There will therefore be great agreement on the definition of luxury tourism where the comparison refers to the super-rich, for example, the over 2000 billionaires, whose combined total assets are estimated at around 8 trillion US dollars (cf. Forbes 2018). This class of the super-rich owns houses and apartments in metropolises and popular destinations, for example in the Caribbean, Monaco, Monte Carlo, St. Moritz, or Aspen; private planes, helicopters, and yachts belong to their means of transport (cf. Beaverstock 2012; Beaverstock and Faulconbridge 2014). The trips of the super-rich, but above all also of other cultural, political, or economic elites, are often well documented in social media and thus also present to a large number of "followers." These "Carbon Elites" (Gössling et al. 2009) are highly mobile and sometimes spend several weeks of the year in the air.

However, as the discussion at the beginning of this section has shown, luxury in the biophysical sense is implicit in almost every journey and defined by "excess consumption." This include pool facilities in regions where water is scarce; elaborate bungalows in island states that have to import building materials, consumer goods, and food; elaborate buffets in countries whose populations are experiencing food shortages; or alcoholic beverages offered for sale tax-free in airports and flown from continent to continent despite their heavy weight. Increased consumption on a finite planet is always a form of luxury, no matter if it is about space, energy, or other resources.

The extent to which this increased consumption is significant depends on the form of travel. A bicycle tour with a tent and food supply would be an example of a trip that can hardly be considered a form of luxury, neither from a social nor a biophysical perspective. A cruise would be the exact opposite. Even the journey to get to the ship, often by plane, is very energy-intensive. The ship passage also costs large amounts of fuel; on board, exclusive meals and buffets generate negative environmental effects through elaborately produced (protein-rich) food and corresponding amounts of waste. In almost every aspect a cruise ship voyage is associated with large resource consumption. However, these can only be described in relative terms, i.e., as consumption that is greater than in everyday life at home.

Whether luxury tourism is sustainable or not can currently only be measured and compared on the basis of one aspect: the emission of greenhouse gases. Since limiting climate change to a maximum of 2 °C is an international policy goal, this maximum acceptable temperature change can be translated into a quantity of greenhouse gas emissions that may still be emitted in the future. It is *highly unlikely* that the 2 °C target will be achieved if more than 1000 Gt of CO_2 were emitted (cf. IPCC 2014). At current emission levels, this quantity would be exhausted within 20 years. It is safe to say that only very drastic emission reductions can lead to a stabilization of the climate system (cf. Davis et al. 2018).

Globally "remaining" emission quantities can thus be broken down to the quantities that are sustainable per year, taking into account a growing world population and an annual reduction in total emissions to a zero level in 2050. The IPCC (2014) has presented numerous scenarios in this direction which, to put it simply, mean that everyone would have to reduce their emissions from the current 4.5 t CO_2 per capita per year to a global average of around 2.5 t CO_2 per capita per year in 2030—within a decade. However, an average citizen in Germany currently emits around 9 t CO_2 per capita per year (cf. UBA 2018), of which at least 6.5 t CO_2 would have to be saved by 2030. Currently, leisure travel contributes around 415 kg CO_2 per capita per year to total emissions (cf. Gössling et al. 2017). Unless other areas of life such as housing, transport, food, and general consumption are drastically restricted, leisure travel by Germans is already a biophysical luxury, since it accounts for around 20% of emissions that would still be justifiable by 2030.

Individual trips can be of central importance for overall emissions. Eijgelaar et al. (2010) calculated, for example, that a single flight in combination with a cruise can lead to emissions in excess of 9 tons of CO_2, which corresponds to the total amount emitted by an average German per year. Research for Germany also shows that the 15% of the most energy-intensive vacations, i.e., cruises and long-haul trips, account for 70% of total holiday emissions (cf. Gössling et al. 2017). It is also scientifically proven that a small class of frequent flyers contributes particularly strongly to total emissions. Individuals who frequently make first-class business trips emit a multiple of the average number (cf. Gössling et al. 2009).

Sustainable Luxury Tourism?

As the previous sections have illustrated, sustainable luxury tourism on closer examination is an oxymoron. In a world approaching or exceeding numerous critical biophysical limits, any waste of resources will prevent the achievement of the Sustainable Development Goals and any additional ton of CO_2 emitted will prevent the achievement of the climate goals. In view of these facts, certain tourism products no longer have any legitimacy from a biophysical perspective: these include cruises as well as long-haul flights. This biophysical luxury tourism calls into question precisely those development goals that certain actors present as the merits of tourism, i.e., the economic development of peripheral states by holidaymakers. The particularly strong storms in the Caribbean in 2017 allow conclusions to be drawn about what can be expected in a world of continued climate change.

Will it be possible to solve the problem of overconsumption in tourism? A prerequisite for this would be that resources in tourism are used more efficiently and are produced sustainably. Technically this would be possible (Davis et al. 2018), but only under the precondition of dramatically higher energy prices. There are currently no initiatives in this direction. In 2018, for example, Germany tacitly abandoned its self-imposed climate targets (cf. FAZ 2018). The fact that these were missed is not surprising at all, as there were no apparent efforts to enforce emission reductions in the transport sector or other fast-growing economic sectors (cf. Gössling and Metzler 2017). It is not likely that this will change in the near future in the current political climate. The current premise is therefore that unsustainable resource use, climate change, and population growth will lead to the future of tourism as a whole being called into question. Luxury tourism contributes disproportionately to this development.

2 Digitalization and Luxury

Marc Aeberhard

Luxury or Curse?

Technical topics such as new achievements, improvements, quantitative and qualitative improvements in the performance of the digitalization of hotel service providers or in distribution technology (Channel Management (CM), Online Travel Agent (OTA), Reservation Technology, etc.) are currently a guarantee for maximum attention and full audiences at tourism, hospitality or other conferences, trade fairs or events. Even more accentuated are the demonstrations of the latest robot technologies and the development of artificial intelligence. But are these two subject areas suitable for enriching the luxury discussion?

To come right to the point: No. They do not. In no possible way.

Digital Infrastructure Becomes a Matter of Course

At the end of the Victorian era, new technologies such as electric light, hot and cold running water, and elevators were presented to a more or less astonished audience with about the same amount of fuss as today's IT companies present the latest smartphone generations, and for a brief moment, the attention of the developers belongs to their products. Especially at the end of the nineteenth century, grand hotels all over the world caused a furore with these architectural and technical achievements, but in the meantime, they have become as self-evident as indispensable. And the same fate befalls digitalization in hotels: the basic basics include IDD telephony services, solid broadband performance for the Internet, W-LAN, and interactive in-room entertainment technology. What 15 years ago still belonged to the realm of science fiction has now become part of everyday life.

Artificial Intelligence Destroys Jobs and Social Interaction

Engineers and technology companies are now working with great success on the introduction of smart home technologies, which means that the accessibility and thus control of technical devices has become location-independent and thus reduces one of the central functions of a hotel: that of the front office. The online services in combination with smart technology already allow the guest today—and to an increasing extent—not only to book and check-in to his room but also to prepare it or have it prepared according to his needs (room temperature/shade level, minibar offer, pillow type, etc.). The front office as a brain and control center in its function as a control organ is thus overridden and massively restricted by these new individualized, location-independent access functions. But with all the chicness of these options—they are and remain only technical gimmicks, which increase lead times and preparation times, dynamize the technical versatility in the hotel, minimize the interaction of the front office and thus eliminate a (potential) source of error. But all these achievements remain in essence purely technical improvements in the sense of original Tayloristic thinking (cf. Ulich 1991, p. 7 ff.) which above all do one thing: reduce human interaction, optimize, accelerate, and de-soul processes (see Hacker 1986, p. 88 ff.).

Particular attention is currently being paid to the rapid development of robot technology and the associated development of artificial intelligence. Cute little white high-tech plastic figurines roll around on tables or in lobby halls, jingle with their camera eyes, and more or less understandably buzz any standardized sentences. The delegated crowd is feeling well and believes to have invented a completely new form of hotel business. But here too, the highest degree of caution is required: There is no doubt that the use of semi-intelligent or even intelligent machines can make sense in the field of highly repetitive work (cf. Ulich 1991, p. 226 et seq.). Especially in the field of housekeeping and public area cleaning, the nightly use of self-controlled vacuum cleaners, window cleaning robots, etc. is a blessing and a great relief for people and operations. The same is conceivable in canteen kitchens, where robots independently take over the stewarding area, not only wash dishes more cleanly but also examine them for defects and then stack them neatly and clear them away. An intelligent drone is also conceivable, which is capable of automatically replacing broken light bulbs or removing cobwebs in the atriums of hotels at lofty heights.

But all these more or less funny and future-oriented possibilities have nothing to do with the modern understanding of luxury. They are only technical aids that simplify work processes or provide the guest with a new level of technical infrastructure as a matter of course.

True Luxury Is Immaterial

Luxury in its modern definition, and studies at the ITB 2015 and 2016 have shown this, is immaterial. The modern understanding of luxury is based on six essential pillars. These include space, time, customized and personalized service concepts, security (especially privacy and data security), health, and exclusivity (cf. Ehlen and Scherhag 2018, p. 138 ff.). Conversely, this means that the concept of luxury, which was based on material criteria and has been valid for decades, is completely outdated. Especially in the saturated economies of the western model in Europe and partly in North America, the concept of material luxury was democratized. On the one hand, this is the result of a massive increase in the purchasing power potential of almost all strata of the population, and on the other hand a consequence of an increasing degree of urbanization. The latter leads to an unbundling of the habitat of modern man and the natural resources that form the basis of this very habitat. In other words, when milk no longer comes from the cow but from the supermarket shelf, the decoupling of cause and effect is already well advanced in the economic equilibrium of the planet.

In this understanding, it becomes clear that neither digitalization nor artificial intelligence contributes to an improved understanding of luxury or a special experience of luxury. On the contrary, the achievements described above contribute ad extremum to the fact that humans can be completely eliminated in many areas (in certain low-budget hotel chains it is already common to have a zero-human interaction policy).

Wanted: The True Values

But the homo sapiens is and remains a social being. So much so that a lack of social interaction or false social integration—let alone insufficient social competence—can lead to misconduct, physical and mental illness (cf. Ulich 1991, p. 298 ff.), and social stigmatization. Technological innovation will not be able to meet this original need, that Maslow addresses at various levels in his pyramid (cf. Maslow 2012, p. 370 et seq.). On the contrary, people increasingly long for embedding, security, respect, attention, and appreciation (cf. Littek et al. 1982, p. 252 et seq.). These are all basic instinctive values which are inseparably linked to being human, which no machine in the world, no matter how advanced, can replace, and which ultimately form the basis of healthy social interaction. It is remarkable that right in the heart of Silicon Valley, a digital-free zone has been created that requires all (electronic) devices, whatever they may be, to be surrendered upon entry. More and more people are using the Out of Office Assistant, which automatically replies to e-mail messages while they are away on vacation, and which indicates that they are now taking a digital time-out. The new hype is called digital detox. Programs such as hotels in radio holes or abstinence in the monastery are currently experiencing a huge influx and not unexpectedly. Burnout therapies first of all focus on curing addiction-like behaviors and are all based on first unbundling people from their supposedly inseparable digital willingness and availability to communicate (cf. Meckel 2010, p. 12 f.). Holiday destinations at the furthest corners of civilization and beyond, where man is (once again) unattainable, are today among the most exclusive destinations there are. This demonstrates a further dimension of the new understanding of luxury. According to this, exclusivity means the limited availability of an offer, which—strictly in accordance with market laws—can therefore be placed particularly successfully on the market (examples are: Safari Lodges in southern Africa beyond the famous trails, diving and snorkeling excursions in the Pulau archipelago, historical haciendas on the Yucatan Peninsula, temporary monastery offers in the Surselva, etc.). In a nutshell: Man longs again for the true values of being human and finds fulfillment in the holistic balance of his being (cf. Kaltenmark 1969, p. 22 f.). The sheer infinity of spa offers worldwide impressively documents man's need for comprehensive well-being. However, it should be mentioned at this point, but not further discussed, that an intensive intellectual process is required in order to reach the point of inner balance. It would go far beyond the scope of this chapter. The fact remains, however, that the availability of digital technology and/or artificial intelligence completely counteracts this human aspiration or at best can catalyze it as a stimulant for precisely this renunciation.

Data Security and Paparazzi-Free Zones Are Essential

Recent political and economic developments such as Wikileaks, hacked large computer systems, sabotaged elections, influence on opinion forming through social media, the misconduct of Cambridge Analytica, etc. point to another dimension of the new understanding of luxury that has become extremely virulent: data security. The need for security, the integrity of life and limb, is as old as mankind and plays a central, even existential role throughout human civilization (see Lenski 1977, p. 46 et seq.). But it is not only about the construction of palisades or forts, electric surveillance cameras, or trained guard dogs, but also about virtual security. The examples described show how sensitive and vulnerable the economic and thus also the political, cultural, and social surrounding system has become. The top-end guest, often High Networth Individuals (HNI) and by definition an opinion leader, is thus exposed to a new uncertainty that can cause devastating damage in the event of misuse.

This clearly shows that the well-intentioned investments in a huge, complex, and sophisticated MICE[1] apparatus and a sophisticated digital structure in many hotels, conference centers, and similar facilities completely miss the goals, expectations, and needs of the top-end clientele. Hidden luxury (cf. Aeberhard 2018) is thus becoming a particularly popular (booking) criterion in this area in particular: important meetings no longer take place in hotels (however chic and renowned they may be) or in visible public spaces, but in secret, remote vacation homes, villas, chalets, yachts, or small hotels. Lawyers and their clients discuss sensitive information on "business walks" in the forest and (really) delicate correspondence is written by hand and, as in old times, physically transmitted by a trustworthy messenger. The rich and famous from Hollywood and the top echelons of politics, society, and business demand paparazzi-free zones and are often only connected to the world via an extremely limited, trust code-secured channel and remain physically and digitally hidden.

Summary

The question posed at the beginning, whether and to what extent digitalization and artificial intelligence contribute to the understanding of luxury, is thus unmistakably and clearly negated for the top-end segment. We are looking for: genuine security and privacy, genuine values and feelings, genuine people, genuine safety, honesty, and trust. J.J. Rousseau is more modern than ever: back to the roots.

3   Demographic Change and Luxury

Jörg Meurer

Age Is En Vogue

Recently, a top manager of a tourism company was asked about the consequences of demographic change for his tour operator business: "How old do you think you will become?" The admittedly very personal question amazed him at first. As an analyst,

---

[1]Meetings, Incentives, Conferences, and Events.

however, he quickly had an answer ready: "The average life expectancy of a man in Germany, and I can do 5 years more. So, late 70s."

This answer reveals at least two things: it indicates that we like to underestimate how old we actually get (life expectancy of men in Germany is currently already 78 years), but it also reveals an important trend: getting old, getting older than "the others," is becoming increasingly valuable for more and more people. Getting old is not only a "scourge," it is also increasingly proof of a good, well-managed life. A statement that certainly does not (yet) apply to the breadth of the population, but to a minority that is privileged because of education, wealth, or ancestry.

This puts us right in the middle of the topic: demographic change and luxury. Because the concept of luxury represented here does not only include *what* we see as luxury and *how* we consume it (category view), but also *who* buys luxury and *why* (target group view): And here we are talking about a minority of Western societies—albeit growing in size—who begin to live, enjoy and fill their old age like no generation before them—and thus gain a completely new, exciting relevance as a target group for luxury and premium brands. Just a little example: In Scandinavia, Steinway & Sons, which manufactures the best and most expensive grand pianos in the world, are increasingly seeing purchases from people over 70 years of age because they (re)start to play an instrument as they get older. It is always been there, but today it is becoming a trend.

The development presented and advocated here—and this too is worth mentioning—is quite contrary to current opinion, which emphasizes the negative consequences of an aging society. The great Frank Schirrmacher, for example, did this almost 15 years ago in his book "Das Methusalem Komplott" (The Methuselah Conspiracy), rich in words and images (cf. Schirrmacher 2004). Nevertheless: Every trend generates a countertrend. So, it seems plausible that the aging of a society leads to more and more people wanting to and being able to age differently and better.

Age is therefore en vogue and everyone is talking about demographic change. But far more is happening besides the fact that we are getting older on average and our view of the last decades of life is beginning to change fundamentally. At the same time, at the other end of the age pyramid, we have an energetic discussion about the two youngest generations. Has a generation ever been discussed more intensively, examined, evaluated as the so-called Millennials—and does the next generation, the generation Z born as Digital Natives become even more demanding, even more egocentric than its predecessor generation—or does it reflect back on other values?

The constant extension of our lifetime results in a unique social and historical situation. For the first time in human history, we are experiencing a situation in which five generations of consumers live side by side (with interesting interactions) (cf. Herhoffer and Meurer). And it is these consumer generations that enable a very special, differentiated view of the connection between demographic change and luxury.

Accordingly, this contribution begins with the definition of the term "demographic change" and an assessment of the current discussion. From there and with a focus on the relevant corporate and management perspective, it opens up the two fundamentally different perspectives on target groups. Then, with a focus on the phenomenon of old age, we describe in detail the current five generations of consumers; always with regards to their diverse relationship to luxury in general and to the travel experience in particular.

Demographic Change

From a statistical and methodological point of view, demographic change is actually an amazingly unexciting construct. In the narrower sense, it describes the development of exactly three variables of population development:

1. The fertility or birth rate, i.e., the development of births per woman and in absolute terms over time
2. The mortality or death rate, that is, the mortality of men and women
3. The balance of migration, i.e., the balance of immigration and emigration and spatial mobility

As unexciting as the statistical variables per se are, so dramatic is what happens when you look at them in interplay, not just over a few years, but over decades. Then a picture is revealed that we all know: the profound change in the so-called population pyramid, which shows the age structure of a population on the basis of the frequency distribution of age in life years. Three findings impressively illustrate which changes are currently taking place in Germany and which will take place in the future (cf. Heckel 2017):

4. Every year the average life expectancy increases by 3 months. Every day we gain an average of 6 h of life! It is worth to think about this a second time.
5. A baby born in 2010 has a 50% chance to live to be 100 years old. Probably one of the reasons why Population Europe, a network of European research institutes on demographic change, recently organized the touring exhibition "How to get to 100—and enjoy it!"
6. At the beginning of the twentieth century, only 8% of people were older than 60 years; by 2050 the figure is expected to increase to one-third, quadrupling.

In short: on average, we are getting older and older and the number of older people is constantly increasing.

Of course, demographic change is not a German phenomenon. All European countries are more or less affected by the phenomenon of (over)aging. On a global level there is another phenomenon that comes into play. The European countries are shrinking (for Germany, the forecasts are between stagnation and a reduction of up to 7 million people; however, France and Great Britain, for example, will grow significantly). Whereas in 1960, 13.3% of the world's population lived in Europe, by 2050 only every 50th (!) person will still have a European passport (cf. Habekuß 2017a, b). It is admittedly a big perspective, but in the long run demographic change also means that, detached from social stratification, luxury markets outside Europe will simply be dramatically larger. China is already giving a lively outlook on this today.

The actual and possible consequences of demographic change are manifold and we encounter them every day: pension security and intergenerational contracts, the state of emergency in nursing care and the explosion of health care costs, distortions in the labor market and a shortage of skilled workers, threats to the competitiveness of Germany as a business location, and the sustainability debate, to name but a few. Demographic change influences the development of a society, an economy, comprehensively.

For the relationship between demographic change and luxury, two perspectives on the topic appear to be particularly relevant: Which changes result from the demographic change for the current and potential luxury consumers (target group view) and which consequences result from the dynamics in the target group change for the understanding of luxury and the luxury travel behavior (category view).

The Two Basic Target Group Perspectives and the Five Consumption Generations

Until about 3 years ago, the existence of different generations of consumers played no or only a subordinate role in discussions about strategy, brand management, product development, marketing or sales for luxury companies and brands. From the age pyramid point of view, there was a horizontal (age-neutral) perspective on target groups that were either classically segmented—for example, demographically, psychographically, or value-based—or one worked with luxury milieus (Sigma, Sinus, GFK/Roper).

Of course, even at that time it was clear to companies that their customers belonged to a certain age group and that a customer base had to be rejuvenated, for example through new products (which, for example, Rolls-Royce succeeded in doing with the Wraith by an average of almost 10 years). But existing and new customers—and this makes the decisive difference—were not regarded as members of a particular consumer generation; with potentially exciting implications for product development, marketing, and sales.

With the discussion about Millennials and Generation Z already mentioned in the introduction, this perspective slowly but surely began to change fundamentally. The main drivers here are likely to be, on the one hand, intensive media coverage and, on the other hand, a significantly increasing presence of the generational phenomenon on the desks of decision-makers in companies, fueled by market research studies and articles.

KEYLENS and INLUX have taken this development as an opportunity and comprehensively empirically examined the five generations of consumers with special reference to the subject of luxury (cf. Herhoffer and Meurer 2018). The so-called upper segments, the upper third of the population, were surveyed representatively. This upper third represents significantly more than *the rich* (the so-called High Net Worth Individuals account for less than 1.5% of the population in Germany, for example). Today, however, luxury products are bought by a much larger proportion of people.

An overview of the five generations of consumption is shown in Fig. 1.

The Five Generations at a Glance

The Silent Generation of those born shortly before, during, and immediately after the Second World War was initially traumatized by war and flight trauma and, in connection with the traditional role and gender images, literally "speechless and quiet." Although the Silent Generation also rebelled against German culture with Rock'n'Roll and later with Jetset airs and graces, the main concern for our (grand) parents was to regain lost security in the self-created prosperity of the years of the economic miracle. They succeeded in this on a material level, their children were born into a prosperous world of economic miracle years.

The so-called Baby Boomers (born 1947–1966) continued the economic miracle trend consistently: "Working their way up" was the credo of a generation for whom cars, beautiful clothing, the first trip to Italy, and perhaps even a small house of their own were life goals. But also this generation broke—albeit not in its entirety, but represented by a few—with its parents and the hitherto untouched establishment: Dutschke, Ohnesorg and Co. unmasked "under the gowns a musty smell of 1000 years abounds" and instigated the 68 revolution. Women discovered emancipation.

The next Generation X (1967–1982) was the first to differentiate itself on a broad front and to break away from the previous generations. In his book "Generation X," Douglas Coupland juxtaposes a lifestyle that has been ground down by social and economic constraints with a philosophy called "Lessness," which does not measure the value of life by the accumulation of status symbols. Thus Generation X also goes down in the history of sociology as the generation of "exhibitionist modesty."

The Millennials (Generation Y, 1983–1994) continue the differentiation and individualization course of the Generation X and lead it, so to speak, into the egocentric absurd. They enjoy themselves and their lives to the fullest and can already be labeled today as the most narcissistic and hedonistic of all consumer generations—with all opportunities for companies as the ideal consumer generation.

It is a necessary question, at least from a social context, whether Generation Z (1995–2010) will continue this trend of the previous generations linearly. But as it seems, it does not: with the Generation Z the pendulum surprisingly swings back toward traditional values such as "good manners," "possession only when I deserve it."

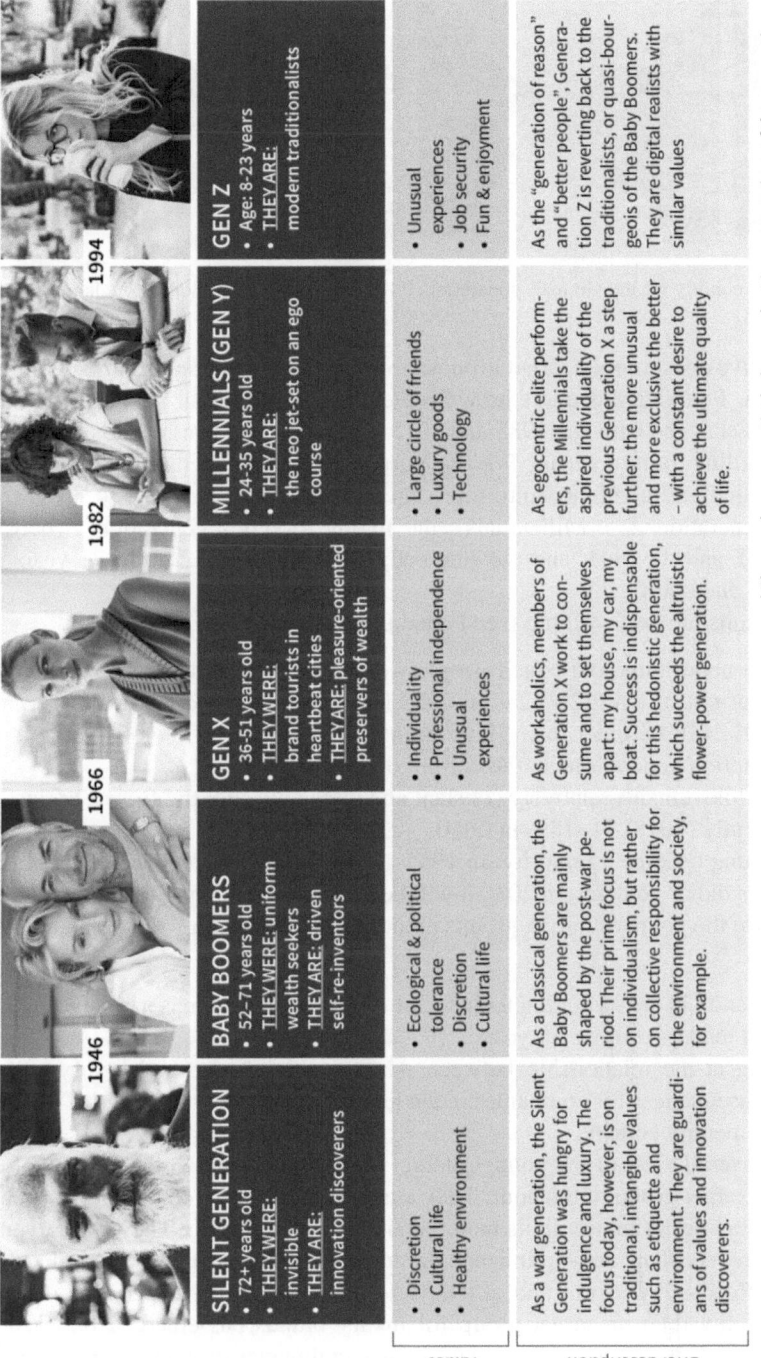

	SILENT GENERATION	BABY BOOMERS	GEN X	MILLENNIALS (GEN Y)	GEN Z
	1946	1966	1982	1994	
	• 72+ years old • THEY WERE: invisible • THEY ARE: innovation discoverers	• 52–71 years old • THEY WERE: uniform wealth seekers • THEY ARE: driven self-re-inventors	• 36–51 years old • THEY WERE: brand tourists in heartbeat cities • THEY ARE: pleasure-oriented preservers of wealth	• 24–35 years old • THEY ARE: the neo jet-set on an ego course	• Age: 8–23 years • THEY ARE: modern traditionalists
Values[1]	• Discretion • Cultural life • Healthy environment	• Ecological & political tolerance • Discretion • Cultural life	• Individuality • Professional independence • Unusual experiences	• Large circle of friends • Luxury goods • Technology	• Unusual experiences • Job security • Fun & enjoyment
Brief description	As a war generation, the Silent Generation was hungry for indulgence and luxury. The focus today, however, is on traditional, intangible values such as etiquette and environment. They are guardians of values and innovation discoverers.	As a classical generation, the Baby Boomers are mainly shaped by the post-war period. Their prime focus is not on individualism, but rather on collective responsibility for the environment and society, for example.	As workaholics, members of Generation X work to consume and to set themselves apart: my house, my car, my boat. Success is indispensable for this hedonistic generation, which succeeds the altruistic flower-power generation.	As egocentric elite performers, the Millennials take the aspired individuality of the previous Generation X a step further: the more unusual and more exclusive the better – with a constant desire to achieve the ultimate quality of life.	As the "generation of reason" and "better people", Generation Z is reverting back to the traditionalists, or quasi-bourgeois of the Baby Boomers. They are digital realists with similar values

[1] These values are particularly prevalent in a comparison of the generations.

Fig. 1 Generation tableau. Source: Herhoffer and Meurer (2018)

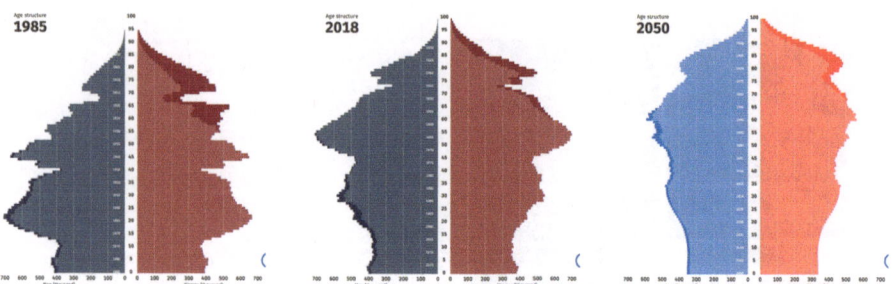

Fig. 2 Age cohorts for the German population 1985, 2018, and 2050. Source: Federal Statistical Office (2018)

These five consumption target groups represent the current age cohorts within the population. Figure 2 illustrates how the weight or distribution between the target groups has changed between 1985 and 2018 and how it will continue to change until 2050 according to the forecast. For this purpose, the five generations of consumers are simplified and combined into three age cohorts: the young (up to 35 years—today's generation Z + Millennials), the middle (today's generation Z between 36 and 51), and the older and old (the baby boomers from 52 up to 71 years and the silent generation).

The results are very striking (see Federal Statistical Office 2018):

- In 1985 about three out of ten people were old (29%), today it is already more than four out of ten (42%) and in 2050 it will be every second person, because then 50% will be older than 51 years (and thus belong to one of the two "old" consumption generations in the current definition).
- The middle remains relatively constant and loses only slightly: from 22% in 1985 to currently (2018) 21–18% in 2050.
- The young become a minority: in 1985 almost half of the population (49%) was 35 years old or younger, in 2018 it will be 37%. By 2050, only about one in three (32%) of the youngest target groups in the current definition will belong to one of the two.

Young and old are switching their social weights—and there is a big exclamation mark to be made!—during 65 years.

Looking at the relationship between demographic change and luxury and the current consequences for companies in the luxury industry, two findings of the study should be mentioned here.

In the overall view of the younger target groups Z, Y, and X, the millennials are the luxury target group of the hour. They attach importance to contemporary quality consumption, desire luxury in its material and immaterial form. They want to enjoy life but are constant self-optimizers: trend-oriented, enjoyment- and education-oriented. Ready to perform and at the same time leisure and experience-oriented. Here, a highly relevant target group for luxury brands has emerged between the mid-20s and mid-30s, for which many luxury companies or even entire industries today have not yet created any offers for. The best example is the cruise industry, where offers for target groups beyond the 50s, i.e., baby boomers and silent generations, dominate in the luxury sector today. With Ruby Cruises, the inventor of "lean luxury," the Ruby Hotels, is setting out to multiply the successful hotel concept for the young and young-at-heart postmodern avant-garde on the water.

But by far the most important finding for the connection between demographic change and luxury concerns the fundamental changes surrounding the phenomenon of old age.

7. A new, highly relevant phase of life is created by the significant change in a lifetime. Whereas in the past we spoke of three phases in a person's life, today a fourth phase is actually added. Sociologists and physicians speak of the so-called worth living years of a human being—and these increase steadily after (for the wealthy often earlier) retirement from the occupation. It is not seldom the case that a person spends a period of 20 or more years in the early/mid-60s of active life today before the actual last phase of life begins; characterized by increasing immobility, physical decay, illness, and finally death. As a 40- or 50-year-old today, imagine what you have done in the last 20 years—then it becomes impressively clear what this additional phase of life offers in terms of possibilities.

8. In particular, target groups with an affinity for luxury are beginning to live this lifetime in a significantly different way: Health and self-optimization are becoming key trends. In view of longer life expectancy, older people are increasingly focusing on preventive medicine to prevent illness and increase viability. All generations take responsibility for their health and leave this neither to chance nor to the sole command of physicians. Overall, all generations see health—and not just wellness for the well-being—as the key to a better quality of life. As a result, new nutrition options are gaining ground, for example. The trend goes from gourmet to functional and superfood. In society, it is already part of good manners to deal with nutrition. A healthy, trained body is a luxury subject and serves social exclusion—even in old age.

9. To be self-optimized means to present a good image to the outside world, but also to have made a certain career as a human being in heart and intellectual education. To think and feel self-respectful, to act empathetically. The goal is to be yourself a desirable luxury subject. The indicators for self-optimization are one's own perception and the feedback in the peer groups, i.e., social reference systems.

10. Thus, a new target group, the so-called SAYAHs, emerges—a target group objectively older in years, but one, two, or even three decades younger in their heart, head, and often in their physique. These *Silver Agers Young at Heart* are increasingly beginning to conquer advertising communication as well. Lancôme separated from Isabella Rossellini as the brand's face at the end of the 1990s. Justification: Women over 40 were no longer considered desirable. Twenty years later, the brand signed the Italian again. Just one example out of many that shows: Age is becoming more attractive. Our image of age is literally changing.

In addition, of course, there is also the question of how these young old people see luxury, what understanding they have of luxury, what significance they attach to it.

Luxury Understanding of Today's Consumer Generations and Luxury Travel Experience

Generations are not static clusters that move through time as stable cohorts. Especially the rapid technical progress has caused even the Silent Generation to change massively. In the sense of reciprocal learning, for the first time in human history, the elderly also orient themselves toward the young and not only vice versa; their thinking and acting are also becoming increasingly digital.

In this millennium, the developments in information and communication technology have a fundamental influence on information, decision-making, and purchasing behavior, both in a generational and cross-generational way. Rarely have the orientation, skills, and behavior of a generation influenced previous generations, but also companies, as in the case of Generation Z and the Millennials.

Moreover—this teaches us the long-term view of the phenomenon of luxury— our understanding of luxury is renewed every 10–20 years and can thus be reversed from one decade to the next. What the parent generation finds desirable is usually no longer desirable for the next generation. Generations thus shape the respective understanding of the luxury of a time (epoch).

Here, the study shows an astonishingly homogeneous picture: Fig. 3 illustrates that the understanding of luxury has completely changed in the last 10 years,

Personal meaning of luxury

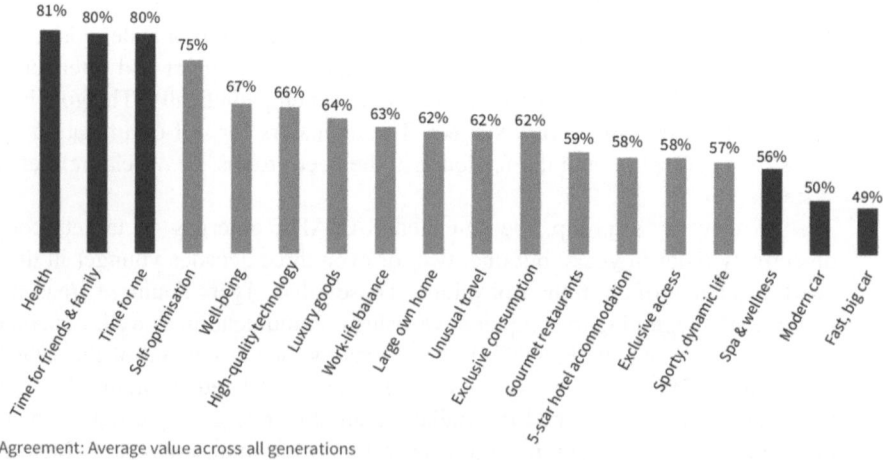

Agreement: Average value across all generations

Fig. 3 Cross-generational significance and understanding of luxury

independent of generations, and has experienced a massive shift toward "immateriality/experience orientation."

- When asked about their personal significance and their understanding of luxury, "high-quality technology" and "luxury goods"—hardware or classic luxury goods only appear in sixth and seventh place in the cross-generational ranking.
- "Immateriality" in the understanding of luxury achieves a clear precision in the direction of "health," "time," "self-optimization." The top 5 values for importance and understanding of luxury of all five generations pay tribute to exactly this trend.
- Only then does classic luxury follow (luxury goods, large property of one's own, exclusive consumption). Classical luxury thus naturally continues to have its right to exist.
- The "car" or "spa & wellness" are at the bottom of the ranking. The classic luxury manifestos of recent decades are losing importance, at least in highly developed economies such as Germany.

This change in the understanding of luxury is of course of the highest relevance for the tourism industry, which essentially serves the desire for unique experiences and what is at the top of Maslow's pyramid of needs: self-determination, self-realization, transcendence (deliberately interpreted by Ritz Carlton on their new cruise ships as "personal transformation"). In this respect, an abstract of the current tourist needs of the five generations of consumers will be presented on the basis of the results of the tourism study (see also Fig. 4): (cf. Keylens Management Consultants 2018):

- *Generation Z: squaring the circle—maximum variety with simultaneous relaxation*
  The "always on" Generation Z carries its hectic, globally oriented, still untidy life one to one into the demands of the vacation: For no other generation are variety, surprise, and extraordinary experiences as strong of a driver as for those under 24 (63%). At the same time, they also have the second-highest value for recreation and relaxation (67%). "Everything can—nothing must" is the general requirement for product developers and marketers.
- *Millennials: from expedition cruise to glamping*
  In addition to their high affinity to classical five-star luxury, the Millennials are basically eager to discover. Sixty-seven percent of this generation have the desire for extraordinary experiences. Although rest and relaxation play an important role for 62% of the respondents as a balance to stressful everyday life, for the Millennials the vacation is also a source of inspiration through individual and strong once-in-a-lifetime experiences. In retrospect, they condense into personal stories that create lasting memories and thus enrich life.
- *Generation X: zero-error recovery, instantly*
  The preferences of Generation X are similar, but with 70% it has the highest percentages of all generations in regards to recovery and relaxation. It is the generation of workaholics under pressure from the double burden of professional

Attitudes to travel and preferences

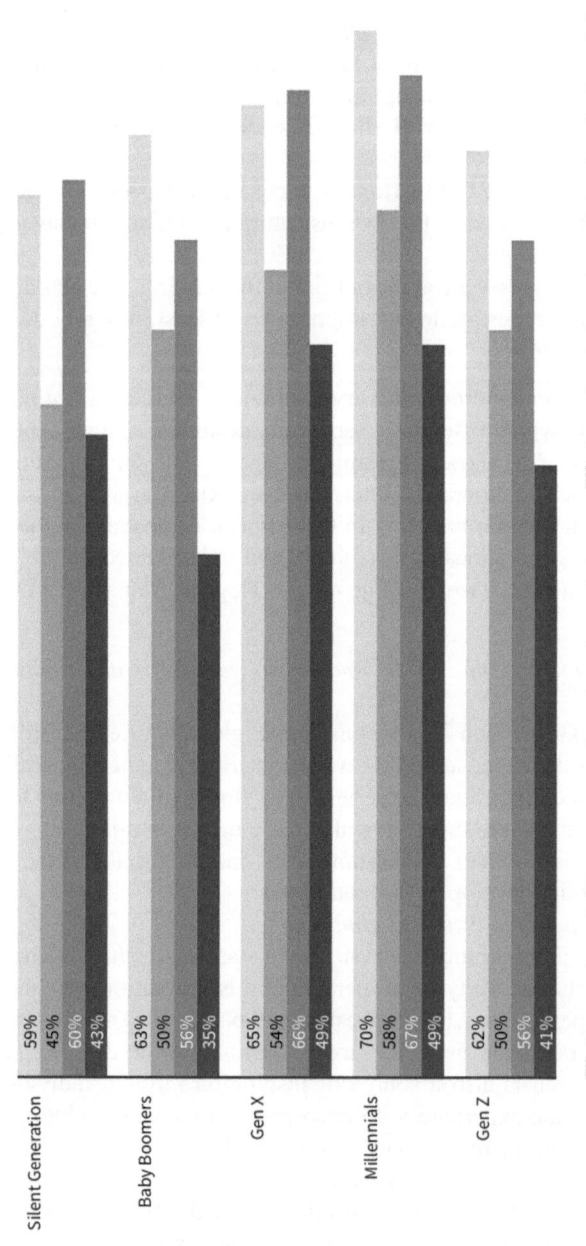

Agreement by generation

- Unusual, one-off experiences (Antarctic cruise, luxury lodge on a reserve, 5-star camping/glamping) epitomise luxury for me.

- For me, a luxury holiday also means sometimes deliberately renouncing comfort (e.g. spending a night in a Tibetan monastery or exclusive walking tours with nights spent under the stars)

- For me, a luxury holiday means a travel experience in a top 5-star hotel (service, location, design/architecture, catering, hotel's history)

- For me, a luxury holiday means variety during the trip: 5-star hotel, a designer Airbnb, but also a few days backpacking can complement each other well in a holiday

Fig. 4 Comparison of travel attitudes and preferences of consumption generations

commitment and family. Generation X defines vacations as a time of regeneration and reflection together with the partner and the family. It is therefore not surprising that this generation, with 48% of all respondents, is still the most familiar with classic stationary expert advice in travel agencies: "zero error recovery" is required from the very first moment.

- *Baby Boomers: to be self-sufficient on vacation*
  For the Baby Boomers, deep experiences, reflection and a balanced approach in a highly comfortable environment are at the top of their personal bucket list. This target group is not convinced by the many exciting alternatives to a classic hotel vacation and the absence of comfort that comes with them. Actually, almost 25% of the respondents actively rejected this option. The Baby Boomers have already seen and experienced a lot—and therefore no longer have the feeling of missing anything. On vacation, they are sufficient in themselves.
- *Silent Generation: be recognized and return*
  The five-star hotel impresses this target group the least. For two-thirds of respondents of the 72+ generation, recognition and familiarity with accommodation and surroundings are decisive. Those who have won their favor can be optimistic that they will return regularly and will be happy to do so—attention!—by booking online and directly. It is worthwhile to look after this well-off clientele as personally as possible with highly professional staff. And last but not least, to invest in an outstanding CRM system.

In the overall view of the results, it becomes clear to what extent demographic change influences luxury—whether in terms of target groups or categories. However, it is also clear that companies and brands serving the luxury segment, in contrast to many others, can see a great opportunity in demographic change.

4   Health and Luxury

Mario Krause

Life Invites You

Luxury Has many Faces

Material luxury manifests itself in exclusive features that set the owner of the luxury good apart from the general public, and it serves as an attribute of success or as a status symbol. Luxury is defined here as a distinguishing criterion between social strata.

We all know such attributes as jewels, handbags, watches, cars, or big real estate, which are viewed as business cards of success and as signs of prestige. Beauty can also be seen as a symbol of luxury because beautiful women did not have to work hard in the fields and were therefore not subject to normal wear and tear. In the past, there was also the opinion that dark tinted skin was less luxurious than white skin. The field workers were tanned while the ladies of the society were equipped with a rather light complexion. Today, this assessment has largely reversed, which already indicates that the faces of luxury can be subject to a temporal change.

In addition to this rather extrinsic motivation, *immaterial luxury* rather seeks an attitude to life in order to express someone's personality. Luxury goods have a certain identification function, they should correspond to individual values. Possibilities of self-realization, the further development of techniques or personality, cultural experiences, enjoyment in various areas, or spiritual experiences can be attributes of this immaterial luxury. The individual is often more concerned with being and feeling than with having and possessing.

At this point, health can be paired with well-being and longevity. On the one hand, health is a value that almost every individual strives for, but cannot always achieve. As a result of this shortage, health is becoming an individual luxury good in addition to a universal basic right. The individual even likes to show this attribute as a status symbol in the form of immaculate beauty. A worn skin due to work, environmental pollutants, wrong lifestyle, or inadequate nutrition symbolizes a deficiency and such a deficiency is only compatible with luxury to a certain extent.

In addition to this more externalized form, luxury in the context of health naturally also means access to pleasure, well-being, performance, success, and longer life expectancy. Thus, luxury also has a very strong personal and hedonistic component.

Material and immaterial luxury have the feeling of meaning and power in common. Both the material insignia of a Ferrari and the feeling of spiritual enlightenment give the respective owner a unique selling proposition, which accommodates his aspiration for an evolutionary (competitive) advantage.

There are *cultural differences* in the definition of luxury. In countries such as Russia, for example, luxury is still defined more by external attributes, while in Germany luxury is shown more cautiously to the outside world. It is interesting to note that health has a very high value through all cultures and is described by many people as the most precious and important thing.

Health as a Value and Luxury Good

There is a general consensus that everyone has the fundamental right to health. In reality, access to health services is very unevenly distributed. Even in highly developed industrial nations, there is a glaring imbalance because special services or innovations are often not refinanced by the social systems. Health thus becomes a luxury good, which also decides whether the individual dies sooner or later and whether he may spend his lifetime with activity, enjoyment, and well-being.

Even if, for example, the health system in Germany suggests that every individual has access to all health resources, the reality is different. Experience has shown that patients with statutory health insurance, for example, have to accept much longer waiting times for the same examinations or treatments than patients with private health insurance. Although the free choice of doctor is also documented in the German health system, it can de facto not be realized, since chief physicians, for example, do not personally treat the regularly insured. For people who can afford to pay a multiple of the regular rates, not only the offers of the top specialists open up, but also a service atmosphere with a luxurious interior and a completely different personnel key. The medical offers with a rather hotel-like service are increasing worldwide.

Offers for Health Versus Offers Against Illness

People around the globe are looking for services and solutions to improve existing health limitations, recharge their used batteries, maintain their health, and live as long as possible.

The vast majority of these services continue to be offered by the traditional health care providers, i.e., doctors or clinics. These can rely on their reputation, the good reputation generally attributed to medical work.

As far as the treatment of acute health disorders is concerned, medicine has earned a reputation for being able to solve these well and safely.

Due to increasing life expectancy, however, more and more chronic or functional complaints are occurring today. There are no equivalent solutions for these yet.

Life expectancy has risen sharply in the industrialized countries over the last four decades. We are used to mountain biking, barefoot surfing, and other fun activities for people over the age of 60 or 70. Society is adapting its offers to the aging society. But not all people are able to participate in it health-wise. Getting older often means getting sicker.

While in the past medicine was able to make a living by taking away the horrors of accidents and acute illnesses and thus making a significant contribution to longer life expectancy, it is more difficult to treat chronic illnesses which have a long history of development and therefore cannot be cured by a single targeted intervention.

The increase in age requires a new strategy due to the associated increase in chronic and functional complaints. All chronic illnesses have their own special development history, some of whose origins date back several decades and are individually influenced by lifestyle factors. In order to influence the development of such diseases or simply the aging processes associated with them, a systemic approach is required which demands the cooperation of the affected person. At this point, illness, health, longevity, or quality of life become a controllable luxury good.

The Dream of Eternal Youth and Immortality

The desire to live long and have the vitality of a 40-year-old is ancient, it is the dream of eternal youth. Whether the ancient Greeks, the Middle Ages, or the twenty-first century—poets, artists, philosophers, scientists, or doctors discuss or interpret the pros and cons of such a state. Or are simply looking for new ways to get one step closer to the immortality of the gods.

In addition to the usual and accepted use of drugs and operations in the therapy of already existing diseases, there are also more extreme approaches among thought leaders such as the transhumanism movement, which is willing to take the step to cyborg. In order to maintain health and performance, transhumanists rely in particular on new and future technologies such as nanotechnology, biotechnology, regenerative medicine, brain–computer interfaces, or uploading human consciousness into digital memories.

Even if today we are still far from finding a cyborg as a work colleague, there are already functioning applications for interfaces between computers and humans, such as brain pacemakers or implants in the inner ear.

Paradigms Are Changing

In the knowledge that chronic diseases usually have a long history of development, preventive and health-improving measures are becoming increasingly important.

Nutrition, exercise, stress management, and personal attitudes play decisive roles in the development or avoidance/delay of illness. Health thus becomes a lifestyle product. We can clearly see this in the fact that health services such as yoga retreats, mindfulness seminars, or medical SPA weeks have increased in recent decades (Fig. 5).

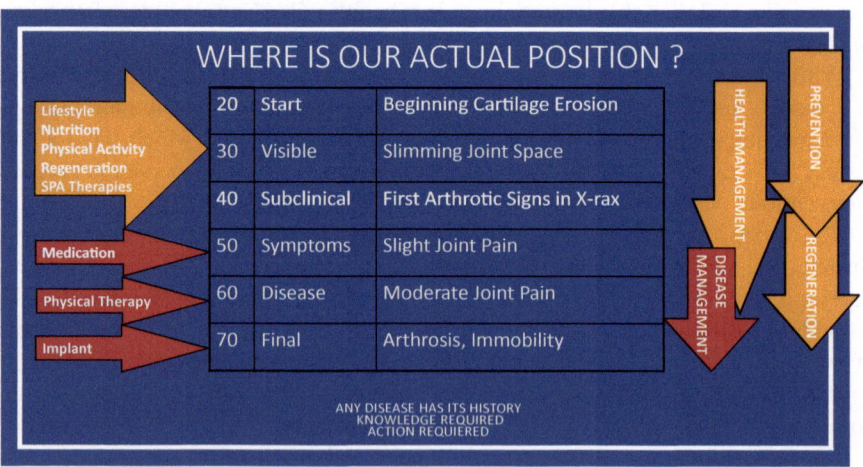

Fig. 5 Continuum health versus disease

A new branch of the bioscience, the so-called systems biology, is aiding the increasingly holistic understanding. It tries to decipher functional processes in organisms very precisely. The understanding that health and health maintenance obey individual rules and that therefore individualized strategies are more successful is gaining more and more acceptance. Health services will therefore in future be increasingly tailored to the body and soul of the individual.

In addition to the systems biology understanding of health, more and more people are opening up to the integration of spiritual health and seeking healing as an extension of the more organic-molecular health.

This development is supported by studies of the Centenarians, who have identified spiritual values such as the meaningfulness of one's own life and a sense of connection with other people or a divine principle as a cornerstone of a long and healthy life.

Blue Zones and the Teachings of 100-Year-Olds

Geographic and demographic areas in which people live particularly long are referred to as "blue zones." These regions include Sardinia, the Okinawa Islands, Loma Linda in California, the Nicoya Peninsula in Costa Rica, and Icaria (Greece).

Residents of these places produce a high rate of centenarians, suffer from a fraction of the diseases that normally kill people in other parts of the developed world, and enjoy healthier life years.

A 2009 study on the island of Ikaria revealed the place with the highest proportion of 90-year-olds on the planet—almost one in three people made it to their 90s. In addition, Ikarians have about 20% lower cancer rates, 50% lower rates of heart disease and almost no dementia.

In a study with almost 3000 people from the blue zones, it was found that the outbreak of the disease occurred decades later in the lives of centenarians than for their younger colleagues. This means that centenarians not only live longer, they are also healthy longer.

Scientific research has shown that there are some basic factors behind long life:

- A predominantly vegetarian diet
- Regular activity with low intensity
- A commitment to the family
- A feeling of faith
- A purpose in life

Interestingly, soft factors such as the meaning of life (7 extra years), the sense of attachment and faith (14 extra years) have resulted in the most significant improvements in lifetime.

The genetic causes connecting 100-year-olds have now been deciphered in many details. Medicine can use these scientific findings to accompany people on their way to a long, fulfilled, happy life.

Here, too, a paradigm shift is helpful. Technical treatments and consultations should offer space for a holistic and spiritual approach.

Deceleration: Mindfulness

Constant stress—be it caused by permanent traffic jams on the way to work, an unhappy relationship, or excessive demands at work—has a direct effect on our body. The changes are measurable and we understand stress can cause a variety of health problems. Chronic stress is an arsonist that triggers inflammations and thus makes the heart, blood vessels, or our brain sick. Stress even often leads people to prefer unhealthy lifestyles such as smoking or drinking alcohol and to act in ways that increase their risk of falling ill.

A positive attitude to life and laughter has been shown to reduce stress hormones, lower inflammatory reactions, and improve cholesterol levels.

Regular meditation, i.e., the practice of inward thinking and deep breathing, has been shown to reduce risk factors for heart diseases such as high blood pressure.

Physical activity such as walking or exercising releases the body's own happiness hormones and not only relieves stress but also protects against heart disease by lowering blood pressure, strengthening the heart muscle, and maintaining a healthy weight.

Regenerative time-outs are increasingly becoming the small luxury moments with which many stress-afflicted people reward themselves. In former times it was rather lipstick and something sweet as a reward. Today it is the hour of yoga, a floating experience, or a mindfulness meditation.

Complex Knowledge Experienced Simply

Every human being is unique and works millions of small miracles every second. Our bodies are made up of about 50,000 billion cells; 3 billion base pairs on our DNA helix mark our individuality. Every second, our brain absorbs over 100,000 pieces of information—but only the tip of the iceberg becomes conscious. Change is constantly taking place: Within a few days, some body cells renew themselves; whole organs can regenerate and heal. Everything works—according to a fascinating construction plan that only nature knows all about. And so the human being holds infinitely many potentials and talents. Much more than he can ever exhaust.

Systems biology and with it personalized medicine are investigating cellular functions, molecular traces in the metabolism or the genetic fingerprint, and thus are creating an ever better blueprint of the human organism. With this knowledge, it is possible to derive a biologic that enables us to understand health and disease processes better and better and even to anticipate future human developments.

Preventive genetic investigations allow us to identify our Achilles heel and develop suitable strategies to protect or improve these weak points. It is difficult to calculate the required knowledge in terabytes or yottabytes. In any case, it overwhelms the individual to verify the arithmetic operations and assumptions. With the increasing flood of knowledge and data, such a great complexity is given that a feeling of powerlessness and paralysis emerges, which quickly nips any necessary impulse for change in the bud. Simple thought processes are energetically more economical for our brain and so we seem to prefer simple facts intuitively. However, the greater the complexity of the interpretation, the more important it becomes to simply present the results.

Simplicity also Becomes a Luxury Good

Imagine that you would be given the opportunity to decipher your genetic blueprint, which is the basis for how you respond to your environment. Imagine you could determine which foods would protect your genes from harmful influences and how to optimize their function even further according to your needs. Imagine that only a single blood test or a single hair would be needed to create a report for you, several hundred pages long, which will allow you to put the right food in your supermarket shopping cart. Future scenario? Not in the least. This is already possible today.

This master plan would accompany you all your life, every step of the way. In every supermarket, at every restaurant visit, your cerebrum would forbid your stomach to choose the tempting offers for a short-term pleasure gain. A permanent conflict between knowledge and need, a permanent stress factor.

But would not you find this life to be an extreme measure of control and renunciation? Where would your personal freedom be at that moment, where would your happiness be? Surely there will be a few people who enjoy having exactly this control cut to them, because it would fit exactly to their psychic basic structure and their neurotransmitter profile.

Luxury, which for us also means excess and exuberance, probably looks different for most of us.

Luxury as a Benefit for Health

When knowledge alone becomes a burden, intelligent steps are needed to transport the implementation into a pleasurable world of experience. Luxury in this example could look like this:

After you have been visited by a friendly medical professional at home, who has gently taken a biological sample from you, you will travel to your chosen holiday destination after a few weeks. For example, you will visit an island in the Pacific Ocean, where you will be expected by a "Personal Health Scout" in addition to a dreamlike backdrop and a tropical climate. In the meantime, your biological sample has been analyzed and a plan tailored to your needs has been drawn up. Over the next few days, your Personal Health Scout will tell you step by step how to make the best provision for your life, which is really important for you in order to achieve your life goals. With all your senses, you will learn what your health-promoting foods do to you. Cardamom, coriander, cumin, turmeric—sniff the spices of the Orient. Mango, papaya—you can pick them yourself in the garden, touch them, smell them, taste them. A specially trained kitchen team conjures up indulgent gourmet menus every day from your elixir of life. Not only do you love your personalized food courses, you are also feeding each of your body cells exactly the nutrients and supplies they need to make your organism scream "yes" to life. Happiness hormones and immune modulators provide you with the best. While you watch the sunset on the veranda, listen to the strong surf, and drink a tea of passion flowers you have collected, you experience this unique, so-special moment where you have understood, are at one with yourself and the universe, grateful and ready for further invitations of life. Once you have understood how easy it is to implement and how much pleasure is found in the new attitude toward life, even a sustainable implementation suddenly becomes possible. It is just fun to indulge in the new experiences. And of course, nothing speaks against refreshing your training with new inspirations at another powerful place on this planet.

Luxury Is a Magical Moment

Luxury lives on emotions, is emotion itself.

When I ask people what really matters to them, I hear from everyone: It is the moments, the memories that cannot be repeated and that have left a lasting impression.

The first car, whether a Renault R4 or a Jaguar Daimler Double Six, it is the special moment, this snapshot of one's own feelings that makes it so precious. A moment of special esteem, of achieving a goal, of being honored with a medal, or simply of a moment of recognition and respect.

It is perhaps this feeling of uniqueness that simply cannot be repeated.

Magical moments are also full of emotions. Here life can take a new direction in just a few seconds. Unexpectedly and without notice. A magical moment is often a turning point, a change of direction. Dramatic on the outside or quiet on the inside, both is possible. It is a moment of being touched. It is a deep experience.

Maybe it is the breath of a buffalo in winter that gets lost in the sunset. Maybe the smell of freshly cut grass, two dogs playing in the park, or an older couple walking down the beach holding hands. It is this one moment that grabs us and triggers something. Our eyes become wet with emotion, a shiver of goosebumps runs through our body or our heart beats faster with joy. We are touched, we feel connected.

Perhaps it is also the view from the top of the mountain, at eye level with the golden eagle, that draws its circles in the sky. The knowledge of having made it. To feel oneself in this state of fulfillment and bliss, to feel one's strength. Knowing that the ascent went over some dangerous passages, the mastered crises of exhaustion behind us.

Life has given us an invitation to experience, to feel. An invitation to forget the program of survival, to leave behind the fear of tomorrow, and to dedicate ourselves to life itself.

Anyone who has experienced such moments will never forget them and they will always be strength and inspiration to accept the next invitation to feel new or familiar again.

Health and Well-Being Are Becoming a New Luxury Trend

Luxury is to use the possibilities that life offers you.

This requires knowledge of these possibilities and the resources to translate this knowledge into action. Resources are not limited to economic resources, but also include personal resources. This is where health comes into play, mentally, physically, emotionally, and energetically.

Health is an essential resource for being able to use one's own potential. Health is much more than the absence of illness.

Health requires harmony with oneself mentally, emotionally, and physically, as well as harmony with one's environment and personal history, dreams, and goals.

And it is here that health becomes a true luxury. What could be a more precious moment than to consciously perceive oneself at this very moment and to feel deep within oneself the connection to other living beings and nature? This moment of pausing and continuing at the same time.

This does not always require a trip to a Pacific island. This would overstretch the resources of every society and our ecological system. Health begins in each of us with a conscious attitude and requires continuous development. Health is therefore rather a journey to oneself. The stations of the journey have to be chosen well and the right travel companions have to be chosen.

Experience Health: Experience Healing

It is time to rethink health and make it tangible. Too much today we are still dealing with illness, with restrictions, losses and try to make people healthier from the wrong end.

Health begins with a positive thought, a lifestyle. It is not a "no" to illness, but a "yes" to life.

Health and healing need an appropriate environment. If we understand that fear and stress make us ill and block healing, then we need not only knowledge but also spaces, places, people, and a respectful communication in order to create it.

Doctors may then become creative healing artists who make health tangible for the protagonists of the individual life story in areas such as enjoyment of life, self-awareness, meaning, growth, connectedness, and spirituality. I am sure this will give all participants of this wonderful journey new perspectives and many magical moments.

Literature

Aeberhard M (2018) Hidden luxury. ITB, Berlin

Beaverstock JV (2012) The privileged world city: private banking, wealth management and the bespoke servicing of the global super-rich. In: Derudder B, Hoyler M, Taylor PJ, Witlox F (eds) International handbook of globalization and world cities. Edward Elgar, Cheltenham, pp 378–389

Beaverstock JV, Faulconbridge JR (2014) Wealth segmentation and the mobilities of the super-rich. In: Birtchnell T, Caletrío J (eds) Elite mobilities. Routledge, London, pp 40–61

Daily GC (1997) Nature's services. Island Press, Washington, DC

Davis SJ, Lewis NS, Shaner M, Aggarwal S, Arent D, Azevedo IL et al (2018) Net-zero emissions energy systems. Science 360(6396):eaas9793

Diamond J (2005) Collapse: how societies choose to fail or succeed. Penguin, New York

Ehlen T, Scherhag K (eds) (2018) Current challenges in the hotel industry. Erich Schmidt, Berlin

Eijgelaar E, Thaper C, Peeters P (2010) Antarctic cruise tourism: the paradoxes of ambassadorship, "last chance tourism" and greenhouse gas emissions. J Sustain Tour 18(3):337–354

FAZ (2018) Germany clearly misses climate targets for 2020. Frankfurter Allgemeine, 9 June 2018. http://www.faz.net/aktuell/wissen/erde-klima/deutschland-verfehlt-selbst-gesetzte-klimaziele-fuer-2020-deutlich-15631595.html. Accessed 26 Nov 2018

Federal Statistical Office (2018) Population pyramid. www.service.destatis.de/bevoelkerungspyramide/#!y=2032. Accessed 10 Nov 2018

Fitzgerald FS (1925) The great gatsby. Simon & Schuster, New York

Forbes (2018) The world's billionaires. Zugegriffen. https://www.forbes.com/billionaires/list/#ver sion:static. Accessed 26 Nov 2018

Gössling S, Metzler D (2017) Germany's climate policy: facing an automobile dilemma. Energy Policy 105:418–428

Gössling S, Ceron JP, Dubois G, Hall CM (2009) Hypermobile travellers. Earthscan, London

Gössling S, Lohmann M, Grimm B, Scott D (2017) Leisure travel distribution patterns of Germans: insights for climate policy. Case Studies on Transport Policy 5(4):596–603

Habekuß F (2017a) Europe's demographic future. Federal Agency for Civic Education. http://www.bpb.de/politik/innenpolitik/demografischer-wandel/196906/europas-demografische-zukunft. Accessed 11 Oct 2018

Habekuß F (2017b) Germany's population in comparison. Federal Agency for Civic Education. http://www.bpb.de/politik/innenpolitik/demografischer-wandel/196909/deutschland-im-vergleich. Accessed 11 Oct 2018

Hacker W (1986) Industrial psychology. Hans Huber, Bern

Heckel M (2017) Impact of demographic change on the economy and labour. Federal Agency for Civic Education. http://www.bpb.de/politik/innenpolitik/demografischer-wandel/195360/wirtschaft-und-arbeit. Accessed 11 Oct 2018

Hennig C (1997) Travellust. Tourists, tourism and holiday culture. Suhrkamp, Frankfurt am Main

Herhoffer P-A, Meurer J (2018) Consumer generations 2018. KEYLENS Management Consultants, Munich

IPCC (2014) Climate change 2014: mitigation of climate change. In: Edenhofer O, Pichs-Madruga-R, Sokona Y, Farahani E, Kadner S, Seyboth K, Adler A, Baum I, Brunner S, Eickemeier P, Kriemann B, Savolainen J, Schlömer S, von Stechow C, Zwickel T, Minx JC (eds) Contribution of working group III to the fifth assessment report of the intergovernmental panel on climate change. Cambridge University Press, Cambridge

Jackson JB, Kirby MX, Berger WH, Bjorndal KA, Botsford LW, Bourque BJ et al (2001) Historical overfishing and the recent collapse of coastal ecosystems. Science 293(5530):629–637

Kaltenmark M (1969) Lao Tsu and Taosim. Stanford University Press, Stanford, CA

Latouche S (1993) In the wake of the affluent society: an exploration of post-development. Zed Books, London

Lenski G (1977) Power and privilege. Suhrkamp, Frankfurt am Main

Lenton TM (2011) Early warning of climate tipping points. Nat Clim Chang 1(4):201

Littek W, Rammert W, Wachtler G (1982) Introduction to the sociology of work and industry. Campus, Frankfurt am Main

Löfgren O (2002) On holiday: a history of vacationing. University of California Press, Berkeley

Maslow AH (2012) A theory of human motivation. Start Publishing, New York

Meckel M (2010) Letter to my life. Rowohlt, Hamburg

Meurer J, Herhoffer P-A, Field Turquoise K (2018) Consumer generations 2018—tourism sector report. KEYLENS Management Consultants, Munich

Peeters P, Gossling S, Becken S (2006) Innovation towards tourism sustainability: climate change and aviation. Int J Innov Sustain Dev 1(3):184–200

Pinstrup-Andersen P (2009) Food security: definition and measurement. Food Sec 1(1):5–7

Pons (2018) Luxury. https://de.pons.com/übersetzung/latein-deutsch/luxus Accessed 18 July 2018

Pyke GH (1981) Optimal travel speeds of animals. Am Nat 118:475–487

Rockström J, Steffen W, Noone K, Persson Å, Chapin FS III, Lambin EF et al (2009) A safe operating space for humanity. Nature 461(7263):472

Schirrmacher F (2004) The methuselah plot. Blessing, Munich

Schmücker D, Grimm B, Wagner P (2015) Travel analysis 2015: summary of findings—demand structure and trends in the German holiday market. Research Association Holiday and Travel e.V. (FUR), Kiel

Skole D, Tucker C (1993) Tropical deforestation and habitat fragmentation in the Amazon: satellite data from 1978 to 1988. Science 260(5116):1905–1910

Steffen W, Grinevald J, Crutzen P, McNeill J (2011) The Anthropocene: conceptual and historical perspectives. Philos Trans A Math Phys Eng Sci 369(1938):842–867

Steitz E (1993) The evolution of man. E. Schweizerbart'sche publishing bookstore, Stuttgart

UBA (2018) Greenhouse gas emissions. https://www.umweltbundesamt.de/themen/klima-energie/treibhausgas-emissionen. Accessed 26 Nov 2018

Ulich E (1991) Industrial psychology. C. E. Poeschel, Zurich

UN (2015) Sustainable development goals. Zugegriffen. https://www.un.org/sustainabledevelopment/sustainable-development-goals/. Accessed 26 Nov 2018

Wackernagel M, Rees W (1998) Our ecological footprint: reducing human impact on the earth. New Society Publishers, Gabriola Island

Wolf ER (1982) Europe and the people without history. University of California Press, Berkeley

Marc Aeberhard founded Luxury Hotel & Spa Management Ltd. in Zurich in 2004 and has been acting as Managing Director ever since. The company has access to a global network of travel trade partners, lifestyle and travel media and (U)HNWI and works closely with public relations and sales and marketing agencies in Frankfurt, Munich, Paris, Dubai, Milan, New York, Hong Kong, and London. Furthermore, the native Swiss is a member of the consulting networks of the Gerson Lehmann Group, USA and Hotellerie Suisse, Bern. He also takes on an active role in the "Luxury" task force of the management of ITB Berlin, Germany. As author and co-author of various specialist publications, his name can be found regularly. He also holds guest lectures in Berlin, Istanbul, Lausanne, Lucerne, Munich, Singapore, Stuttgart, Thun, Vienna, Worms, Zurich, etc. The graduate hotelier graduated with distinction from the Ecole Hôtelière de Lausanne (EHL) and previously completed his studies in business administration as lic.rer.pol. (MBA) at the University of Bern. The luxury hotelier has more than 20 years of experience in the fields of hotel opening, management, and renovation/refurbishment in Abu Dhabi, Germany, France, Maldives, Morocco, Seychelles, Sri

Lanka, Switzerland, Thailand, Ukraine, and Cyprus of small hotels in high and top end. Many of the hotels have been awarded international prizes. All projects are based on the definition of New Luxury and work according to the principles of Triple Bottomline.

Stefan Gössling is a geographer and biologist (University of Münster) and holds a doctorate in human ecology from the University of Lund (Sweden). He holds chairs at Lund University and Linné University (Kalmar, Sweden). He is also a research coordinator at the Western Norway Research Institute (Vestlandsforsking).

Mario Krause has been seducing people to better health for many years. As an expert in holistic prevention, stress medicine, and personalized medicine, he regularly speaks at international conferences. He is the founder of THE KRAUSE, a concept specialized in luxury medicine, and director of the German Center for Individualized Prevention and Performance Improvement.

Jörg Meurer is a strategy consultant and specialist for brand management, marketing, and communication. As a consultant, he focused early on growth strategies, strategic change, and transformation processes as well as value-oriented marketing/sales programs. Jörg Meurer has been a Managing Partner and shareholder of KEYLENS Management Consultants since 2007. Prior to that, he worked for Deutsche Lufthansa AG and Roland Berger Strategy Consultants. Among other things, he is a founding member of the Luxury.Brand.Lifestyle Forum of the German Brands Association.

Analysis, Design, and Future Perspectives of Luxury Features

Hannes Gurzki, Philipp Schmid, Daniel Schönbächler, David M. Woisetschläger, and Verena Zaugg-Faszl

1 "Me to Me" Perspective

Daniel Schönbächler

There are different models to arrange the needs of people. The pyramid of needs of the US-American psychologist Abraham Maslow (1908–1970), which has become famous, places biological needs at the basis: Breathing (clean air), warmth (clothing), drinking, eating, sleeping. The needs for security follow: accommodation, health, order. Building on this, there are social needs: the desire for friendship, partnership, love, care, communication. Then the needs for recognition and appreciation are next. The top of the pyramid forms the need for self-realization. Maslow does not see the pyramid strictly as a ranking, because the priority of needs is individual and variable depending on the situation (cf. o. V. 2018a).

In economics, needs are differentiated according to the urgency of their fulfillment: The most urgent needs for every human being are the existential needs that are

H. Gurzki (✉)
European School of Management and Technology, Berlin, Germany

P. Schmid (✉)
avintas:schmid, Leuk-Stadt, Switzerland
e-mail: info@avintas.ch

D. Schönbächler
Kloster Disentis, Disentis, Switzerland

D. M. Woisetschläger
TU Braunschweig, Braunschweig, Germany
e-mail: d.woisetschlaeger@tu-braunschweig.de

V. Zaugg-Faszl (✉)
St. Niklausen, Switzerland
e-mail: zauggverena@b.uewin.ch

© Springer Nature Switzerland AG 2020
R. Conrady et al. (eds.), *Luxury Tourism*, Tourism, Hospitality & Event Management, https://doi.org/10.1007/978-3-030-59893-8_8

also necessary when in need: sufficient food and water, air, clothing, living space and accommodation, employment and medical care, security and partnership—and, depending on the cultural environment, also the need for pleasure satisfaction and fun. Cultural needs, like the desire for esthetics, creative expression, and education are lifted from these basic needs.

The so-called luxury needs are distinguished from the aforementioned basic needs: the desire for luxurious services (jewelry, cars, travel), even if in other places they promote misery, suffering, and perhaps environmental pollution. There is no limit to desire. Luxury needs can superimpose and abuse basic needs: I want not only an apartment, but a luxury apartment, need not only food, but exquisite banquets. This inevitably leads to competition and competitive behavior.

Individualization

In view of the question posed by this book, it is necessary to determine the luxury features more precisely, namely in the field of tourism. Luxury originally means waste, abundance, that is, what is not essential to life. One could certainly do without it—and many contemporaries are forced to do without it. Luxury is expensive, and not everyone can afford it. Renunciation is only possible where there is a hierarchy of values. It is easier to do without something if I "trade in" another, for me higher value in its place. For renunciation for renunciation's sake is not a desirable good. Luxury is only attractive if it promises added value. This added value is usually based on the need for recognition and appreciation or in the area of satisfaction, i.e., in the emotional area. Luxury is elitist:

Whoever has it stands out from those who cannot afford it. We can observe a continuing democratization of luxury in our time. In the past, luxury items were only available to the upper class, but today people up to the lower middle class are earning enough financial means to be able to afford something special from time to time (cf. Kappler 2012, p. 4). As a result, the "ranking list" of luxury goods is shifting.

The gradation of "being able to afford it" and the gradation of willingness to renounce are individually different. Every individual lives and acts essentially on the basis of his genetically determined "start program." Every human being is guided by the three brain dimensions: the brain stem, the limbic system, and the cerebrum. So, he has instincts, feelings, and thinking skills. But in every human being one of these systems has from the beginning of life a slight, but serious preference. This causes us to distinguish between head people, emotional people, and people that rely on their gut feeling. Apart from having a lot in common, there are typical differences: people that rely on their gut have the primary need for space and autonomy; emotional people must primarily be loved and recognized; and head people must recognize and understand the world. This is coupled with other characteristic distinguishing features: Some people are directed toward the outside world and are therefore dependent on it, others focus on their inner being and primarily seek to carry through their own concepts. The relationship to time is also characteristic: Forced to act, some go resolutely forward into the future, others retreat and reflect on the experiences of the past, others remain in the present moment and often stand blocked like the mouse in front of the cat (cf. the critical overview in Wikipedia 2018a).[1]

This rough sketch suggests that our character is like a filter for our perceptions, feelings, thinking, and acting. The character determines our predispositions, from which we can deviate if necessary, but which in the existential always form the basis of our life. How education, milieu, and life experiences contribute to our individualization are merely updated adaptations, behind which our "start program" still retains its validity.

If we want to gain clarity about our needs, desires, and last but not least our behavioral habits, we have to become aware of our character. Self-knowledge is a spiritual task: I tend to say: "Spirituality means to redeem the word that God spoke when He called me into being."

Self-Determination

So, if a person is "implanted" with his character at the time of conception, he will soon begin—already prenatally—to establish himself in his environment, which at first happens completely unconsciously. Life requires adaptation and demarcation. Man learns most through imitation. It is fascinating to watch a baby fit in, adopt behaviors, and then gradually take initiatives of its own.

The standard is first of all the so-called "superego," the reference person who conveys values and determines what is permitted and desirable and what is forbidden and not tolerated. With the gradual socialization, further instances and role models are added. Normally the developing person has the feeling that he himself determines his life. To a certain extent, this is the case: Every person makes his own world, nobody has the right to blame someone else for his life. Frustration tolerance is an indispensable element of personality development—it used to be called humility.

Thus, relatively uncritically one takes over first the standards of the origin family. In further development, it comes then inevitably to the confrontation with other concepts. The original learning may become questionable, new possibilities are perceived, and more or less tried out. Most people today also come into contact with other cultures, we have become global. Traveling to other continents is no longer considered a luxury. In all this, however, the desire for self-determination remains; it is an essential part of the human being.

---

[1] As an introduction to the Enneagram "Rohr, Richard/Ebert, Andreas: The Enneagram. The nine faces of the soul. Munich: Claudius 48th edition 2017" is still suitable. Also recommended: Gallen, Marie-Anne/Neidhardt, Hans: Das Enneagramm unserer Beziehungen. Complications, interactions, developments. (rororo 19616). Reinbek b. Hamburg: Rowohlt 14th edition 2014—In my seminars on personality development the personal profile is verified by means of psychokinesiology.

So, we believe we have the regiment over our needs. We believe we know what luxury means to us and how much we are prepared to give up. But it is not so easy to get clarity about your own motivation, since we constantly measure our behavior against what others do.

Transcendence and Self-Development

Maslow's pyramid of needs rises from the broad base of biological needs, through increasingly specific emotional and cognitive needs, to the peak of transcendence needs. "What lies outside or beyond a realm of possible experience, especially the realm of normal sensory perception, and is not dependent on it, is considered transcendent". The notion of "transcending" refers above all to a transcendence of the finite world of experience on its divine basis (Wikipedia 2018b). The "divine reason" is the point of reference that gives meaning to earthly existence, even when all other references to meaning have been lost and everything seems to have become meaningless.

The Austrian neurologist and psychiatrist Viktor Frankl (1905–1997) was in such a situation when he was taken by the National Socialists to the concentration camps of Theresienstadt, Auschwitz and Türkheim near Dachau. He said to himself that if I stop believing in a higher meaning in this inhumane, devilish situation, I will not get out of it alive. He survived and in 1946 wrote the field report "Ein Psychologe erlebt das Konzentrationslager" (A psychologist experiences the concentration camp), in the new edition 2009 under the title "... I'd still say yes to life. A psychologist experiences the concentration camp" (Frankl 2015). The Greek word for meaning is "Logos," the form of therapy in which the human being is led to basic trust in the meaning of existence has been given the name "Logotherapy": "Logotherapy is based on the assumption that every human being is originally a sense-oriented being and possesses his or her own sense of value". If he does not live his values and if his will to make sense is frustrated in the long run, then the human being experiences an inner emptiness, into which then different negative experiences proliferate: neurotic disturbances, depressions, boredom, unwillingness to work, fears and phobias, disorientation, nihilistic thoughts or also excessive urge for recognition, forced striving for pleasure, paralyzing inertia, addictions, etc. The state of inner emptiness, which a person can hardly endure, is countered by the striving for meaning and fullness of meaning (o. V. 2018b). So, there are exposed phases in life in which I notice: Now it depends on whether I realize the meaning of my life or miss it. In such situations, all other needs are put into perspective, and thoughts of luxury goods are completely distant. I know how to set my priorities right.

Peace and Inner Reflection

But the reference to transcendence does not simply fall into our laps when we find ourselves in an extreme situation. Rather, it must be practiced until it becomes a matter of course. Viktor Frankl did not learn to believe in a last meaning of life in the concentration camp. Certainly, faith is not feasible, some have a greater affinity for it than others. Faith is a gift. But there is also a grateful handling of the gift—and a decisive practice.

Spiritual masters such as Anthony de Mello (1931–1987) provoke us with the statement: "Spirituality means to wake up. Most people are sleeping without knowing it. They were born asleep, they live asleep, they marry in sleep, educate their children in sleep and die in sleep without ever having woken up. They never understand the charm and beauty of what we call 'human life' (. . .). Most people tell you they want to get out of kindergarten, but don't believe them. You really don't want to believe them! All they want is for their broken toys to be repaired: 'I want my wife back. I want my money back, my prestige, my success! That's all they want: their toys back" (de Mello 1991, pp. 9–12). And the experienced spiritual individuals tell us: "The first step to wake up is to be honest enough and admit that you don't want it." For it is well known that awakening is unpleasant and annoying. "Or let's be more specific: we don't want to be unconditionally happy. I am ready to be happy, provided I have this and that and who knows what else" (de Mello 1991, p. 12).

In order to be able to devote ourselves to the spiritual "business," we need competent support. The old Egyptian monks were convinced of this, they went to the experienced "Altvater" (old father) or to the "Amma" in the desert to ask for "a word of advice." The social developments of our time are pushing us back on this path. Silence and tranquility are the main requirements. In a time in which noise and hustle and bustle tyrannize us everywhere and we are always and everywhere "online" and available, rest and free time become the most urgent luxury goods. Tourism and the leisure industry are now also aware of this luxury niche. Having free time, peace, inner reflection are what we would like to have today, but do not have. So who can afford this luxury?

Experience Instead of Adventure

Where should one find peace and inner reflection, if not in the monasteries? Benedictine monasteries have been open to hospitality from the very beginning. Benedict of Nursia (around 480–547) dedicates in his Rule, which became the basis of Western monasticism, a separate detailed chapter to the "reception of guests" (Benedictus de Nursia 1990, Regula Benedicti, Chap. 53). The motivation for hospitality is a spiritual one: "All strangers who come shall be received as Christ; for he will say, 'I was a stranger, and you received me'" (Regula Benedicti,

Chap. 53,1). Benedict thus describes hospitality from the point of view of the monastic community.

However, he also knows the opposite view, that people come to the monastery because they expect something. As a rule, the term "boys and young people" was often used for those entrusted to the monks for education (Regula Benedicti, Chap. 59). The admission of candidates (novices) was regulated (Regula Benedicti, Chap. 58), but also the admission of priests (Regula Benedicti, Chap. 60) or of a foreign monk who comes from afar and wants to stay as a guest in the monastery. There he said: "If he is satisfied with the way of life he finds there and does not bring confusion into the monastery by making unreasonable demands, but is content without further ado with what he finds, one takes him in, and he stays as long as he wants" (Regula Benedicti, Chap. 61).

In today's formerly large abbeys of our cultural circle, there is usually a lack of offspring. The people of our time find it increasingly difficult to commit themselves bindingly and definitively to the entry into a religious community. Nevertheless, the attraction of the places of "tranquility and inner contemplation" remains evident. New forms of affiliation evolved. Many monasteries have a circle of friends (Oblates) who maintain regular contact with the monastic community. Sociologically, this model is called "center and periphery" (although one can of course ask oneself: what does the periphery do if the center becomes weaker and weaker?).

In 1962, Abbot Emmanuel Heufelder (1898–1982) in the Bavarian Abbey of Niederaltaich initiated the idea to invite people to a "temporary monastery" for a few weeks (see 2011–2017). Unfortunately, this model has become more and more defused into a tourist offer: "We invite you to a monastery—a time-out that leads you into the world of the monastery, a world of inner contemplation and peace" (o. V. 2018c). You are invited to "participate in our life and daily routine—according to your possibilities and needs" (o. V. 2018c). In line with the market, the model is being transformed on relevant websites into a wellness spirituality: "The vacation type 'temporary monastery' is becoming more and more popular (...). In the seclusion of a monastery you can relax and meditate. Here one finds the peace, which one looks for in vain in the everyday life. 'Temporary monastery' is a wonderful alternative to vacations where you rush from one place of interest to the next" (o. V. 2018c). The stress timeout agency "offers you a wide range of trips during which you can personally get to know the advantages of the temporary monastery model. With us you can book several-day monastery breaks throughout the year. Let us take you to another world—a world in which you, and not the needs of others, are at the center of attention. After your monastery trip you will notice that your state of mind has noticeably improved. A stay in the monastery is always worthwhile" (2018c). In addition, an introduction to yoga may be recommended or beautiful walks and cultural sights can be promoted: "If you walk along the corridors of the monastery complex of your choice, you will feel that it was a good idea to book a temporary monastery trip. There are no demanding bosses or work colleagues here to bother you. There are no noisy photocopiers in the corridors—instead you can see old paintings and other works of art that bear witness to the monastery's long tradition. The ubiquitous peace and quiet is good for you and lays like a protective coat on your strained nerve costume" (o. V. 2018c).

Such canvassing moves far beyond what would be the most radical luxury of "rest and reflection", the renunciation of all the usual luxury tourism in order to face the radical challenge of answering the questions: Who am I, where do I come from, where am I going? Anyone seeking undisturbed well-being or experiences in the monastery is in the wrong place. But perhaps one experiences "abundance and emptiness" (Steindl-Rast 2011). To stand before God with empty hands and let Him fill them. "Spirituality" is not wellness, but the hard work of finding out who you really are, of stopping lying to yourself. I recommended a suitable abbey to a friend who decided to stay in a monastery after leaving a demanding job, and I knew that there was also spiritual guidance there. When he arrived, the father led him to his room. When asked about his Internet connection, he received the answer: "Do you want to stay at a monastery—or do you want to leave right away?" He confessed to me that this question instantly got him thinking. The stay in the monastery did not become an adventure for him, but rather an experience he no longer wanted to miss. The abbey has become a kind of second home for him. "Peace and inner reflection" is indeed the most radical luxury we can afford!

2   Perspective "Me and the Other"

Hannes Gurzki and David M. Woisetschläger

Introduction

Throughout history, luxury has always been a social phenomenon (Gurzki and Woisetschläger 2017). The ability of luxury goods to signal social status makes luxury stand apart from other goods (Albrecht et al. 2013a). Status is fundamental to all social structures and brings individuals social benefits such as power, esteem, and recognition (Bourdieu 1984). According to Bourdieu (1984), individuals use their economic (financial resources), social (such as relationships, affiliations), and cultural resources (such as skills, taste, knowledge, cultural objects, or degrees) to build symbolic capital through which they influence the practices that define taste and thereby create the codes that create status (Gurzki 2018).[2]

What has changed over the course of history is how the status is constructed. While in ancient times, luxury goods reflected the status of the powerful, they have

---

[2]This chapter is based on the thesis of Gurzki, H. (2019). The Creation of the Extraordinary—Principles of luxury. Dissertation. Wiesbaden: Springer.

become a way for individuals to create and maintain their social status (Kapferer and Bastien 2009). In ancient times, status was externally given through birth, and ascendance and social structures were relatively structure and fixed. With the enlightenment, the given order was challenged, and individuals took the agency to create and manifest their social positions. With the increasing emergence of a consumer society, the way how social status was visible signaled also changed from the inheritance of heirlooms to goods traded in the marketplace. Consumers have become agents to manage their ascribed status and as many authors argue have thereby contributed to the democratization of luxury. Today the status system has become elaborated with a multiplicity of different codes. One such example of the refinement of the code system is the emergence of so-called inconspicuous consumption, which favors subtle signals such as design cues that can only be interpreted by people in the know (connoisseurs) as opposed to conspicuous signals such as a high price or expensive materials which have been the focus of study before. With an increasing growth of experiences, their meaning and performances become new ways in status codes are enacted through consumption rituals and tastes. Today social media provides the space to share the stories, build symbolic capital, and construct status (Berry 1994; Kapferer and Bastien 2009).

Luxury goods are part of a communication system in which they visibly symbolize wealth and power, social relationships, and group membership (Anderson et al. 2015). Two main motives drive this behavior: Affiliation and the want to belong to an aspirational target group and distinction and the want to dissociate oneself from non-aspirational groups or highlighting one's uniqueness within the group (Albrecht et al. 2013a; Gurzki 2018). The status luxury goods bring to their owners is thus closely related to exclusivity, meaning having the power to exclude someone or being unwilling to admit outsiders to the club.[3] This chapter will provide models to understand the social dynamics of luxury with a focus on the processes that construct status to derive design characteristics of luxury today and in the future.

Luxury as Status Signal

All social interactions express and reproduce status relations (Goffman 1951; Mason 1985). Individuals are social actors who permanently engage in the interpretation and creation of positional signs (Goffman 1951). They (consciously and unconsciously) compare their abilities with their social environment, resulting in a continuous process of upward and downward comparisons (Arbore and Estes 2013; Festinger 1954). Particularly for groups that are perceived as desirable, individuals have a high pressure to conform to their rules and group norms (Festinger 1954). According to their adherence to the norms, they either gain or lose status, which can

---

[3]https://www.merriam-webster.com/dictionary/exclusive; https://www.etymonline.com/word/exclusive

even lead to exclusion from the group in the case of non-conformance (Festinger 1954). The importance of "the other" and the social environment has received large prominence in academic studies. One such example is the study of Lasaleta et al. (2014) who showed that feelings of social exclusion increase the desire for money to restore the social connectedness, whereas feelings of social connectedness weaken this desire.

Theories of conspicuous consumption have long ago highlighted the social nature and importance of consumer goods in the creation of status (Veblen 1899). Most prominently, Veblen (1899) proposed that members of the "leisure class" signal their wealth by consuming expensive and conspicuous luxuries to affirm their class membership. But not only the price of the good was thought to be a status signal. Also, the time invested to understand how to correctly consume complex products with their rituals and practices signals to other members of the leisure class their affiliation (Veblen 1899). This led Simmel (1957) to propose that status symbols are created by the highest class then continually trickle-down to the other classes as an explanation of how signals spread. Simmel (1957, p. 541) proposed that "Fashion is a form of imitation and so of social equalization but, paradoxically, in changing incessantly, it differentiates one time from another and one social stratum from another." Signals emerge in a social process and change over time as symbols are reinterpreted or new symbols are invented (Goffman 1951).

Vigneron and Johnson (2004, p. 486) claimed that for luxury goods "[...] the simple use or display of a particular branded product brings esteem to the owner, apart from any functional utility." Luxury goods are status symbols as they visibly mark the social categories (Goffman 1951). "By definition, necessities are possessed by virtually everyone while luxuries have a degree of exclusivity" as Bearden and Etzel (1982, p. 184) states. Kapferer (2009) even proposes that through its acceptance by the leading classes, anything that can become a social signifier and luxury. Status and its benefits such as preferential treatment is endowed on the individual by its social environment (Goffman 1951). Today, many different flexible social arrangements or communities exist that provide sources for identification through their membership that have challenged previous conceptualizations such as social class (Cova and Cova 2002; Holt 1998). Even communities dedicated to consumption or particular brands have emerged which share a symbolic system of values, codes, rituals, and beliefs (Schouten and McAlexander 1995). As Holt (1998, p. 4) points out: "Consumption is a particular status game" in which status relations are expressed through the choice of consumer goods. Consumption has become an important way of expressing and constructing identity and status (Berger and Heath 2007). Through their expression of tastes, consumers can negotiate the dominant codes for status construction (Üstüner and Thompson 2012). These tastes also dictate which consumption objects are valued and prescribe how they should be consumed (Holt 1998). By going through several learning stages, from an initial experimentation, to conformity, to mastery and internalization of the community's rules and principles, individuals can become members of the group (Schouten and McAlexander 1995).

Often signaling theory has been used in the literature to understand and explain the social processes underlying the construction of status through luxury consumption (Gurzki 2018). Signaling theory proposes that an individual has an unobservable quality that can be signaled (such as economic, social, or cultural capital), that signaling the quality would bring benefits to the sender (such recognition and power) as well as to the receiver who correctly identifies the signal (credibility of signal) (Bird and Smith 2005). To be a valuable signal, the signal needs to be costly (requiring personal investment such as money or time) and hard to achieve for someone who does not possess the unobservable quality (Bird and Smith 2005). Therefore, codes are generated based on social practices that give meaning to the signal. The social group familiar to these codes can code and decode the signal clearly thus decipher the signal correctly (Bird and Smith 2005). McCracken (1990) provided an example of how these codes work and are shaped by their social environment: Before the sixteenth century, in the Elizabethan, patina that is the visible age of objects was a marker of status as this aging signaled that the object and the status it embeds has been passed on over generations. However, the emergence of trade, the liberation, and the increasing introduction of marketable products came to challenge this code. The risk of status forgery in an increasingly anonymous society from someone just purchasing antiquities was seen as too high and hence patina lost its signaling credibility. It thus gradually became replaced by novelty, as only the rich people could afford to buy all novelties on the market (McCracken 1990). A phenomenon similar to the "status forgery" that drivers the market for counterfeits today (Nia and Zaichkowsky 2000).

A related view comes from the field of evolutionary psychology that also with a strong interest analyzes the social function of luxury goods. This view proposes that luxury goods generate an advantage for survival as they can signal reproductive value which is of high importance in a mating context (Nelissen and Meijers 2011). Zahavi (1975) provided the example of a peacock to illustrate the costliness of the signal: The colorful tail of the peacock has little function other than to advertise its qualities as a relationship partner. It makes the peacock easily identifiable by predators and thus carries a high cost and risk. The fact that the peacock still advertises his colorful tail and is still alive must mean that the tail signals strength and reproductive value (Zahavi 1975). Bird and Smith (2005) extended this argument by proposing that different practices such as artistic elaboration, religious rituals, or body modifications are quite similar as they also can signal resourcefulness, personal qualities, or commitment to a community. And research has found ample evidence for the link between mating and luxury consumption. For example, in a series of experiments, Griskevicius et al. (2007) found that men with a mating goal in mind consume more conspicuously compared to men without a mating goal. Contrarily for women, the mating goal had no effect on conspicuous consumption but on benevolence. Men only displayed benevolent behavior if it could be publicly displayed. Nelissen and Meijers (2011) demonstrated the possession of luxury products can bring social benefits, such as increased compliance with requests or a higher preference for job candidates as they increase the individual's perceived status.

Signaling Theory in Practice

The social environment and particularly reference groups play a strong role in luxury consumption. Particularly for public and visible luxuries, their impact on individual consumers is high (Bearden and Etzel 1982). The study of Han et al. (2010) illustrated very well how the association and dissociation with certain reference groups work using luxury goods as status signals. They propose that consumers can be segmented into four groups depending on their need for status and wealth maturity (Fig. 1). Patricians, consumers with a high wealth maturity and low need for status, only want to affiliate themselves with other patricians. Thus, they use mainly inconspicuous signals that only their peers can decipher. Parvenus, consumers with have a high wealth maturity and a high need for status, want to associate themselves with higher-wealth groups, but also clearly dissociate themselves from consumers with lower wealth levels. Thus, they choose more conspicuous signals

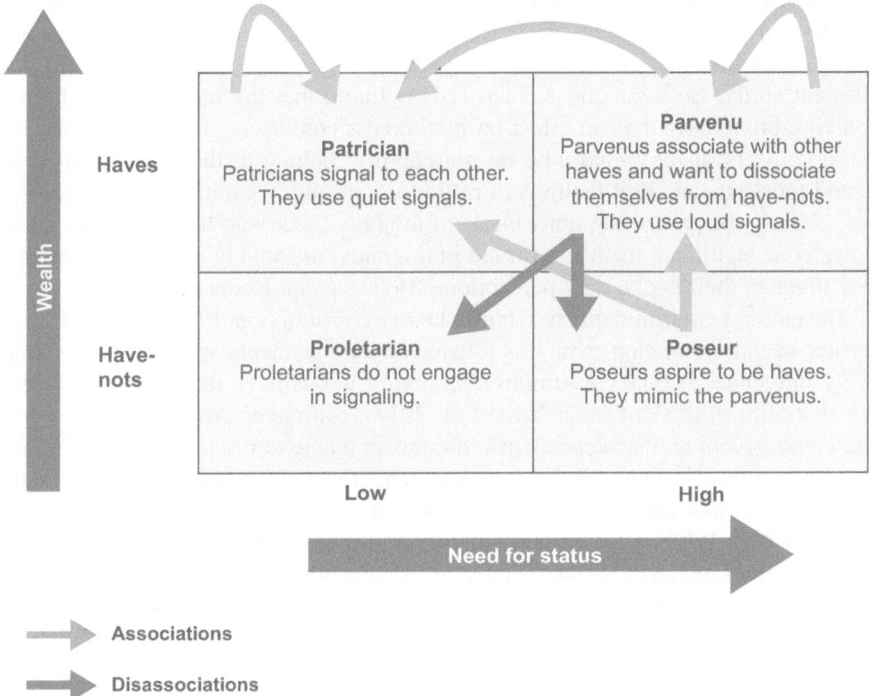

Fig. 1 Consumer taxonomy based on the need for status and wealth maturity. Source: Han et al. (2010)

that are clearly visible and more broadly recognized, such as well-known and conspicuous luxury products. Poseurs, consumers with a low wealth level and high need for status, strive to belong to the higher wealth level groups. In particular, they want to emulate and mimic the parvenus as they use consumption codes that are more broadly recognized and thus more clearly signal the status they aspire. However, in lack of financial resources, they often buy counterfeit products, that if identified as status forgery by the aspirational community can lead to their illegitimacy and exclusion. Lastly, proletarians are low in both wealth and need for status and thus do not actively take part in luxury consumption, but still serve as dissociative reference groups for the other segments (see Han et al. 2010).

To understand the social codes of luxury, the poseur group is particularly interesting to study. The study of Han et al. (2010) looked at counterfeit bags to identify which bags are more heavily counterfeited and found that particularly the less expensive and more affordable and more conspicuous bags are used, as these are the ones most popular with the parvenu segment, the poseurs want to identify with. Indeed, for most luxury brand handbag designs the conspicuousness decreases with a higher price for which patricians are willing to pay a price premium for the subtler design. Interestingly, Chanel is one of the exceptions where the higher-priced products are more conspicuously branded than the lower priced ones. Interestingly, in their study, while patricians correctly identified the price, non-patricians presumed that the quiet Chanel bag would be the more expensive one. This highlights the imitative social aspects of luxury consumption as well as the role luxury goods play as loud (conspicuous) and quiet (inconspicuous) signals (see Han et al. 2010). In a different study, Bellezza and Keinan (2014) found that the membership claim of non-core brand users has an effect on the brand's core users. The study found that non-core users of the brand who do not claim to belong to the brand community (brand tourists) can elicit feelings of pride for core users. Contrary, non-core users who claim to be part of the inner circle of the brand, but who the core users do not consider as legitimate members (brand immigrants) negatively impact core users as they threaten their exclusivity perceptions (Bellezza and Keinan 2014).

The cultural environment also shapes luxury consumption. For example, a strong vertical social orientation, which is a focus on achievements and hierarchies, positively influences luxury consumption as luxury goods have the ability to recreate social stratification (Yi-Cheon Yim et al. 2014). Moreover, research has proposed that cultures with an interdependent self-concept that assumes a strong interconnectedness between individuals in a society tend to place a stronger role in status consumption and public luxuries (Kastanakis and Balabanis 2012; Wong and Ahuvia 1998). While a stronger normative influence of the group leads to conformance motives to group norms (bandwagon consumption), the consumer's need for uniqueness more strongly emphasizes differentiation motives (snob consumption) (Kastanakis et al. 2014). Learned codes shape consumption choices. Research has demonstrated how individuals with high cultural capital (HCCs) differ in their tastes, their corresponding consumption choices, and their status codes compared to individuals with low cultural capital (LCCs) in the USA (Holt 1998). For example, individuals with high cultural capital prefer esthetics over practicality and utilitarian value (Holt 1998). Moreover, their interpretation of luxury differs: While HCCs emphasize metaphysical aspects such as creativity, their subjective appreciation, and the refinement of their own tastes, LCCs emphasize extravagant and material aspects of the luxury object (Holt 1998). Also, the way they construct their identity through consumption is different: While HCCs view consumption as a pursuit of individuality through a unique and individual style, LCCs have a more collective orientation that emphasizes the belonging to a local group. Consequently, HCCs value the authenticity of objects and build their connoisseurship to find unique consumption experiences that are removed from the commoditized marketplace. HCCs have a stronger global orientation and their cosmopolitan tastes become a dominant status code (Holt 1998). In their leisure time, HCCs prefer meaningful and self-actualizing experiences such as creation or learning which provide a sense of achievement over the intrinsic enjoyment of the activity itself and its social nature (Holt 1998). Extending this model to less industrialized countries (LICs), Üstüner and Holt (2010) found that local consumers adapt the global HCC practices to their local environment to embrace the imagined lifestyle. While the local HCCs seek to emulate the behavior on a global level, local LCCs focus on the local environment. For example, focusing on the experience and receiving superior treatment in local shops rather than looking for value for money are used by LCCs to associate themselves with the upper class and to dissociate themselves from the lower classes (Üstüner and Holt 2010). Yet, Üstüner and Holt (2010) also observed the difficulties in emulating the unconscious aspects of behaviors and practices. While this invites normative discussions, this goes beyond the scope of this chapter. The point we want to make here is to illustrate the complexities of the mechanisms that govern the construction of status through luxury consumption.

Implications for Luxury Tourism

What does this mean for luxury tourism? Luxury will remain about me. Personalization, intense positive emotions, and pleasure have been and will be core to luxury services (Albrecht et al. 2013b). The individual is continuing to be at the center of the luxury experience, no matter in which luxury sector. For example, luxury fashion is seeing a rise in mix-and-match and emerging trends such as luxury casual and streetwear (such as the rise of brands like Supreme and their collaborations). Often these trends are driven by Millennials, but the Millennial mindset has spread across generations. Collaborations with smaller up-and-coming brands from other luxury sectors such as fashion could be an interesting way for luxury tourism players to enrich their story and proposition.

Luxury will remain about the other. Luxury is a social phenomenon deeply interlinked with the social environment and the construction of status (Gurzki 2018). Exclusivity is core to luxury and to the creation of desire. Even if most people do not engage in a luxury travel experience only for the sake of getting others recognition, the social context still plays a role in making consumers feel special and has received substantial evidence in research (as outlined above). Moreover, qualification fuels desire. If something is readily available and does not require any investment from be (be it time or money) cannot be worth a lot. So goes the heuristic that is deeply ingrained into our decision-making processes. But the status codes are evolving. While the underlying mechanics described above are likely to remain the same, the codes are getting more diverse. With an increasing sophistication of the consumption sign system, codes become more subtle, and recently the popularity of inconspicuous consumption has spurred (for example, Berger and Ward 2010; Eckhardt et al. 2015). While this is not a new phenomenon but has been known as for centuries (for example, Kroen 2004), it provides insights into the complexity of today's code systems. Some luxury brands have consciously decided to only use subtle design cues rather than a visible brand logo, such as Bottega Veneta and their claim "when your own initials are enough."[4] Luxury codes are not only the choice of brand and logo but also the consumption constellations (which products are consumed together) or knowledge of consumption rituals and practices (McCracken 1990; Solomon 1983). These codes provide the "invisible ink" that glues together luxury brand communities (McCracken 1990, p. 34). The influence of digital, particularly through social media the ways for status construction are changing. Particularly younger generations who desire luxury experiences can use social media to construct their status within their peer groups. Understanding their network of influence and codes can help luxury marketers to better tailor their offerings and stories. Younger generations have a strong potential for the luxury travel market and are eager to learn about authentic and meaningful stories and share their learnings.

And luxury will be more about me and the other. Even altruism can be seen as luxury consumption and is reaching new heights. Here lies a great potential for designing experiences for a community of like-minded luxury enthusiasts, be it Millennials of culture lovers. One such idea to play with this trend is to transform hotels into places of everything for everyone (Skift 2018). Such new *experiencescapes* with interactive showrooms, immersive entertainment options, or even coworking spaces could redefine and innovate the new luxury experience. Particularly for global travelers, this home away from home concept and the idea to get to know more like-minded people that share the same ideas and dreams could prove to be a strong proposition to bring the world of today Millennials and beyond to the next level. Luxury will continue to evolve and needs to evolve to reach the next level of refinement that is needed to redefine the extraordinary. Thus, innovation is crucial for luxury brands, particularly in a global context. Fashion styles spread quickly and being the first to identify a trend brings status in itself as well as satisfying the curiosity of luxury travelers.

[4]https://www.bottegaveneta.com/us/unisex/when-your-own-initials-are-enough_grd30416

3 Perspective "I and the Environment": Architecture and Interior Design

Verena Zaugg-Faszl

Space as space and not as place

Introduction

What makes a space a space? A simple question that can be answered very practically and almost pragmatically: a useful, enclosed area. But space is much more.

In architecture in general and for interior design in particular, a space is more than just a three-dimensionally limited usable area. On the one hand, a space can consist of space-limiting elements, but empty spaces, air spaces, and openings can also create space. As an architect or interior designer, you quickly become familiar with the effects of spatial conditions on people. First semester of architectural studies, first spatial exercise: You have an A3-sized area that can only be divided into a spatially exciting structure with 1/50 scale wall panels. There is no universal solution for this task, yet it shows much of the basic understanding of space sequence, tension, and conditions. If you only put straight walls in your model and offer one or the other freestanding wall panel, you will almost certainly have to rebuild your model and rethink your work. Anyone who already considers the principles of the relationship of excitement—boredom, narrowness—width or light—shadow in this task comes very close to the ideal result.

The inspiring model in this exercise is the Barcelona Pavilion, designed by the German architect Mies van der Rohe.

Created as a symbol of German craftsmanship, the pavilion is characterized above all by an open floor plan, in which an enclosed volume is not subdivided into individual, self-contained subspaces, but is perceived as a single, large, and flowing floor plan with reference to the outside space. This spatial principle shows that the preoccupation with space-expanding measures, such as transparency and vastness or the connection with the natural space of the environment, creates interior spaces that are constantly changing due to factors which man cannot influence and which bring excitement to a space. This approach contradicts the approach of the space as an enclosed area, at least when transparent enclosures are regarded as open transitions.

The actual significance and importance of this first architectural exercise can only be recognized much later in one's studies, if not much later in one's work as an architect or interior designer. In this simple exercise, one usually encounters the term "void" for the first time. "Voids" are empty spaces between structures, in the universe, and in architecture. An example of the use of such voids is the Jewish Museum in Berlin, designed by the architect Daniel Libeskind. These voids create areas in the museum that are not accessible to the visitor but are optically perceived and visible. Only through conscious omission, through conscious emptiness one creates excitement, which in turn creates feelings. Feelings for space. And if an architecture or interior design evokes feelings and emotions, then it is successful and sets a benchmark. In order to achieve this, a limited space is not always required, as the example of the Memorial to the Murdered Jews of Europe (Holocaust Memorial) in Berlin shows.

Spatial Sensations

A space that is right for the human sensation has more to offer than just size. The truth of a harmonious space lies in its overall relationship between surfaces, colors, light, and shadows as well as olfactory perception. Spatial luxury can be described absolutely objectively as a large space or very subjectively as a space in which the guest feels comfortable and secure. Large houses, large entrance halls have always been the representation of wealth, power, status, and luxury, but the length of stay in such halls is usually rather short, since size is not always equated with well-being. Abstractly, we seek the vastness of nature, the generosity, and emptiness in which we can feel comfortable and regenerate, but would feel uncomfortable and lost in the same situation knowing that we are in an enclosed space at that moment. The room would be too big for us, too empty, and too wasteful. In a small room that lets in only little daylight and whose interior design is no more than bare walls, we feel constricted and oppressed. Associations evoke our sensations. If we have the same room, the same lighting conditions, but design the room with warm tonality, many will perceive the room as a pleasant, homely retreat and feel safe and comfortable.

This shows that human perception of spaces can be manipulated with design in such a way that we can create a positive or negative experience with one and the same space with the help of subconsciously evoked associations, no matter to what extent.

The spatial effect on humans can be so strongly influenced by the change of space-limiting area that we can even lose our balance or, in extreme cases, our orientation in space. Depending on which surface, floor, walls, ceiling we focus on, we can create absolutely different spatial effects. A bright ceiling lifts the room, a dark one pushes it down. A strongly accentuated sidewall lets us tend to one side, a light floor with dark walls causes discomfort because we lack the visual support. Gimmicks of this kind can give a different expression to one and the same space, and when the factor of dimensional change is added, i.e., when we bend, arch, tilt, or lift a wall or ceiling on one side, the human sense of orientation can be tricked, and on the emotional level, unconscious discomfort or experience that makes the space unforgettable arises again. Again, the Jewish Museum with the Garden of Exile should be mentioned here as an example. The deliberately chosen sloping position of the floor between meter-high steles creates disorientation and discomfort for the visitor but at the same time an unforgettable and impressive experience (Fig. 2).

Fig. 2 Libeskind: JMBerlin, Garden of Exile. Source: Stiftung Jüdisches Museum Berlin (o. J.)

In addition to colors and lighting conditions, the sound in a room is an essential aspect, which is unfortunately too often neglected in planning, but which has a significant influence on the room and the perception of the room. With a spatial limitation, we create an echo in the room, with which we hardly ever have to deal with in nature. The ambience, the views, the light—everything can be so harmonious, but if the sound in the room is not right, the room will have difficulties feeling pleasant. It is therefore important that acoustically relevant side effects such as reverberation, roaring, and dullness are corrected at the planning stage of a room by placing walls slightly at an angle, and taking the surfaces of furniture, floor texture, or walls with textiles of all kinds, cushions, curtains, wallpaper, voiles and similar measures into consideration. Depending on the use of the space, the echo is decisive for the success of the architecture and can be thought through to perfection and also be esthetically designed, as the examples of the KKL in Lucerne (Jean Nouvel) and the Elbcphilharmonie in Hamburg (Herzog & De Meuron) show.

In order to describe a room with words and to be able to express a feeling of well-being or just the opposite, the sensitivity with regard to the approaches mentioned poses a challenge for many users. We can usually assess logical and structural processes very quickly and easily. But what would happen if, for example, we were to turn the usual processes upside down in a hotel room? Imagine you enter the room through the large window front, right next to the bed and wardrobe, the cloakroom is at the end of the room. Probably no guest would want to stay very long in this hotel. Discomfort arises when a person is torn out of a familiar process and pattern. This does not mean, however, that such new concepts cannot also lead to success if this approach is consistently applied and if commitment and handling are chosen very consciously.

Can a guest also distinguish between a good, harmonious room and a room that is not rhythmically perfect? Yes. Architectural perfection, the coherence, and harmony of a room are perceived subconsciously. If this perception of the subconscious is disturbed, the problem usually lies in expressing this discrepancy in words. How often do we hear the sentence: "Something is wrong." You do not feel well. In the interior, it is easier to expose something as uncomfortable or inconsistent. Many people feel this imbalance of a space but cannot assign it. An example of this is a restaurant, in which an architect installed an additional element under the existing ceiling, which made the area below even deeper. This element, which had no relation to the other spatial lines and probably only served as a decorative cladding for a technical installation, now has a disturbing, false, and oppressive effect on both the guest and the host. A space is created in which we do not like to spend time in. The balance is upset. Our gut is rebelling. This element has created a proportional imbalance between the floor and ceiling surfaces of the room, and the natural opposite pole, which strives upwards from the floor, is missing in the room to balance the weight of the element. With a simple element, you can counteract and restore the balance in this space.

In the hotel and gastronomy industry, the guest usually has a forced superficial perception of a room. Floor, walls, ceiling, size, light, and materials can and are very quickly assessed and perceived as a whole. I like the colors, the materials appeal to me. But there is more behind them, and beauty, or in our case ambience, is also in the eye of the beholder. For this to happen, a balanced relationship between all these elements is necessary. The associated spatial experience is created from an interior design point of view. Size ratios, exciting and eventful floor plans, varied rooms. Views from outside, inside, and looking through are elements that provide variety in a room.

Architecture is often inspired by nature. Interior architecture is not—classically speaking—the view from outside, from free space to or into the building, but also the view from the inside of the building to the outside space has to deal strongly with the elements of nature. We create openings, bring nature into the interior, direct our views to them, and find technical possibilities to reproduce, for example, natural light almost completely. It is not for nothing that naturalness is a top priority and natural materials, colors and shapes create harmony. Man has always lived in a symbiotic relationship with nature and has been expressing this for centuries by drawing elements of nature into his living spaces (ornamental plants, wallpapers with plant motifs, photo wallpapers with palm beaches, etc.), by moving his living quarters outside (patios, open bathrooms integrated into the garden, garden seats, etc.), by adapting architecture to the circumstances of nature or integrating architecture into nature (treehouse hotels). The conversion of many seventeenth- and eighteenth-century orangeries into cafés and restaurants shows that such interiors, which were created as oases, still function as such—albeit with different uses—in today's zeitgeist.

The example of the golden section shows that people have been preoccupied with the conditions that nature has dictated since antiquity.

The golden ratio is a natural constant that has shaped our image of beauty as a harmonious sense of composition and is mathematically proven (cf. Beutelspachter and Petri 2013). As unconsciously as many use the golden ratio, as unconsciously we perceive it and perceive a corresponding division as pleasing and perfect. The golden ratio can be found again and again in nature and shows through its ratio of approx. 1:1.6 an astonishingly harmonic inequality for our perception. In design, we can consciously use this as a guideline for division and distribution in order to create harmonic compositions. But even nature sometimes offers us unbelievable forms which seem absolutely identical and for this reason almost unnatural, as Pyrite crystals show us.

Luxury in a Spatial Context

The change of generations also changed our needs and our demands. Man defines himself in a new and different way. This change consequently influences the interior design. How do we still create spaces that are timeless in constant change, and what does this timelessness actually mean, apart from freeing ourselves from trends?

"Space" has not changed per se as a concept or as a place of being. What has changed is human perception and everything around it, the totality. The human being no longer lives on a small scale, the human being lives in context, inside–outside, city–country, narrow–wide, mass–individual. The question is how this change affects our built space. Our living space is getting smaller, more and more people live in a smaller space, we are building more and more upwards. To occupy airspace is the epitome of luxury and prestige in eastern cultures, and despite this higher aspiration, the buildings are still generous inside. Despite the minimalization of our living space, we seek the vastness of nature as a contrasting program and want to get away from our everyday limitations. New urban planning concepts for future generations bring compressed private living areas, but large areas for the community, community living, and activities. But it is only with the expanse, the peace, and this feeling of freedom that we experience true spatial luxury, which a built space cannot offer in the same quality. As much as we appreciate the minimalized living space in urban everyday life, the more we need freedom in our leisure time. Selfness and awareness are omnipresent today and must be incorporated into the new spatial concepts.

Since we can never create the luxury that nature gives us in a spatial, built context, we focus our interior design on the world we can create. Hotels need to focus much more on their target audience in order to hit the nerve of that audience with the correct design statement. In this statement, interior design finds great significance. The generations have different demands and hoteliers, who want to please everyone, will not be able to hold themselves up sustainably in the long run. Never before have there been so many studies on the sinus milieu (cf. publisuisse SA 2018) of our society, so many classes and so many changes in demand and so rapid changes in our needs and behavior and a change in values. Although the more urban a hotel is, the less differentiation there is between the various target groups, but even in the hippest city, you must not lose sight of your positioning. We are living a fast life. *In* today, *out* tomorrow.

Interior architecture and spatial design can create something luxurious, not just by stringing together noble and expensive materials, but above all by focusing on people and their well-being. In different cultures, the feeling of luxury is character- ized by big differences, and thus the question arises who one wants to please or who one is aiming at. Is it the noble materials or is it the large window front with a view of the outside that brings nature into the room and creates an incomparable experience? Is it the gold-embroidered silk fabric or is it the simple, genuine Berber carpet that creates luxury? For the operator, luxury in the hotel business also means turning a blind eye when it comes to the operational side. Natural materials, highly pigmented colors with a matt surface, glossy metallic highlights, etc. are maintenance-intensive, prone to damage, or have a limited service life. Everything that is a headache for housekeeping today belongs in this segment and cannot be replaced by even the best photorealistic replication. The emotional effect of the feel of a natural material and the charisma it leaves in the room are the epitome of luxury in interior design and trigger emotions. In a world that has a very saturated market and where good design is no longer necessarily luxury, it is important to define yourself through emotions.

The term "Stealth Luxury" (cf. Pogorelova 2016; o. V. 2018d) is shaping our new generation. Post-material, without bling and glamor. The interior design approach to this new luxury is based on material understatement combined with high quality. High-quality furniture, excellent details, and the already mentioned natural and genuine materials are in the foreground. The combination of spatial generosity and the successful staging make up the new luxury. Colors are a trendy theme. Over the centuries, color palettes and fashion trends have always been subject to increasingly rapid change. While the 1970s pleaded for bright orange and brown, the 1990s for dark blue and yellow, and the 2000s for brown, anthracite, and cream, today it is petrol and purple, with the clear knowledge that this wave is already being replaced by colorful flowers in tangy green, lime with pink and stabilo colors. And a new trend is already emerging: gold on facades in combination with concrete, black and dark brown. For the feel-good factor, not only a successful mixture of tonalities is necessary, but also the outburst and the disturbance factor in an overall picture. In a minimalist environment, the guest unintentionally becomes a disruptive factor, which in turn is perceived as disruptive by the guest himself. The feel-good factor is often lost in a minimalist, perfectly reduced world.

Only the interesting things that stand out remain in positive memory, the monot- onous classic quickly leaves our thoughts again. Whereas in the past it was en vogue to build everything cleanly into a hotel room and to hide it behind large cupboard fronts, the new guiding principle is the casual handling of the elements that make up a room. Sometimes a hotel room is decorated with useless things: When casually placed, old suitcases, bicycles, and chairs, which can be used for nothing more than to look at, create the design of the rooms and thus give us the feeling of luxury. The luxury of having space-wasting space. But it is precisely these elements, which do not fit into the classic-practical hotel business and constitute a new type of design, that make room and design concepts exciting and unique. But only the combination of the right technical implementation and the skillful use of material, color, and light can turn the individual parts into a successful whole. How do we create color effects that are exciting but not agitating? A lot of color is compatible if the design language remains soft and harmonious. Different forms, on the other hand, do not tolerate many colors. If one outweighs the other, the sway toward technical sleek discomfort can go very fast, as can the over-stimulation of the senses, which is created by a too lavish staging of elements in space. Here, too, the focus must once again be strongly on the target group and the design of the rooms must be adapted accordingly.

Concept and Spatial Context

Do we still build for eternity in this fast-paced life, do we build for several years, or do we only stage for a few months in order to constantly reinvent ourselves? Where does the overlap take place between the naturalness we seek and the artificial world we are offered? Today more than ever before, clear positioning plays a central role in the hotel industry in order to clearly differentiate oneself and to spin one's own guidelines. If this positioning is well thought out and sustainable, it lays the foundation for a concept and begins to tell a story that is the basis of a presentation. All areas are based on this thread and can be spun together to form a coherent, harmonious whole. However, positioning is not a trend to be found on social platforms. Positioning is the deep engagement with the own brand, the own values and goals. The cornerstone of a whole strategy and an even more important cornerstone of implementation. In the conceptual, we always look for the unique, the distinguishing feature. A good conceptual implementation of the positioning is based on the basic values of the hosts, the surroundings, and the house and not on the mainstream design served to us on a silver platter by the social media world. However, it will hardly be possible to completely detach oneself from trends and tendencies in design, and so every concept, no matter how much it bears the signature of a designer, will always also carry a hint of a trend. For the general understanding, the acceptance and the desire for good design social media does excellent work, but this representation of trends is not to be confused with skillfully developed interior design. How does the trend load to which we are subject affect the sustainability of a concept? An artificially created or searched story will never achieve the depth and authenticity that is really necessary for a unique concept. But the confrontation, the search for this conceptual straw needs courage and patience. Two factors that are not always present today. Often the omnipresent seems too close and too common for it to arouse our interest. And this can be exactly the salt in the soup, which makes a concept unique, because it is real and lived. A house that has a story can tell it and underline its uniqueness by being diverse and having experienced a lot (such as castles, converted water towers, unused prisons, mills, greenhouses, etc.). But that is exactly what makes it interesting, because we wonder what it wants to tell us. And with this narrative, we experience something new, something unique, which in turn defines the new kind of luxury.

Sustainability in interior architecture is more than just concentrating on materials. The term "sustainable" (or in its meaning as lasting or permanent for a longer period of time) already plays an important role in the first conceptual idea. The decision as to which service life I will give my interior is decisive for the entire further planning. A strong conceptual basis makes a story, a space, and a behavior real and true, tangible, and comprehensible. The implementation of this concept in all areas spins the red thread that makes the guest experience complete and a place authentic. A consistent attitude to this self-created concept basis makes life easier in many respects, but sometimes also a little more complex. To stand behind one's concept, to live it, and to have internalized it is important in order to appear authentic. Preaching water and drinking wine is synonymous in interior design with the presentation of an ecological, sustainable menu and the laying of vinyl straps delivered by a container ship from the eastern market. The examination of one's own possibilities is an important factor for the decisions that are made in spatial design. We want real materials, but we have to ask ourselves whether we can cope with the effort involved or whether only the thought of the naturalness of the materials triggers a pleasant feeling. One could almost call today's techniques for the reproduction of wooden floors or linen fabrics, for example, a false and fake luxury, since they are so deceptively genuine that even the expert has to look twice.

But interior design sustainability also means that we take things from where they come from. Supporting the local economy, locating excellent products from the neighborhood, and exploring the opportunities and capabilities of the local market are real luxuries. More and more suppliers and producers are supporting the local and regional markets, and it is precisely these markets where the selection of furniture, textiles, and products should start. The idea of "everything fast" and "everything cheap" gives way slowly but steadily to the idea of quality and less (products) is more (quality).

A long life span also means thinking far ahead of everything that my room has to cover with its equipment, even in the distant future. However, this also means that the main focus is no longer on short-lived trends, but on much longer product cycles. This heralds an actual deceleration. Deceleration is a big topic anyway in a world flooded with stimuli, characterized by hectic, stress, and restlessness, and opens up many possibilities for the hotel industry. Often, however, there are technical issues involved and they have to clarify whether and how investments made today can also be used in the future and whether they bring the right benefits. Technical innovations are a rapidly developing field that is outdated within a very short time. Even if the technical trend in the hotel industry is back to simplicity, we can guess what will be regarded as "simple" in a few years and should be prepared accordingly. The fact is that although only simple toggle switches, plug-in plugs, plug-out plugs and sockets are desired in the front-of-house areas, these often require high-tech back-of-house installations.

However, conceptual sustainability means finding a basis that will not be out in a few years' time and will only be oriented to the current zeitgeist. Certain trends, such as the generosity of the lobby, which is used more than just for arrival and departure, will continue. The lobby becomes a multifunctional living room. Lounges become zones with meet-sit-work-dine-and-live functions. So, the future guest will need power and WiFi, not in the corner under the side table, but omnipresent.

Today's society is living in rapid and constant change. Topics such as burn-out, electrosmog, sensory overload, or overtourism are terms that are increasingly becoming the focus of hotel operators and thus also of interior designers and architects. The multitude of Barefoot luxury hotels shows that the consciously chosen isolation, the conscious turning away, the conscious renunciation of all-round availability of technology, and sensory stimulation opens up a new luxury dimension. The luxury to rest: Zen Buddhist monks in Japan, but also monastic communities in Europe, have been living this life of luxe de la simplicité for well over 1000 years.

Space as Space and Not as Place

Man will always need space, but in the future, our thinking of space as a place will change. The focus on the human being as the measure of all things is becoming more and more important in the narrower space and only with this measure will we be able to deal with reduced space conditions properly. In addition, intelligent and human-oriented formulas are necessary for learning how living space should be created. The new megatrends show it. It is about:

- Sustainable use of natural resources
- Empathic integration of nature into man-made structures
- A holistic approach to the choice of materials, design in space and time
- The consideration of all senses in order to break the elegance of pure esthetics that has existed for far too long
- The creation of a generous sense of space for people with an urban character
- and much more

Even though scientific methods and analyses have enabled us to get to know people's needs better and better, as well as the characteristics of a building and the sense of beauty, the desire for the perfect home remains an asymptotic approach to what we lost when we began to live in closed spaces.

4   Perspective "I and the Environment": Natural Space

When the Intensity of Nature Experiences Becomes a True Luxury

Philipp Schmid

> In nature we feel so well because it has no judgement over us. (Friedrich Wilhelm Nietzsche, German philosopher, 1844–1900)

When it comes to leisure activities in general and vacations in particular, we want to enjoy fulfilling, happy, eventful, unforgettable, in short: perfect days. We aim to maximize hours, days, and hopefully weeks during which everything is just right for us. So, the question arises: what are the ingredients for such a day, such a vacation? How do we design such a time window? What diverse activities ultimately make it so special for us? Why are nature experiences the ultimate luxury for such days? What experiences make this luxury so special? What is going on in our bodies?

The prerequisite is that we first switch off completely and shut down. We take the necessary time for relaxation, for special, active experiences, for example in and with nature, which we enjoy with all our senses. We treat ourselves to peace, let ourselves drift, discover, experience, taste, and feel. What happens in our bodies?

Release of Happiness Hormones

Six different happiness hormones are continuously produced in our body: dopamine, serotonin, endorphins, norepinephrine, phenethylamine, and oxytocin. Through thinking, eating, talking, and acting, we have a direct influence on the amount and type of happiness hormones that are released into our body. Happiness hormones, the popular scientific term for messenger substances for some hormones and neurotransmitters, are vital for our survival, because without them we can neither think clearly nor feel emotions. This enables us to make decisions, deal constructively with stress, and deal with bad news and bad days. They evoke well-being and happiness and for this reason are also called "endogenous drugs."

An active day with exercise, be it on the mountain or in nature in general, is therefore an effective and quick way to stimulate the biochemical processes for the production of happiness hormones.

Depending on the intensity of the movement, our brain begins to release dopamine after only a few minutes, which makes us more alert, concentrated, and focused: we simply feel better. After the activity, the dopamine level slowly decreases again, parallel to this the level of its opponent, serotonin, increases, which is involved in various functions such as the control of the sleep–wake rhythm and body temperature. In addition, it controls the appetite as the so-called feel-good hormone alleviates pain sensitivity and otherwise counteracts stress very effectively.

By the way, bright sunlight and fresh air also help our bodies to boost the production of serotonin. Another side effect of sunlight: it improves the release of the sleep hormone melatonin. The sunlight during the day helps us sleep deeper and more restful at night and it helps us get up out of bed better and more rested in the morning. Finally, oxytocin is often referred to as the "cuddle hormone." It promotes the feeling of solidarity and makes it easier for us to trust people.

However, endorphins, which reduce anxiety and help against depression, are regarded as the messenger of happiness par excellence. They are also produced during exercise in general and sport in particular. By the way, even simple laughter promotes their production.

Nature Is Luxury and Lifestyle: Timeless!

As a patient teacher, nature provides answers to our various questions. We find in it the already mentioned and much-needed tranquility. We throw off ballast and begin to feel ourselves again in the mirror of nature. Our entire body regenerates through physical activity. These activities can be completely different and we come to an inner connection of spirit and soul. We recharge our batteries and strengthen our personal senses and sharpen our perception if we really take the necessary time to do so.

The healing power of nature is therefore an effective medicine against our various diseases of civilization. Every single one of us can tap into their potential. We are not talking about the healing power of plants, but about the spiritual power of experiences with physical exercise outside in nature.

We take our technical gadgets offline and listen to nature. We put the sensations for the purifying effect of nature, for the fascination of the four seasons, and for the messages of the plant world online. Embedded in nature we let ourselves drift, we relax and strengthen our senses. Be it through the changing sunlight, through beautiful mountain peaks, through wild glaciers or crystal-clear water, through colorful flower meadows or colorful autumn forests, through the contemplative silence of a winter landscape and much more. We experience the perfection of nature. We experience harmony and peace through balance. The (untouched) nature with its partly spectacular and magical phenomena needs neither flavor enhancers nor colorings. Nature as a source of ideas and energy makes us fitter, makes us more positive, more confident, more satisfied, more efficient, and therefore happier.

Trekking in the Himalaya or a Canoe Trip in the Canadian Yukon?

The sensation of nature and the physical strain associated with it, depending on the activity, is individual for each person. Each of us should take the time to find out what is best for him: an action-packed adventure in exotic places around the world, a Greenland expedition, a trek in the Himalaya, the Andes, or Patagonia? Or rather a canoe trip with fishing on the Canadian Yukon River? Hiking, mountain biking, or climbing in the Alps? A relaxed summer hike in Iceland? Snowshoeing in the wilderness of Lapland or a wine walk in the hills of Tuscany? Or would you prefer a wilderness safari in the endless vastness of Namibia? A desert trip, a journey on a pilgrimage path, or on the Pacific Coast Trail? Swimming and diving in coral reefs such as the Great Barrier Reef? For some of us, activities in tropical rainforests are the highest of feelings.

Whether we prefer to enjoy the warm bonfire outside in the evening and spend the night in a tent, relax in our sleeping bag under the stars, or exchange experiences with locals in a simple, secluded mountain hut, for example at a regular's table, or have a massage with a sauna in a 5-star hotel, is ultimately very individual. The possibilities are inexhaustible. And certainly different for each of us in the respective phase of life and the current personal state.

In the future, the spa experience, in particular, will be less about pampering treatments and more about life-changing experiences. The term spa also extends to sporting challenges in nature with inspirations for our minds. Hiking as a physical activity also has a positive effect on us. According to sports physiologists, it stimulates the formation of additional vascular structures such as capillaries and mitochondria and reduces the release of stress hormones. Or even better: It supports the release of the already mentioned endorphins.

A Glass of Wine Never Tasted Better

After an intensive, sensual, passive, or active nature experience, the use of the wellness facility or a massage for our soaked and sweaty body takes on a completely new dimension. The sauna and a cleansing, cooling shower never have a more relaxing effect. Never is a book more entertaining. Hardly ever are culinary experiences (preferably with regional specialties) more intense or a bottle of wine tastes better than after an intense experience in nature. The satisfaction after the physical activity in nature increases the enjoyment of the food, the wine, and the whole holiday feeling. This kind of (natural) wellness is a luxury for our mind and extremely helpful for our mental feeling. Because we can let our soul dangle and get inspiration for everyday life through a new balance.

Luxury Is to Maximize Time in and with Nature

Experiences in and with nature increase the quality of our vacation enormously, the perfect days accumulate. The challenge is and remains to find the time in our everyday life to win even more perfect days for us with such nature experiences. So let us take the necessary time out to afford the great luxury, to undertake relaxing tours in nature, and to be active under the motto "away from everyday life, into nature." Let us follow this timeless luxury. Let us make friends with nature. And let us not forget: There is no bad weather, only a wrong attitude to it.

Literature

Albrecht C-M, Backhaus C, Gurzki H, Woisetschläger DM (2013a) Drivers of brand extension success: what really matters for luxury brands. Psychol Mark 30(8):647–659

Albrecht C-M, Backhaus C, Gurzki H, Woisetschläger DM (2013b) Value creation for luxury brands through brand extensions: an investigation of forward and reciprocal effects. Market ZFP 35(2):91–108

Anderson C, Hildreth J, Howland L (2015) Is the desire for status a fundamental human motive? A review of the empirical literature. Psychol Bull 141(3):574–601

Arbore A, Estes Z (2013) Loyalty program structure and consumers' perceptions of status: feeling special in a grocery store? J Retail Consum Serv 20(5):439–444

Bearden W, Etzel M (1982) Reference group influence on product and brand purchase decisions. J Consum Res 9(2):183–194

Bellezza S, Keinan A (2014) Brand tourists: how non-core users enhance the brand image by eliciting pride. J Consum Res 41(2):397–417

Benedictus de Nursia (1990) The rule of St. Benedict, 15th ed. Edited on behalf of the Salzburg Abbots' Conference. Beuroner Kunstverlag, Beuron

Berger J, Heath C (2007) Where consumers diverge from others: identity signaling and product domains. J Consum Res 34(2):121–134

Berger J, Ward M (2010) Subtle signals of inconspicuous consumption. J Consum Res 37 (4):555–569

Berry CJ (1994) The idea of luxury: a conceptual and historical investigation. Cambridge University Press, Cambridge

Beutelspachter A, Petri B (2013) The golden ratio. Springer Spectrum, Wiesbaden

Bird R, Smith E (2005) Signaling theory, strategic interaction, and symbolic capital. Curr Anthropol 46(2):221–248

Bourdieu P (1984) Distinction: a social critique of the judgement of taste (Richard N, Trans.). Harvard University Press, Cambridge, MA

Cova B, Cova V (2002) Tribal marketing: the tribalisation of society and its impact on the conduct of marketing. Eur J Mark 36(5/6):595–620

de Mello A (1991) The crux of the matter. Wake up and be happy. Herder, Freiburg

Eckhardt GM, Belk RW, Wilson JAJ (2015) The rise of inconspicuous consumption. J Mark Manag 31(7–8):807–826

Festinger L (1954) A theory of social comparison processes. Hum Relat 7(2):117–140

Frankl V (2015) . . .still say yes to life. A psychologist experiences the concentration camp, 7th edn. Kösel, Munich

Goffman E (1951) Symbols of class status. Br J Sociol 2(4):294–304

Griskevicius V, Tybur JM, Sundie JM, Cialdini RB, Miller GF, Kenrick DT (2007) Blatant benevolence and conspicuous consumption: when romantic motives elicit strategic costly signals. J Pers Soc Psychol 93(1):85–102

Grütter JK (2015) Light and colour. In: Fundamentals of architecture perception. Springer Vieweg, Wiesbaden

Gurzki H (2018) The creation of the extraordinary—principles of luxury. Dissertation. TU Braunschweig, Braunschweig

Gurzki H, Woisetschläger DM (2017) Mapping the luxury research landscape: a bibliometric citation analysis. J Bus Res 77:147–166

Han YJ, Nunes JC, Drèze X (2010) Signaling status with luxury goods: the role of brand prominence. J Mark 74(4):15–30

Holt DB (1998) Does cultural capital structure American consumption? J Consum Res 25(1):1–25

Ilg A (2017) ReGen villages. https://experimentselbstversorgung.net/regen-villages-utopie-oekodorf/. Accessed 13 Dec 2018

Jewish Museum Berlin Foundation (n.d.) The Libeskind building. https://www.jmberlin.de/libeskind-bau. Accessed 3 Dec 2018

Kapferer J-N, Bastien V (2009) The luxury strategy: break the rules of marketing to build luxury brands. Kogan Page Publishers, London

Kappler A (2012) Luxese—luxury and asceticism, the new meaning of luxury. In: Facets (Kappler Management AG), February, p 4

Kastanakis MN, Balabanis G (2012) Between the mass and the class: antecedents of the "bandwagon" luxury consumption behavior. J Bus Res 65(10):1399–1407

Kastanakis MN, Kastanakis MN, Balabanis G, Balabanis G (2014) Explaining variation in conspicuous luxury consumption: an individual differences' perspective. J Bus Res 67 (10):2147–2154

Kroen S (2004) A political history of the consumer. Hist J 47(3):709–736

Lasaleta JD, Sedikides C, Vohs KD (2014) Nostalgia weakens the desire for money. J Consum Res 41(3):713–729

Mason R (1985) Ethics and the supply of status goods. J Bus Ethics 4(6):457–464

McCracken G (1990) Culture and consumption: new approaches to the symbolic character of consumer goods and activities. Indiana University Press, Bloomington

Nelissen RMA, Meijers MHC (2011) Social benefits of luxury brands as costly signals of wealth and status. Evol Hum Behav 32(5):343–355

Nia A, Zaichkowsky J (2000) Do counterfeits devalue the ownership of luxury brands? J Prod Brand Manag 9(7):485–497

o. V. (2011–2017) Emmanuel Maria Heufelder (1898–1982). http://www.abtei-niederaltaich.de/oekumene/abt-emmanuel-heufelder-preis/emmanuel-heufelder/ Accessed 8 May 2018

o. V. (2018a) The Maslow pyramid of needs. https://www.lpb-bw.de/fileadmin/Abteilung_III/jugend/pdf/ws_beteiligung_dings/2017/ws6_17/maslowsche_beduerfnispyramide.pdf. Accessed 8 May 2018

o. V. (2018b) What is logotherapy and who helps. http://www.logotherapie.de/was-ist-und-wem-hilft-logotherapie.html. Accessed 8 May 2018

o. V. (2018c) Kloster auf Zeit: total relaxation for body and soul. https://www.stress-auszeit.ch/kloster-auf-zeit/. Accessed 8 May 2018

o. V. (2018d) The future of the department store. https://www.zukunftsinstitut.de/artikel/stealth-luxury-subtiler-statuskonsum/. Accessed 6 Sept 2018

Pogorelova K (2016) Luxury becomes immaterial. https://www.zukunftsinstitut.de/artikel/der-luxus-wird-immateriell-infografik. Accessed 6 Sept 2018

publisuisse SA (2018) The 10 Swiss Sinus Milieus. Bern. https://blog.zhdk.ch/loop/files/2011/11/sinus_broschuere_d.pdf

Schouten J, McAlexander J (1995) Subcultures of consumption: an ethnography of the new bikers. J Consum Res 22(1):43–61

Simmel G (1957) Fashion. Am J Sociol 62(6):541–558

Skift (2018) Travel Megatrends 2018

Solomon M (1983) The role of products as social stimuli: a symbolic interactionism perspective. J Consum Res 10(3):319–329

Steindl-Rast D (2011) Fullness and nothing. Awakening to life from within, 5th Aufl. Herder Spektrum 5653. Herder, Freiburg

Üstüner T, Holt DB (2010) Toward a theory of status consumption in less industrialized countries. J Consum Res 37(1):37–56

Üstüner T, Thompson CJ (2012) How marketplace performances produce interdependent status games and contested forms of symbolic capital. J Consum Res 38(5):796–814

Veblen T (1899) The theory of the leisure class: an economic study of institutions. Macmillan, New York

Vigneron F, Johnson LW (2004) Measuring perceptions of brand luxury. J Brand Manag 11 (6):484–506

von Flüe J (2018) Bootshaus am Vierwaldstättersee becomes a temporary hotel room. https://www. luzernerzeitung.ch/zentralschweiz/luzern/bootshaus-am-vierwaldstaettersee-wird-zum-temporaeren-hotelzimmer-ld.1026312. Accessed 30 Nov 2018

Wikipedia (2018a) Enneagram. https://de.wikipedia.org/wiki/Enneagramm#Typentests. Accessed 8 May 2018

Wikipedia (2018b) Transcendence. https://de.wikipedia.org/wiki/Transzendenz

Wong N, Ahuvia A (1998) Personal taste and family face: luxury consumption in Confucian and Western societies. Psychol Mark 15(5):423–441

Yi-Cheon Yim M, Sauer LP, Williams J, Lee S-J, Macrury I (2014) Drivers of attitudes toward luxury brands. Int Mark Rev 31(4):363–389

Zahavi A (1975) Mate selection—a selection for a handicap. J Theor Biol 53(1):205–214

Hannes Gurzki is a consultant of The Boston Consulting Group and research Fellow of the Technical University of Braunschweig. He is an expert in luxury, marketing, branding, and consumer behavior. He studied business administration and cultural management at the University of Mannheim and holds an MBA from ESSEC Business School. His research appeared in leading academic journals such as the Journal of Business Research or Psychology & Marketing.

Philipp Schmid (Dipl. Natw.) is co-owner of a Swiss catering company and a restaurant as well as the initiator of the Global Forum Wallis. As an entrepreneur and project manager, he is currently involved in setting up Spitzhorli-Entwicklung, Marketing & Invest GmbH and the international start-up company Hazu Inc. Philipp Schmid is a graduate of the Management Seminar for Small and Medium Enterprises at the University of St. Gallen. Since graduating from ETH Zurich with a degree in biotechnology, he has been involved in various companies as a director or member of the executive board and in the founding of various companies. He often undertakes hikes in the mountains and would also like to inspire third parties with his motto of enjoyable nature experiences with the Trekking Friends.

Daniel Schönbächler (Dr. phil. and lic theol.) is the 65th abbot of the Disentis Monastery, one of the oldest and most important Benedictine monasteries north of the Alps. Abbot Daniel has been conducting seminars on the subject of "Personality, Resources and Talents" or "Personality Structures and Communication Patterns" for small groups for many years. The typological model of the enneagram and the method of psychokinesiology in the personality seminars help to recognize the personality structure and to break down blockades. Individual work is possible in small groups.

David M. Woisetschläger is Professor of Services Management and Director of the Institute of Automotive Management and Industrial Production at TU Braunschweig University. His research interests are in the fields of Brand Management, Customer Relationship Management, Sales, and Sponsorship. He is also a consultant for new service development and the analysis of customer survey and behavioral data with an industry focus in the automobile and telecommunications industries. His research is published in leading international journals such as the Journal of Marketing, Journal of the Academy of Marketing Science, Journal of Business Research, and Journal of Retailing.

Verena Zaugg-Faszl (Bac. Interior Architecture) is an interior architect at a Swiss interior design and construction management office specializing in the overall conception, design, planning, and execution of hotel and gastronomy conversions and new buildings. Verena Zaugg-Faszl is a graduate of the HTL Ortweinschule and the Darmstadt University of Applied Sciences in the field of architecture and since 2018 a member of VSI.ASAI, which promotes and disseminates the recognition and quality of the profession of interior designers throughout Switzerland.

Case Studies and Best Practice Examples of Luxury Tourism

Marc Aeberhard, Andreas Caminada, Sergio Comino, Dirk Gowin,
Marcus Krall, Brett McDonald, Thomas Reimann, and Marco Walter

M. Aeberhard
Luxury Hotel & Spa Management Ltd., Zürich, Switzerland

A. Caminada (✉)
Genusswerkstatt GmbH Schloss Schauenstein, Fürstenau, Switzerland
e-mail: kontakt@schauenstein.ch; stich@andreascaminada.com

S. Comino
Jesolo International Club Camping, Lido di Jesolo, Italy
e-mail: manager@jesolointernational.it

D. Gowin (✉)
Select Luxury Travel GmbH, Berlin, Germany
e-mail: gowin@select-luxury.travel

M. Krall (✉)
Ocean Independence AG, Düsseldorf, Germany
e-mail: marcus@ocyachts.com

B. McDonald (✉)
Flame of Africa, Kasane, Botswana
e-mail: brett@flameofafrica.com

T. Reimann
thr media, Hamburg, Germany
e-mail: reimann@thr-media.de

M. Walter (✉)
ECOCAMPING Service GmbH, Konstanz, Germany
e-mail: marco.walter@ecocamping.gmbh

© Springer Nature Switzerland AG 2020
R. Conrady et al. (eds.), *Luxury Tourism*, Tourism, Hospitality & Event
Management, https://doi.org/10.1007/978-3-030-59893-8_9

1   Luxury Travel from the Tour Operator's Point of View Using the Destination Oman as an Example

Dirk Gowin

Luxury Tourism: Facts and an Elastic Term

For a long time, luxury tourism was regarded as individual tourism that only the rich could afford. But this has changed with growing incomes and the prosperity of broad sections of the population. Internationally, around 190 billion euros in sales were generated in this growing segment in 2017. And it is no longer just the 20,000 five-star hotels worldwide that characterize a luxury vacation; today it is also the individual, exclusive experience offers at a higher, but quite bearable price.

The luxury tourist travels on average three to four times a year, preferably in spring, autumn, and winter, for 10–12 days and spends between 5000 and 12,000 euros per person. And at the "Marriott President Wilson" in Geneva, you can spend a night in what is probably the most expensive suite in the world for as much as 80,000 US dollars. And the Germans' favorite luxury expense, even before jewelry and watches, are vacations and traveling (see Statista 2018; Buck and Ruetz 2018), with the desire for very special and long-term, meticulously planned, unique travel experiences such as cruises, cultural tourism or slow travel, which serves as a magic wand against everyday stress, generally increasing.

Luxury Tourism in Transition: Longing for Simplicity

The author's experiences after almost 30 years of passionate work in luxury tourism and the results of various customer surveys, analyses, and studies prove the paradigm shift of luxury customers in terms of vacations. At Select Luxury Travel it can be observed that more and more wealthy and so-called insider tourists, especially those of Generation X (36–55 years of age), are interested in discovering destinations that have retained their authenticity and are ecologically sustainable. So, a new type of travelers, those who want to live like the locals. Looking for the unique, preciously simple experience, for time to breathe in nature and for cultural and culinary pleasure—in other words, always looking for special experiences in a peaceful paradise. More and more ultra-rich people, too, appreciate the kick of a special, original ambience without showing off and pomp, and prefer smaller boutique hotels that offer the comfort of five-star hotels, resulting in a fundamental reorientation of offers in the hotel sector. All this places special demands on the travel professional, who has to meet the demanding needs optimally. Travel in the luxury segment is predominantly planned and booked face-to-face with travel experts—a gratifying development for tour operators.

Oman Tourism: Rapidly on the Upswing

Away from mass tourism, the Sultanate of Oman (with roughly the same area as the Federal Republic of Germany) and the Middle East in general rank in the top third of the "bucket list" of luxury travelers in terms of demand and this mostly as a second trip (another trip in addition to the main vacation) with an average travel duration of 8–10 days and a share of about 22–24%, almost on a par with Southeast Asia, South Africa, the Indian Ocean, and the Caribbean—followed by Western Europe, the Pacific, and North America.

In addition to oil and gas production, fishing, agriculture, and trade, Oman also focuses on soft tourism. Thanks to the petrodollar, people can invest heavily: The future program of the Ministry of Tourism of Oman envisages the provision of around 3000 new hotel rooms by 2020. By 2040, 5 million guests are to be accommodated and 500,000 workers (mainly Omanis) are to find employment in the tourism sector (German-Omani Society). Oman aims to grow in the luxury segment and especially in the areas of family, study, sports, and business travel (MICE). The Sultanate does not want to become a mainstream destination and regulates the inbound via consistently higher to very high pricing.

Expansion of the Ultra-Luxury Segment

In Oman, five-star luxury tourism has almost doubled within a decade since 2008. Higher growth in this segment is to be promoted with ambitious projects. One of the most important current investment projects is "The Wave Muscat" (Al Mouji Muscat)—it is labeled as an "integrated tourism project" and is a combination of housing for the Omani population and hotel resorts for tourists. This new 2.5 million m^2 district on the Gulf of Oman has a 6-km-long beach, an artificial reef, the luxury hotels Kempinski and Fairmont, luxurious villas, apartments, townhouses, lifestyle shopping malls, marina, 18-hole golf course, parks, and playgrounds to meet the needs of rich travelers. Majid Al-Futtaim Investments from Dubai ("Mall of the Emirates" project) and two government-owned companies of the Sultanate of Oman have designed this ultra-luxury resort to catch up in the internationally prospering luxury tourism segment, looking at the successful United Arab Emirates and to keep the younger generation of Omanis in the country.

The simplified entry policy with Visa on Arrival directly at the airport makes it particularly easy for tourists to visit Oman. A high-level infrastructure is in place. A well-thought-out and planned traffic system with roads (1000 km of comfortable motorway from the capital Maskat to Salalah), airports of the renowned airline

Oman Air, luxury vehicles, and a very good hotel infrastructure in the high-end sector make Oman an attractive travel destination.

An overview of the current development of inbound and outbound tourism in the period 2009–2017 with a breakdown of the number of travelers by destination to Oman can be found in Figs. 1, 2, and 3.

Tourism in Oman: Traditional, Diverse, Cosmopolitan

Tourism in the Sultanate of Oman (about 80 times the size of the neighboring UAE in terms of area) is deliberately cosmopolitan and based on the principle of tolerance traditionally anchored by religion in society, the economy, and foreign policy (75% of the population belongs to the Islamic branch of Ibadism).

Oman focuses on sustainable and high-quality eco-tourism. The country offers the best conditions for this. The people are wealthy and hospitable, crime is virtually nonexistent. An ideal, safe, and almost museum-like destination, with lots of exoticism, great culture and history, varied nature, wildlife, dream beaches, original souks, the scent of incense and spices, wonderful cuisine, warm hospitality, and an attractive range of top luxury hotels for premium travelers. Over 1000 forts, castles, and watchtowers can be found in the desert landscape. They once guarded oases, coasts, harbors, date plantations, and caravan routes. This fascinating variety of landscapes with deserts and oases, around 2000 km of coastline and the impressive Oman Mountains with its highest mountain, Jebel Akhdar (around 3000 m.a.s.l.), which are popular with nature lovers and round-trippers, make it one of the most varied destinations in the world.

Destination Oman: Firmly Anchored in the Program of Tour Operators

Since the mid and late 1990s, Oman has been a permanent feature of luxury tour operators. At that time the Sultanate opened to tourism. The road was rocky—the Gulf War brought losses after 2001, many organizers had canceled Oman again—but for Select Luxury Travel Oman as one of the top luxury destinations in the world has always been an integral part of the program without interruption. With a lot of intuition, thoughtfulness, and creative commitment, Sultan Qabus has managed over the last few years to let tourism in Oman grow slowly and cautiously and to establish it as a new source of income. The Ministry of Tourism was created. It is also worth noting that about half of government revenue is invested in education as the key to success and health, thus benefiting the "ordinary citizen." The requirements demand economical energy consumption, an environmentally friendly construction method with buildings that cannot have more than 12 storeys. High-rise buildings and

Inbound Tourism 2009-2016

Years	(*Number of visitors (000))						Visitors Expenditure (000') R.O						(*Number of Nights spent (000))					
	G.C.C	Other Arabs	Asians	Europeans	Other	Total	G.C.C	Other Arabs	Asians	Europeans	Other	Total	G.C.C	Other Arabs	Asians	Europeans	Other	Total
2009	839	76	249	270	151	1,585	37,489	6,910	10,931	73,767	15,271	######	3,547	335	745	1,115	793	6,535
2010	698	80	291	270	161	1,500	36,777	6,839	10,571	71,379	14,982	######	2,952	355	869	1,117	844	6,137
2011	660	70	258	250	155	1,393	44,142	7,087	19,766	66,518	21,100	######	2,895	292	1,188	1,161	959	6,495
2012	696	103	389	321	204	1,714	46,566	10,425	29,802	85,621	27,856	######	3,582	557	1,781	1,509	1,146	8,575
2013	870	124	435	372	121	1,923	52,000	12,613	33,591	99,684	29,382	######	4,481	669	1,995	1,745	680	9,569
2014	961	130	478	508	148	2,225	58,412	13,402	37,274	######	31,833	######	4,778	828	2,572	2,139	763	11,079
2015	1,208	164	578	515	170	2,634	75,518	17,039	45,912	######	35,107	######	6,368	1,017	2,695	2,219	841	13,139
2016*	1,561	185	651	584	169	3,151	97,868	19,541	52,043	######	36,820	######	8,052	1,296	3,364	2,805	987	16,504

Fig. 1 Inbound tourism 2009–2016. Quelle: Ministry of Tourism, Sultanate of Oman (2016a)

Outbound Tourism 2009-2016

Years	(Number of visitors (000))						Visitors Expenditure (000') R.O						(Number of Nights spent (000))					
	Omanis	Other Arabs	Asians	Europeans	Other	Total	Omanis	Other Arabs	Asians	Europeans	Other	Total	Omanis	Other Arabs	Asians	Europeans	Other	Total
2009	1,438	138	727	118	48	2,470	1,51,541	18,797	52,679	3,429	1,783	2,28,230	2,328	4,471	30,323	752	437	38,311
2010	1,670	155	817	133	54	2,829	1,74,358	21,702	60,917	3,938	2,055	2,62,971	2,703	4,807	34,254	849	486	43,100
2011	1,990	78	1,139	86	48	3,341	1,97,843	6,288	70,730	5,942	5,008	2,85,811	7,013	6,178	36,347	567	458	50,563
2012	2,417	90	1,311	99	56	3,972	2,05,395	7,613	1,11,420	8,382	4,722	3,37,532	7,872	6,935	40,795	636	514	56,751
2013	2,679	93	1,367	103	58	4,301	2,30,612	8,040	1,17,665	8,852	4,986	3,70,155	9,720	8,563	50,373	786	634	70,075
2014	3,078	95	1,391	105	59	4,727	2,69,836	8,330	1,21,904	9,171	5,166	4,14,406	11,165	8,709	51,235	799	645	72,553
2015	3,572	107	1,562	117	66	5,424	3,21,325	9,600	1,40,495	10,569	5,954	4,87,944	12,958	9,782	57,548	897	725	81,910
2016*	3,910	115	1,680	126	71	5,902	3,24,770	9,536	1,39,562	10,499	5,914	4,90,282	14,183	10,523	61,907	965	779	88,359

Fig. 2 Outbound tourism 2009–2016. Quelle: Ministry of Tourism, Sultanate of Oman (2016a)

Number of Tourists to Oman from 2011-2017

Nationlities	2011	2012	2013	2014	*2015	*2016	*2017	Change percent % 2016-2017
Britain	1,02,806	1,21,528	1,33,529	1,39,362	1,41,214	1,37,170	1,43,224	4.4
Germany	38,034	46,652	55,126	59,400	66,044	72,566	1,02,203	40.8
USA	35,984	46,073	53,165	56,962	60,874	57,493	58,598	-5.6
France	30,040	40,105	47,830	50,570	46,520	41,366	49,913	-11.1
Canada	9,128	12,213	17,254	23,333	26,482	26,227	26,747	-1.0
Nertherland	11,605	14,948	16,228	17,480	16,539	15,126	16,649	-8.5
Italy	14,297	19,806	26,063	31,858	33,915	32,280	43,664	-4.8
Sweden	8,329	9,680	10,762	9,154	6,615	4,757	4,794	-28.1
Turkey	6,695	7,566	7,229	6,815	7,940	8,664	8,851	9.1
Ireland	4,712	6,166	7,452	7,723	8,185	8,580	8,667	4.8
Spain	4,087	6,593	9,318	11,117	11,770	9,672	9,964	-17.8
Denmark	4,001	4,786	4,603	4,557	4,473	3,620	3,785	-19.1
Austria	8,223	8,487	8,150	8,543	6,939	7,051	8,338	1.6
Swisserland	6,969	10,592	13,272	14,224	14,790	14,224	19,746	-3.8
Belgium	4,176	6,346	7,316	7,630	7,098	6,343	7,722	-10.6
Russia	2,084	2,847	3,326	3,050	3,426	3,223	4,858	-5.9
Finland	1,368	1,939	2,130	2,313	2,088	1,889	1,927	-9.5
Poland	1,246	2,301	2,817	4,198	6,527	10,754	15,062	64.8
Brazil	1,768	1,881	1,588	1,945	2,290	2,440	3,056	6.6
Greece	1,959	3,264	4,618	5,241	4,429	3,133	3,571	-29.3
Portugal	1,852	2,591	3,360	3,877	4,180	3,272	3,559	-21.7
Norway	2,062	2,820	3,330	3,013	2,407	2,356	2,819	-2.1
Australia	8,213	14,058	15,616	17,002	17,643	16,833	17,441	-4.6
Czech	919	1,301	2,472	3,798	4,281	4,638	6,738	8.3
Lebanon	10,695	15,112	17,602	19,372	22,783	24,175	23,296	6.1
Jordan	10,942	16,116	18,008	19,244	20,945	21,991	23,432	5.0
Egypt	17,606	25,148	28,541	28,263	33,630	36,596	41,385	8.8
Syria	9,256	15,198	17,654	16,701	19,763	22,755	22,107	15.1
Iraq	4,566	5,812	6,430	6,347	6,295	6,982	7,835	10.9
Tunisia	3,537	2,912	2,155	2,374	2,786	2,978	3,351	6.9
Plastin	2,495	3,533	3,819	3,407	3,803	4,066	4,040	6.9
Morocco	1,831	2,732	2,993	3,352	3,661	3,685	4,317	0.7
Algeria	467	1,262	1,518	1,526	1,598	2,090	2,126	30.8
Sudan	4,807	6,711	8,447	8,451	8,703	10,103	11,175	16.1
Yamen	4,125	10,819	12,634	14,039	29,993	43,036	24,734	43.5
India	1,63,451	2,21,623	2,44,786	2,56,210	2,99,022	2,97,628	3,21,161	-0.5
Pakistan	39,994	61,198	67,893	73,055	84,906	88,974	87,090	4.8
Philippine	10,945	20,145	24,897	35,348	54,781	1,13,257	86,852	106.7
Malaysia	5,086	6,452	6,975	9,399	8,104	8,007	7,549	-1.2
Thailand	2,897	3,945	6,612	8,694	11,768	14,455	6,903	22.8
Iran	9,010	11,939	12,685	12,635	15,530	19,717	22,747	27.0
Japan	3,954	6,398	5,508	6,298	6,640	6,205	6,533	-6.6
China	6,431	8,810	9,460	11,839	13,829	16,532	20,021	19.5
Taiwan	406	361	501	505	567	599	703	5.6
Bangladish	811	18,408	20,191	24,422	29,292	29,261	24,904	-0.1
Srilanka	3,630	5,414	6,162	6,249	6,402	6,246	6,882	-2.4
Singapore	2,215	2,665	2,804	2,754	3,087	3,134	3,352	1.5
Indonesia	2,219	7,913	11,506	13,101	15,934	18,542	20,385	16.4
Hangary	911	1,443	1,464	2,097	2,460	1,848	2,089	-24.9
S.Africa	4,894	5,924	5,971	6,544	7,356	7,119	8,225	-3.2
Tanzania	5,053	6,035	6,330	7,051	7,907	8,207	8,320	3.8
Others	17,606	40,401	42,223	44,477	50,234	62,352	75,723	24.1
Total of Tourists by Visas	6,60,397	9,28,972	10,52,323	11,36,919	12,78,448	13,74,217	14,49,133	7.5
Omanis (Non Resident) in Oman	72,258	88,802	97,863	1,11,760	1,30,345	1,41,986	—	8.9
Others G.C.C *	6,62,196	7,89,461	7,72,535	8,49,546	10,62,215	13,11,349	15,90,538	23.5
Cruise Ships Passingers*	-	2,56,721	2,02,159	1,25,375	1,47,269	2,17,153	2,21,813	47.5
Grand Total of Tourists	13,94,851	20,63,956	21,24,880	22,23,600	26,18,277	30,44,705	32,61,484	16.3

Fig. 3 Number of tourists to Oman from 2011 till 2017. Quelle: Ministry of Tourism, Sultanate of Oman (2016a)

skyscrapers made of steel, glass and concrete, mega-buildings such as huge malls and fun parks such as those in the UAE are therefore sought in vain, houses are built consistently in the Omani style.

The Luxury Hotel Industry in Oman

The Intercontinental Muscat, the Al Bustan Palace Hotel, and the Ritz Carlton, which is still a guest house for state visits today, were the first hotels in the premium class. In recent years, substantial investments have been made in the hotel infrastructure. Surrounded by magnificent gardens and spacious parks, the following hotels belong to the five-star category: The Chedi, Six Senses, Shangri-La, Grand Hyatt, Anantara, Alila, Fairmont, Ritz Carlton, Intercontinental, Hilton, Rotana, Marriott, Mövenpick, and the Kempinski.

This creates the best conditions for luxury tourism and also for discerning round trip guests there is now the possibility to enjoy the luxury of five-star hotels in Maskat, on the Musandam Peninsula, in the Hajar Mountains, in the desert, or in the south on the beach of Salalah.

Select Luxury Travel works exclusively and on a long-term basis with local partners. In regular exchange, we learn from each other and create lasting values. Select Luxury Travel realizes trips to Oman on an individual basis. In addition to luxury hotel travel, the agency as a global provider is also geared to the increasing demand for luxurious, emotionally complex, and individualized travel routes (*experiential travel*), the requirements of travelers are competently planned and implemented. Oman sets the standards for an ideal luxury destination of modern character, which Oman Tourism officially describes as follows (Ministry of Tourism, Sultanate of Oman 2016b):

Beauty Has an Address: Oman

Our call to action is used to convey the message: there are moments in Oman that make you realize beauty has an address. This includes tangible and intangible visitor experiences from world-class scenery to the sparkle of meeting enchanting people.

Our Mission

Providing professional high-quality services to diversify our economy and to create jobs by offering the world enriching tourism experiences with Omani personality.

Our Vision

Providing services efficiently and effectively for Oman to become, by 2040, a top-of-mind destination for vacations, discovery, and meetings, attracting international and local tourists annually.

This philosophy means for Select Luxury Travel, as a committed and service-oriented luxury tour operator in Oman, to successfully transform the creative ideas and wishes of its customers into dream vacations. The agency also supports sustainability projects such as the protection of turtles and the avoidance of plastic waste.

2 Destination and Luxury Hotel: Frégate Island Private/ Seychelles

How Much Man Can Nature Take?

Marc Aeberhard

Figure 4 shows a place not far from Mont Signal, the highest elevation on Frégate Island. These granite rocks have towered from the (present-day) Indian Ocean for hundreds of millions of years and created the basis for the unique fauna and flora on the island.

Fig. 4 Best practice: Frégate Island Private, Seychelles

Forgotten by Evolution

In the geological history of Frégate Island lies probably the real fascination and therefore the real treasure that there is for the world to preserve: Located in the archipelago of the Seychelles, about 50 nautical miles north-northeast of Mahé, the main island, Frégate is a small granite island of about 3 km^2 in the middle of the Indian Ocean. When sometime between Carboniferous and Jurassic times, about 180 million years ago, the ancient continent of Gondwana drifted apart and formed the basic structure of today's known continents, some small splinters remained in the Indian Ocean, which did not really want to decide in which direction they wanted to go, and thus simply stayed where they had always been. And while tectonic plate shifts folded, leveled and flooded mountain ranges, plains, oceans, hills and then unfolded them again, the granite islands of the Seychelles remained erratic blocks and small islands, where evolution developed—or rather did not develop—completely independently from the rest of the world. So, it is not surprising that the fauna and flora of these islands (as well as Madagascar) has a multitude of endemic species that do not exist elsewhere, even in a modified or further developed form. This small island of Frégate Island is also a small Garden of Eden, which has survived all the changes of evolution for almost 200 million years. With species like the Magpie Robin, the Giant Tenebrionid Beetle, or the Giant Thousand Feet, testimony is given about animals that exist nowhere else in the world. However, this little paradise was greatly upset when in the late eighteenth-century man set foot on the island and first pirates, then French settlers and later Englishmen cultivated the island and began to clear the primary forest in order to plant cinnamon and coconut plantations in its place. To this day, the island has been traversed by wild citrus trees from the time of the pirates, a good 1000 cinnamon trees and countless coconut palms as well as other invasive neophytes such as bamboo or philodendron, which pose a massive threat to the remaining remains of the primeval primary forest.

Invasive Threats

When the current owner took over the island, a radical restoration of the island, unique in its kind, began. A process that has now lasted almost 20 years, but is far from complete. First of all, all mammals had to clear the island, since they never developed in the millions of years of history on the island, especially the extermination of rodents (mice and rats) required not only a huge effort of man, material, time, and money but was the indispensable basis to prepare the habitat again so that the massively threatened population of endemic fauna, which never developed defense or escape strategies against predators, could unfold again safely. It should be noted that evolution on Frégate Island has never made the developmental stage beyond the status of the dinosaur and thus the fauna remains restricted to insects, a few reptiles, and approaching birds.

Parallel to the extermination of all imported animals, first the containment and later the extermination of introduced plant species began in order to create a natural habitat again for the sensitive primeval plants. In the last 15 years, well over 200,000 rare trees have been reintroduced to the island in the island's own tree nursery. However, the stock of coconut palms in particular is proving to be particularly tough and difficult to control. Although the coconut palm is the epitome of a tropical beach paradise, it is important to emphasize that this tree on Frégate is an extremely threatening weed plant. So much so that a team of almost 20 men stays busy pulling out, collecting, and destroying both fallen coconuts and the rapidly sprouting young shoots.

The renaturation efforts of this small island are an example of how quickly the natural order of a fragile ecosystem can be disrupted by humans and how long it takes afterward before damage can be limited and then repaired in very small steps. The fate of this island serves as an extremely dramatic example for human intervention in much larger ecosystems, which in the past centuries have certainly been unintentional, today perhaps unconscious, but often out of well-calculated economic greed and greed for profit with completely unpredictable consequences (deforestation on Borneo or the Amazon): extreme weather phenomena, unpredictable seasons, soil erosion, expansion of the deserts, increases in storms and drought, etc. are—nota bene always with a considerable time delay—the reactions of nature to such interventions. To upset nature's balance is a means to destroy it. And although nature has an amazing power of self-regeneration, this process would take much more time than modern man allows. And so, it has become the sad truth today that man's restlessness, greed, impatience, and ultimate ignorance and disrespect for the life-creating and life-preserving system—Mother Earth—are increasing.

Small Illustration for the Big Picture

Frégate Island Private may only be a tiny dot on the world map, but its history is exemplary for the great connections that are currently also running through planet Earth and thus, in its strategic orientation as an ultra top-end resort, not only fulfills the usual functions of recreation, excellent service, and good food but also appeals as an example to a UHNWI[1] clientele, which in turn can extrapolate the connections experienced in the small to the global situation and carry out sensitization work.

---

[1]Ultra High Net Worth Inidividual.

New Luxury, Exemplarily Depicted and Lived

From these facts unmistakable realizations are extracted in relation to the new understanding of luxury:

Material ideas of luxury disappear completely into thin air and become insignificant. They may be a pleasant basis for a very wealthy clientele, who are used to not have to do without any luxury in the traditional sense. Therefore, the tangible infrastructure is not worth mentioning. Perfect table and bed linen, crockery, cutlery, and glasses as well as an internationally recognized gourmet kitchen are naturally available.

Rather, the focus is on the dimensions of the new luxury, which is defined exclusively immaterially.

Space and Time

So, it is no coincidence that the dimensions of time and space are of paramount importance. Time is understood here not only as a package of 24-h services provided by employees but also as the experience of time and timelessness, of creation and passing, transience and birth in a (partially) original, recreated ecosystem unique in the world, which rests on ancient roots. Here nature becomes a charismatic authority, in which man degenerates into a small being of evolution. The ultimate message on the island one understands when opening one's senses is: We are not the masters of the universe, we are members of the universe.

The example of Frégate Island also shows, however, that an individual human being needs a lot of space in the modern way of thinking of today's Western civilizations, so that he can satisfy his supposedly necessary needs from the creative power of the earth with all its products and resources. Of course, this poses the big question of whether the world has not become too small for a total population of soon 8 billion people, if we are as zealous as we have been in the past about resource wear and tear, which the planet is unable to compensate for. Frégate Island is one of several vivid examples of what happens when man exaggerates (see also the deforested Dalmatian coast by the Venetians or the vanished civilization of the Easter Islands as a result of overpopulation).

Safety and Security

And suddenly another dimension of the new understanding of luxury comes to the fore: security. It is not so much about the integrity of life and limb by attackers, but more about existential dimensions such as the availability of good air, clean water, and healthy soil from which man creates food. How far away the urban man from western civilizations is from this fine balance of all-natural forces can be seen when hotel guests insist on boat trips or helicopter flights during tropical storms with wind forces of 7 or 8 on the Beaufort scale and do not (any longer) recognize the danger to their own lives. The overestimation of one's own importance in a post-hedonist social system completely lacks humility and even the beginnings of a deeper understanding of the global connections of nature and the forces of nature. This ignorance is life-threatening and here a small island like Frégate Island offers a laboratory setting in which the inclined and interested guest gets a unique opportunity to experience causes and effects.

Exclusiveness

Consequently, these arguments give rise to the question of the ideal size, which is proportionate for such an island. Or to put it another way: How many people can the island take? The scary thing is, it is less than I thought. And so, it is no coincidence that the island offers only 16 units for guests and furthermore only accommodation for the employees. This means that a further dimension of the New Luxury definition is mapped: Exclusivity: "excludere" from Latin means to exclude, to limit and exactly this is the goal of ultra top-end hotels. On the one hand, it is about the artificial shortage of the supply in order to provoke a demand-pull, on the other hand, and much more importantly it is about the understanding that the harmony between man and nature is maintained or created. In the common choice of words "man and nature" lies an almost cynical paradox, because in its understanding it excludes man as part of nature. And it is precisely here that a misunderstanding sets in, which has its roots in the Renaissance but pronounced in the Enlightenment and perverted in industrialization, which continues to shape modern humanity to this day: When man describes himself as an exclusive being in a natural environment, he has indeed—and with fatal consequences—excluded himself. The Eco-Luxury, Barefoot-Luxury hotel industry plays a key role here, as it must/may/can create understanding and consciousness and lead people back into their habitat. And this habitat is not outside, but within nature, man is part of it. Consequently, the aforementioned rebuke to man applies: "Man is not master of the universe, man is member of the universe."

Service Thoughts

This leads to the next dimension of the new understanding of luxury: individual and personalized service: empathy and attention, but also the careful handling, the elegant introduction, and a service culture geared to the explorative curiosity of the guest are the new main aspects of good service. Impeccable labels and perfect service techniques are a prerequisite. The point is to take the guest (literally) by the hand and make the vacation not only a great experience but rather a lasting experience, after which he or she goes home enriched. As a result, Frégate Island invests a great deal of energy and effort, providing a team of qualified rangers and horticulturists who act as knowledge mediators and can turn the green jungle thicket into a unique, perfectly harmonious universe of fauna and flora.

Summary

In summary, it can be said that the demands placed on top-end hotels addressed by the UHNWI must be met by a new catalog of criteria. The profiling of star cuisine, extravagances in equipment, and perfection in service and wine cellar have their limits at some point, and the question remains: What is next? And only at this point does the actual further development of the product mix in the actual sense of the marketing management context begin. This shows that the profiling of a hotel with a clear Unique Selling Proposition (USP—a significant difference in performance due to clearly recognizable unique selling propositions) is of strategic importance in order to gain convincing market shares in a completely saturated hotel market. Today, the successful design of the three pillars of the Triple Bottomline Approach is increasing more and more. So, it is not only about economic profitability, but also about social responsibility, active embedding in society, and ecological sustainability. With this balancing act, it is also important to approach the guest in a holistic approach by touching not only the head but also the heart and stomach. The point is that the guest experiences, recognizes, and accepts the new dimensions of immaterial luxury not only within the framework of an (interchangeable) stay, but also through internalized experiences and personally enriching experiences (Fig. 5).

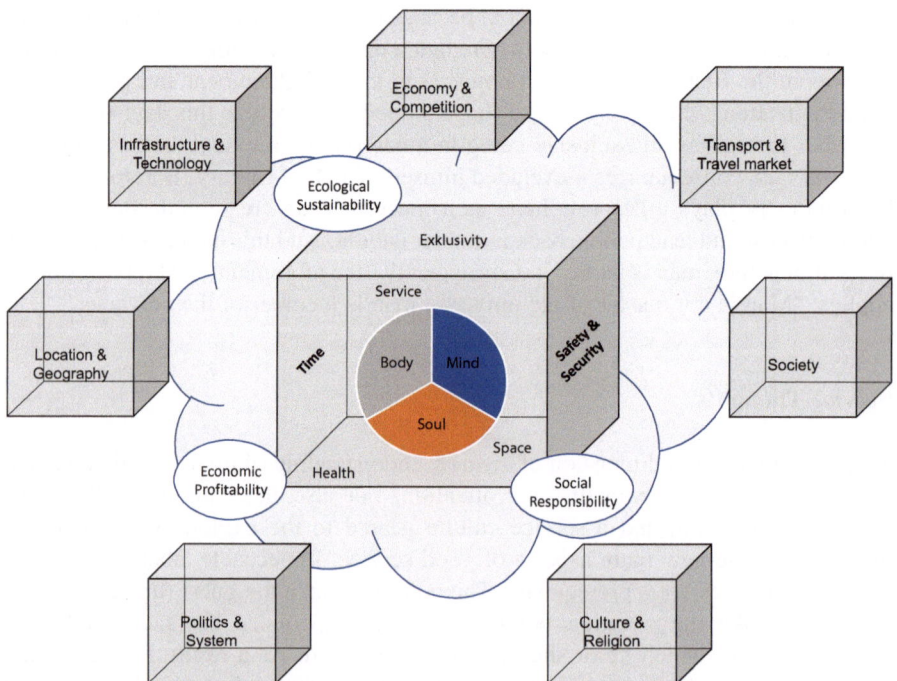

Fig. 5 Strategic influencing factors

3   Hotel industry 1: Monastery Disentis/Switzerland: "Reduce to the Max"

Marc Aeberhard

Obeying the 1400 years old order: grant hospitality!

History

The exact beginnings of the Disentis monastery (Fig. 6) lie in the early Middle Ages and in the dark. Although the founding myth of 614 A.D. gives first clues, which are based on various documents, such as the hymn of praise of St. Gallus, which the two monastery founders Sigisbert and Placidus mention, or a biography from the thirteenth century. However, it is considered certain that the first (Marien) church was built at the beginning of the eighth century. The village developed rapidly, and as early as 810 A.D. under Abbot Ursicin of Churratia the first monastery following the

Fig. 6 Disentis monastery

rules of St. Benedict was opened and has been an important monastery, school complex, and an important economic factor for the Surselva ever since (cf. Schönbächler 1999, p. 4 et seq.)

This makes the Disentis monastery one of the oldest monastery foundations north of the Alps and it has been active for almost 1400 years. The location of the monastery at the crossroads of the important Alpine passes east of the Gotthard massif (Oberalp Pass and Lukmanier Pass), already known in ancient times, explains the outstanding importance and responsibility of the complex over centuries, as a monastery complex, school (boarding school and grammar school to the present day), the production site for agriculture and trade as well as a pilgrims' hostel.

But the monastery also had to go through difficult times. The complex was first devastated by the Saracens in 940, burned down several times (1387, 1514, 1621), and was also plundered by Napoleonic troops in 1799. But despite all the setbacks, even the secularization of the mid-nineteenth century, the monastery kept on getting back on its feet, experiencing a blossoming and economic and political upswing. Its magnificent Baroque façade from the late seventeenth century characterizes the entire valley from afar and thus also sets a clear architectural sign in the (then inhospitable) alpine landscape, but in its magnificent Baroque opulence in the church also unmistakable signals to all people.

Philosophy of a Hotel Business Within Monastery Walls

One of the most important rules of Saint Benedict describes the granting of hospitality. In the deep belief that God can reveal Himself in everyone, no matter how poor, needy, or exhausted he may be, there is a natural conviction that everyone who knocks on the door must be given access, protection, a roof, a meal, and consolation. In this ancient tradition, which is still alive today, monastic congregations become nolens volens the original hoteliers of Europe: the word "hotel" is still derived directly from the Latin "hospes" and "hospice." Hospices were founded and managed practically exclusively by monasteries. This tradition also lives on to this day.

The form has changed, but not the motif why people knock on the door. While for hundreds of years it was mainly physical reasons (protection from storms, wind and weather, snow and cold) that caused people to stop and rest on pilgrim paths, today they are practically the same terms, but they increasingly refer to psychological challenges, dangers, and excessive demands. Today it is economic storms and icy winds that strike people in their daily work in the form of mobbing, intra-company skirmishes, bad company mood, burnout, etc. And again, the monastery becomes a place of psychological shelter for more and more people. The postmodern human being, overstrained by society, seeks meaning, protection, and/or simply security from fear and despair—be it while traveling through, at a seminar, or during a temporary stay in a monastery. Or, as Hanspeter Gschwend writes: "(...) There's a monk slumbering in every person who is constantly in search of perfection and renounces everything that is not necessary. Being a monk is not bound to monastic life, but the monastery is a protective space, in which a life according to strict rules allows concentration on this search. (. . .) What I seek and find for a short time is the protective place, the orderly course of the day and the liberation from all duties and tasks that are part of everyday life in the outside world (. . .)" (Benedictine Monastery Disentis 2014, p. 24 f.).

Unique Selling Proposition

In our world, which is characterized by digitalization, real-time communication, artificial intelligence, multitasking, omnipresent globalization, the "bliss" of people, processes, and structures, etc. the over 1000-year-old basic values, ways of thinking, and rules have an almost anachronistic effect. On closer inspection, however, they are of frightening and impressive topicality.

Beyond material (worldly) desires and motives, the monastery creates a "meaningful" space, which directs the focus of the human being toward being, the senses, and gives the observer the possibility to change his accustomed ways, with the aim of redefining the point of view.

Only a few hotels manage to formulate such a brutally striking USP. In the motto of the Disentis monastery, it is masterfully bundled and summed up: "stabilitas in progressu."

Operational Accents

Regardless of origin or convictions, this place leaves no one untouched. Be it the majestic backdrop, the impressive architecture, the precious baroque church—the first encounter with the monastery complex in Disentis shows man his earthly meaninglessness before the Almighty. Without wasting even a single word, it becomes clear in what proportion man stands to the sublime and the highest, he is tiny, he is insignificant, he is fragile. The charisma of the complex is omnipresent and is additionally reinforced by the monks in this monastery. The encounter with the monks makes it unmistakably clear that here a completely different set of values is shaping the way of thinking and everyday life. It is like leaving your usual coordinate network to immerse yourself in a completely different sphere. Already the first steps behind the almost modest-looking monastery gate reveal that accents are set by the thick walls, massive structures, vaults, small doors, and windows. It is as if they were whispering incessantly: they are protection from outside and security from within. Simple, unpretentious, modest. Hardly any wall decoration, little decoration, completely reduced to the necessary. This place lives from its unique location, its history, and its values. The fact that the choice of the location of a monastery complex in the eighth century was no coincidence can be explained solely by the congenial construction of the very first church complex, which is still preserved in foundations and still today on July 11th (the name day of St. Martin, the patron saint of the monastery), the sunlight shines through a single choir window onto the altar. What used to be a carefully selected patch of earth is still today a place of power of supra-regional importance.

1. *Seminar business*

So, what could be more suitable than holding seminars at a 1400-year-old powerhouse? In addition to offering various seminar and retreat rooms, the monastery has also set up its own adult education facilities. Under the direction of the alto abbot Dr. Daniel Schönbächler and Br. Magnus Bosshard personality seminars are offered, which should help "(...) to better recognize one's own personality structure, to reduce stress and inner blockades, to use one's personal resources sensibly and to broaden one's horizon. The Abbot relies on the typographic model of the enneagram and the method of psycho-kinesiology. Added to this is his wealth of experience from thousands of personality analyses (...)" (Benedictine monastery Disentis 2014, p. 73).

Even though all seminar rooms are equipped with basic technical equipment, the real focus is not on technical gimmicks, but on thoughts, people, interactions, and encounters.

2. *Hotel Accommodation*

In the south wing of the huge monastery, several Klausas have been converted into hotel rooms in recent years (the last stage completed in 2017). Obeying the motto of the order "ora et labora," the design of the 21 hotel rooms (Fig. 7) does without anything superfluous but does not lose any of its quality. The excellent high-quality and solid wooden furniture from local/regional production, the pleasant beds, the functional but impeccable sanitary facilities in the rooms underline once again the first impression: here it has never been about material things and it will never be about that. One of the extraordinary qualities of the monastery hotel is its tremendous tranquility. 1000 years of monastic life, the massive construction, the absence of entertainment offers (or distraction offers) and the location in the heart of the Bündner Alps leads to the fact that in the evening lying in bed you suddenly only hear your own heart beating and nothing else. Nothing.

This place forces the guest to deal with himself. The deliberate omission of any technical installations (from data port to TV) and the rigorous reduction to Holy Scripture and Regula Benedicti on a simple shelf cements the beneficial transformation of coordination from the usual everyday life outside the monastery walls. This is where the motto "reduce to the max" once coined by the company Smart becomes reality.

But the most amazing effect of these rooms is that, despite the simplicity of the furnishings, the strict rules of the order, and the deep silence, every guest at any time and in any place has the feeling of finding coziness, beauty, love, and peace. Or summarized in one word: Security.

Fig. 7 Premises in the
Disentis monastery

3. *Self-Sufficiency*

Since time immemorial the monastery has farmed estates in the Surselva (and beyond). In the Middle Ages, this important monastery complex was entrusted not only with spiritual and sacred tasks but also with secular ones. As a large land and forest owner, the abbey had and still has great economic importance and is one of the largest employers in the region (over 100 employees). Even today—as then, albeit to a much lesser extent, but still in keeping with tradition—the affiliated companies make a considerable contribution to the self-sufficiency of the hotel, boarding school, and monastery. In addition to a nationally significant milk and cheese business, these include a monastery bakery and confectionery run by Brother Gerhard, a carpenter's workshop, and an extensive agricultural and forestry business.

Products from the respective companies, from nut cakes and hand-picked Alpine herb tea mixtures to books and hand-knitted winter socks, are sold directly online or in the monastery's own retail shop (which also serves as a check-in and check-out counter for the hotel).

Target Market Segments

The history, the location, and the infrastructure result in specific market segments for the monastery with its special hotel offer, which are ideally addressed. Here the seminar segment from the German-speaking area is in the foreground. On the one hand, the topic-specific seminar offers which are created and offered within the monastery are captivating, and on the other hand, we have the very special location with its extraordinary charisma. It is therefore hardly surprising that large companies, renowned universities, and colleges, but also small think-tank organizations are regularly guests behind the monastery walls.

But also like at the very beginning pilgrims stop by on the Way of St James to Santiago de Compostela. Thus, the monastery continues its ancient tradition even after 1400 years. But the seeker of meaning sometimes does not travel alone to distant Spain but seeks comfort and pastoral care in the monastery itself. This becomes the destination per se. As already described, the tradition lived over centuries, the stability of values and customs, the regulated everyday life, and the renunciation of worldly goods thus become for many a place of refuge from a world marked by impermanence, stress, and an unimaginable flood of stimuli. Like other monasteries, Disentis offers the possibility of joining the monastery for a limited time.

However, the F&B offer and the kitchen infrastructure are rather limited. Although the newly created café "Stiva" on the ground floor of the south wing offers a limited selection of drinks and snacks, there is no actual restaurant. Nevertheless, the kitchen can also be used for banquet orders in connection with the boarding school and the monastery. In the new parts of the grammar school wing, catering for larger groups in halls is possible, but in relation to the other services and core competencies of the monastery, it is only a subordinate offer.

Sales and Marketing

The commercial marketing of a monastery offer is delicate and follows an extremely narrow tightrope walk. The habits of life, the completely different pace of everyday life in monasteries, the omnipresent spirituality are initially very unusual for an outsider and require a very careful approach. The uniqueness of being allowed to be a guest in an active monastery goes far beyond the usual use of sales and marketing instruments. In contrast to unused monastery complexes, in which the coexistence with the active religious life no longer exists, a thorough empathic examination of the monastery, its personality, and history is required here. You have to do justice to the complexity of the product. A lot of previous knowledge and knowledge about dos and don'ts is necessary, because neither the presentation nor the marketing should be bold. Monasteries are not theme parks à la Disney World, but centuries-old centers of partly secular, but above all spiritual creativity, which must be treated with maximum respect. Therefore, intelligent public relations (media) work is in the foreground and focus. The monastic community, especially today, is torn back and forth as to the extent to which the doors and gates are to be opened to the outside, or to what extent the order remains among its peers. The tempers, tendencies, and interpretations of this oscillate sometimes more in one direction, sometimes more in the other. In any case, it remains a fact that an invasion by the masses would bring the dignity, the aura, and the charisma of this institution to an end and possibly destroy it. Overtourism on a small scale: Here, in particular, it is important to think about monastic life in the narrower sense and about the main mission that a monastery has to fulfill.

The gentle opening of this very special world and way of thinking is a unique opportunity to immerse oneself in seemingly long past centuries and traditions, which, however, are unimaginably topical, modern, and in demand today.

Outlook

The retrospect thus becomes the outlook and obeys the motto of the monastery: "Progresso in Stabilitas." Social developments are characterized by ever faster change, excessive demands, unreliability. At the same time, however, the call for security, traditional values, family, and order in urban centers in the western (secular) world is becoming louder and louder. Here the ancient values of the monastery prove to be colossally contemporary and forward-looking.

The gentle management of the hotel offer in harmony with the monastic everyday life, the outstanding quality of this unpretentious authenticity, the radically lived deceleration, and the unconditional immersion in a spiritual world, in which the encounter with one's own self takes place, allow undreamt-of thoughts and give people who are constantly distracted from something the opportunity to focus on the greatest luxury that exists: the well-being, harmony, and balance of their own self.

There is strength in serenity.

4   Hotel Industry 2: Flame of Africa/Botswana: "The Luxury of Time"

The Luxury of Time: Or Time is a Luxury

Brett McDonald

When I was approached to contribute a chapter to this book, I must say I found it rather intimidating. My concept of luxury is probably very different from others. After 30 years in the travel industry, most of which has been spent providing luxury getaways in some of the remotest parts of Africa, I have come to realize that luxury is most often defined as that commodity that may not easily be available to any particular individual.

For the poor, luxury may come from money. Money to pay for running water, money to buy electricity, money for food. While these may sound like necessities to most, for those who do not have them, they are intrinsically luxuries. The middle-income sect of the world may find luxury in a new family car, or simply the ability to afford an annual holiday. But for those who can be considered "luxury travelers," and by that, I mean someone who might, say, drink their wine out of Riedel Glassware at home, or who find necessity in Egyptian cotton, enjoy the services of chauffeur-driven luxury cars or might even criss-cross the globe in their own private Lear jet, the hotelier would be hard-pressed to match or better these experiences in the quest to provide luxury beyond their "norm."

For those of us in the luxury travel industry, we would be ignorant to believe that if we have provided those things and perhaps a good bottle of wine, we have provided a luxury experience. In my mind, these essentially are the nuts and bolts of our craft, the basic ingredients. Luxury, if someone has the means to purchase the luxury holiday, is essentially the pursuit of what one does not have. It has been my experience of most people across the spectrum, and almost certainly for the well-heeled, the one luxury they do not have enough of, is the luxury of time. So, it is up to us to ensure that we provide them the most unique experience possible in the time that is allotted, because their time is absolutely irreplaceable. Arguably, in their world, money often is easily replaced.

To give a little illustration, I was involved in the opening of the One&Only Hotel in Cape Town in 2009, which was the brainchild of Sol Kersner, well known for his Sun City development. Comprising a seven-storey hotel block designed to imitate an island resort, the complex with two man-made islands overlooks the Waterfront Marina and commands sweeping views of Table Mountain. Completed at a cost of R900 million, it is part of the renowned worldwide One&Only Hotel franchise. Guests enjoy a level of luxury second to none.

I had been tasked with hosting the VIPs that were resident at the hotel. We had many celebrities, whose names I will not mention, attending and they were some of the most well-known and prominent people with few limitations on satiating their heart's desires. I had the opportunity to take them out on a luxury yacht and give them a unique and memorable experience, which for some of them, it was. One particular guest stuck out for me. As a wealthy man, there were likely a few things his heart yearned for. He was staying in an incredibly luxurious suite at the hotel and had access to most anything his heart desired. However, he had at one stage tasted Nando's, Portuguese-style flame-grilled fast-food chicken, and it had obviously appealed, and now it was all he yearned for.

In the hotel, we had a world-celebrated Nobu Restaurant which served Japanese signature cuisine, as well as a Gordon Ramsey restaurant, among other incredible dining options. The dilemma was, how could we bring Nando's Chicken takeaways into this brand new six-star establishment. The image made us all cringe. So we hatched a plan. We had Nando's deliver the boxed chicken to the aquarium which was next door to the hotel and we went out there with a tray and two cloches to collect it. We placed the chicken under the cloches and smuggled them back through the hotel and sent them to the celebrity's room and as requested, he had Nando's Chicken that night.

For him, sitting in his luxury VIP suite, enjoying Nando's Chicken was his idea of luxury. And you know what, I tip my hat to him and say "well done." It is these little unique experiences that count. If he had gone and had a five-star dining experience, it would not have been special. That is what he has all the time. He wanted something different, and we gave him something different.

My mind goes back as well to a time when I hosted one of Asia's richest men, a very successful property developer who shall remain anonymous. He had spoiled my wife and me years before by taking us to dinner where he had José Carreras sing for us, which was something that was mind-blowing. I thought how can I reciprocate when he came out to Africa. How do you give such a man a luxury experience when he has had luxury beyond imagination?

The lodge at which I was hosting him was in the bush but nestled on the banks of a large lake. I decided I would put a line of about 20–30 paraffin lanterns going out into the bush and in the end I placed two deck chairs. The remote setting and expanse of water close-by made for some exceptional night sounds. After dark, I walked this gentleman in the bush along this lighted walkway of lanterns. Unbeknown to him, my staff were walking behind us blowing out the lanterns. All he could see was the lighted pathway ahead, not realizing behind was pitch black. As we got to the two deck chairs, I popped the top of the champagne off with a bush machete, which he found terribly amusing as that was sabrage with a difference. We had a glass of bubbly together while we chatted.

It is interesting in the African bush, that while you speak and make any noise, all the creatures stay quiet. I got him to sit down and we snuffed out the last light and I said to him "You gave me a chance to hear José Carreras sing at dinner accompanied by an incredible little orchestra. Now I am going to give you the Orchestra of Creation." I started pointing out to him, as the creatures began making noise in the now quiet bush, saying "Listen to my soprano. . . listen to my tenors. . ." and slowly the crickets and reed frogs along with the bullfrogs, and the nightjars and even the hippos in the distance, all started putting together an incredible symphony of sounds of the bush. And then I pointed up to the stars. It was one of those dark black-satin nights without a moon, though the Milky Way was out in its full glory. And I just said to him, "Look at this. Our opera house has the most incredible ceiling." And you know, this man just cried!

What I had done there was I had touched his heart. I had given him luxury in my definition of luxury. And my definition of luxury is being able to make sure that you give a person an experience that you cannot just go out and buy. Basically, it has to be a choreography of all sorts of things which encompass all the senses: sight, sound, smell, taste, everything. You need to be able to make sure that you give someone something so unique that it touches the heart and the mind, and if you can do that, then I believe you have achieved something that is truly luxurious.

So, provide the nuts and bolts. Give them the Egyptian cotton and Ridell glassware and if you can, a great bottle of wine. But do not be afraid to let them experiment. Wine is a great example of turning the otherwise ordinary into something special. Everybody talks about DRC, Domaine de la Romanee-Contiand, and how absolutely amazing it is. But I found a little vineyard in South Africa that some of the top wine experts rate very comparable to DRC. So, you do not need to always be flash. Sometimes you can just say "try something unique. Try something special." (and be able to supply it). Explain it is from a little boutique winery and let them experience and taste something that is hard to find elsewhere and does not necessarily have a massive price tag either. In a nutshell, luxury to me is the opportunity to give an experience which they will remember for the rest of their lives, something which touches the heart, mind, and soul.

But what is the one resource all of this requires? Time! These anecdotes provide an insight into the essential ingredient of time. Let me explain. There is nothing more valuable to another human being than someone taking the time to notice them. Giving of one's time is one of the greatest forms of empathy and quite literally has the ability to touch hearts and souls. Sitting quietly with a grieving friend, hugging a grubby-faced child, or listening to someone's deepest fears, there is little to compare. It is a most valuable commodity and when given freely, is priceless. Many African cultures see time as a currency. They may not necessarily have material wealth, but they do have time, and taking the time to sit beneath a tree and discuss the weather, crops, one's family, or the health of your cattle, is priceless. Anything special that is done for a client intrinsically requires time, and taking the time to listen to them bestows a different kind of wealth upon them.

There is a vast difference between how you "pass" your time, and how you "spend" your time!

So now that you can see what I think luxury tourism actually is, let us go more into what the title of this chapter is; the leadership of the luxury hotel or lodge, or whatever you happen to be running. It is up to you to be able to create the breeding ground for the creativity that is needed in order to come up with the time element, and the ingredients necessary for the successful luxury concept I am putting across.

I remember when we built the Zambezi Queen, which was deemed to be a luxurious boat on the Chobe River and has won, among other awards, the "Leading Boutique River Cruise in the World" from the World Travel Awards. Onboard that boat, we noticed that if we got to know a little bit about each person and if each staff member had just a little something that he could mention, the results were most rewarding. This knowledge is invaluable.

So, what we did is lined a staff staircase that linked the galley to the dining room with flip charts. Each flip chart had a cabin number on it and it was up to every staff member to notice something about the client and write it on the relevant chart. So, it could be that Mr. and Mrs. Jones have two children going to Harvard, or that Mr. Jones caught a tigerfish the day before which weighed 5 kg. Mrs. Jones likes a Pink gin and tonic—only a single, mind you—whereas Mr. Jones has fallen in love with the local beer. Whatever, just make a note. It is just music to the ears of your client when you can go up and say "Good morning Mr. Jones, I hear you caught a wonderful tigerfish yesterday. How big was it?" Allowing him to talk a bit about it and actively listening is key. Likewise, maybe it is sundowner time and you go over to Mrs. Jones saying "Here's your Pink gin, and Mr. Jones shall we do that lovely local beer that you enjoyed yesterday?" Whatever, just let them hear their own name, let them realize you have taken a little bit of extra care, and listen and engage them when they talk. They do not need to know that everybody has a cheat-chart going up and down the stairs. It just shows that you have taken that little bit of extra time and attention. In that way, you are going to provide a luxury experience for them.

This is particularly relevant in the disconnected, fast-paced computer-driven world in which we live. Social media and a plethora of applications used to make our lives easier, have distanced us from real human contact. This is especially true of many high-powered businessmen, who by necessity dwell in this realm. Being able to interact one on one with a guest is in itself a form of luxury.

It is well known in our industry that knowing a guest's name and being able to greet somebody by name when you see them is important. The subject of one's name and how the human psyche interprets and responds to it is a subject worthy of a little discussion, as the science is applicable to my concept of luxury. Using functional magnetic resonance imaging, scientists have measured brain activation patterns in response to hearing one's own first name in contrast to hearing the names of others. There are several regions in the left hemisphere that show greater activation to one's own name, as would be expected, as this is the tag by which we have been known and self-identify ourselves with virtually since birth. By the time we reach adulthood, there is a whole social dynamic attached to our name.

The way in which we respond to our name also depends on the tone used by the speaker. For instance, facial expressions and even the number of times it is used indicate how relevant it, and we are, to that person. If overheard in a conversation, our attention is automatically drawn to that conversation and like it or not, we pick up on the feelings they have associated our name with and we will often associate those feelings with that person.

When engaged by a charismatic person, they will use your name often in conversation. Every use of your name is screaming your importance, emphasizing just how special you are, saying "I am listening to you and you alone."

So, like the Egyptian linen, fine glassware, and other comfortable amenities expected, the foregoing examples demonstrate that lanterns along a path, a starry night, or some tasty fast food are also just elements. In Africa, we are lucky that the average luxury destination is imbued with a magic which enhances the tools we have at our disposal. However, in my opinion, the main currency of luxury is time.

So, I believe leadership in a luxury hotel is about instilling in your staff or your crew the concept that everybody is special. Make them feel special. In order to do that, get your staff to touch the hearts and minds of clients by taking the time to go out of their way. It could be a picnic on the beach, it could be something as simple as taking them out to see the night sky. Equip staff with the necessary tools and give them the time to develop their hosting skills. It may be as simple as teaching them a couple of stars, or one or two constellations, just something where they are able to engage clients and inspire them to perhaps do something they have never stopped and done before.

The old expression "stop and smell the roses" is especially applicable in luxury travel.

5 Glamping: Camping, Luxury, Sustainability

Marco Walter, Sergio Comino, and Thomas Reimann

Glamping: Attempt at a Definition

Glamping stands for glamorous camping. However, the term is not yet clearly defined and is sometimes used relatively arbitrarily as a marketing term. For most experts, glamping consists of particularly luxurious or at least innovative rental accommodation on campsites, often paired with upscale service and embedded in an attractive ambience. It is therefore about a nature and camping experience with luxury and comfort at the same time. In this context, factors such as larger, beautifully located parking spaces—and thus more space and privacy—as well as exclusive or unusual furnishing and design concepts also play an important role. A pleasant, natural environment is essential for the glamping experience. The term was first established in the Anglo-American language area and used for safari accommodation in Africa. However, this chapter does not intend to examine luxury models such as lodges in the high-price range (for example: Vumbura Plains in Botswana with rates of up to 2500 dollars per person per night in a double room). For some years, glamping has spread also in the not English-speaking Europe. Many providers today are returning to the beginnings of the former mass phenomenon of camping, which was originally reserved for those who could afford camping equipment, in addition to the military and the tightly organized scouts. Some glamping tour operators are recording high growth rates, which seems to confirm Eicke Wenzel's forecast from 2012 when he said: "The idea of freedom and adventure, proximity to nature combined with luxurious camping offers will prevail. *Glamping is the Tourism Trend 2020*" (Wenzel 2012).

Target Audiences

With the glamping trend, the camping economy is reaching completely new target groups, which were not receptive for normal camping offers so far. People who are uncomfortable with the idea of sleeping on a sleeping mat in a shaky tent and sharing a toilet at night are more likely to try the glamorous form of camping. This mixture of camping freedom and adventure, the intensive nature experience paired with the amenities of a hotel stay brings new, solvent clientele to camping/glamping. The "Jack Wolfskin" generation is looking for a natural experience combined with a certain degree of luxury and comfort, without overburdening nature. The so-called LOHAS families (LOHAS = Lifestyle of Health and Sustainability) are also discovering glamping for themselves. For Silver Agers, glamping is a welcome alternative to classic camping or caravanning, as the service and comfort outweigh the perceived shortcomings of camping holidays and even offer attractive prices, especially in the early and late seasons. For all these target groups who are oriented toward a healthy and sustainable lifestyle without their own camping equipment, glamping offers an almost ideal holiday offer.

Types of Accommodation and Services

The variety of campsite accommodation marketed under the term glamping has been growing for years. A basic distinction must be made between dry accommodation and accommodation with sanitary facilities. Some glamping purists refuse to classify dry accommodation as luxury camping. From the authors' point of view, dry accommodation can fall under glamping if it includes individual sanitary use outside the glamping object. These can include special tree houses as well as safari tents, shepherd wagons, or sleeping barrels (see for example Fig. 8, greened sleeping barrel on Via Claudia Camping). In addition, the categorization into tents and fixed accommodations offers itself, whereby there are also hybrid forms here like the Airlodge or the hybrid lodge Clever of the tour operator Vacanceselect.

But not only the kind of the accommodation plays a role with glamping—also everything around it must match. So, the glamping guest expects more than most regular campers. This attitude calls for a rethink in the direction of hotel-related services, especially on the part of camping companies. This means more service for the glamping guest—such as breakfast service, freshly made beds, and regular or

Fig. 8 Sleeping barrel with a green roof on the Via Claudia Camping (Bavaria)

even daily cleaning of the accommodation. Glamping guests are also receptive to sophisticated and extravagant ideas when it comes to cultural and culinary offerings. The camping industry reacts with exclusive wine tastings, opera evenings, golf tournaments, and high-quality regional dishes in the local restaurant.

Providers and Platforms

How much glamping has established itself as a holiday form can easily be proven with a few clicks on the Internet. A large number of glamping providers and platforms are courting the favor of the new attractive glamping target groups. Glamping is also an integral part of the tourist offer on social media, for example, at Campingdreams.com, Europe's largest camping and glamping community. The Vacanceselect/Selectcamp Group, recently acquired by the French camping group Vacalians, is the European market leader in glamping. Vacanceselect/Selectcamp Group markets more than 1200 glamping objects at more than 100 locations in Europe. Beside Vacanceselect numerous other glamping specialists are active. The authors like to refer here to three examples from the English and German language areas:

glampinghub.com, a global glamping provider with international offers, glamping.info with a broad-based Central and Southern European focus based on user ratings, and Sawday's Canopy & Stars, which offer exclusive, original, and British-spleen glamping spots. Even Airbnb and booking.com now have glamping accommodations in their program.

Sustainability

First of all, the fundamental question arises: Is luxury compatible with a sustainable tourism or is it generally a form of waste that will never be ecologically correct, but can be cosmetically enhanced through greenwashing campaigns?

The authors of this chapter advocate a more modern meaning of luxury, which respects a world with finite resources and which should invest the higher revenues associated with luxury tourism in a high sustainability performance.

A major challenge is therefore to pay particular attention to the sustainability aspects of glamping products. Firstly, because, as with camping, nature is the primary basis of every glamping offer and the protection of nature and the environment is therefore a compulsory task. Because only socially and ecologically conscious luxury tourism will be successful in the long term and be accepted by customers and their social environment. This is partly due to the fact that luxury tourism and in this case glamping appeals to a well-educated clientele with a higher income, which places above-average value on sustainable consumption and is prepared to spend significantly more money on it than the average population.

Since the sustainability of tourism companies is a complex field that also applies to camping and glamping areas, it is advisable to orient oneself toward relevant sustainability labels such as ECOCAMPING, the European eco-label for accommodation companies, or Greenkey Europe.

The key issues that a sustainable glamping provider should address are: Nature-friendly design and maintenance of the site, energy efficiency and conservation of natural resources (water, materials), waste avoidance, environmentally friendly mobility (e.g., through the use of company bicycles and electric vehicles), ecological cleaning, use of regional products and services. In social areas, sustainable glamping providers should strive for accessibility, fair pay and employee involvement, health protection, and regional cooperation, among other things.

Awards

As in any other tourism segment, glamping offers have received a number of awards. A distinction must be made between those that relate significantly to the quality and service of a glamping facility and those that place an additional emphasis on sustainability. In the absence of an internationally clear definition of glamping, the transition from pure camping awards to actual glamping awards can be fluid. For example, the editorial staff of the ADAC's (General German Automobile Club) camping and caravanning guide awards the best campsites in Europe annually, with reference to the range of luxurious rental accommodation. The same applies to the European Prize of the German Camping Club. The European rating and booking platform camping.info also awards the so-called camping.info Award annually, a list of the best, which is drawn up on the basis of the ratings given by users of camping.info. Here, too, you will regularly find businesses with a strong glamping connection. A rather journalistic appreciation of outstanding glamping offers is the "glamping inspector," by the Hamburg travel journalist and glamping expert Thomas Reimann. The "Glamping-Stars by ECOCAMPING" contain the most systematic definition to date, with a strong focus on sustainability aspects. This award was launched in 2016 as part of a consultation process involving key stakeholders and experts. It evaluates glamping offers from the following points of view:

- Environment and nature
- Design, comfort, and privacy
- Cleanliness and maintenance
- Leisure and relaxation
- Sustainability
- Marketing, sales, guest services

A Pioneer in Glamping and Sustainability: Club Camping Jesolo International

- *Description*

  Jesolo International Club Camping is a medium-sized, traditional five-star camp-site that has been in existence for over 60 years. It is located on the Adriatic coast not far from Venice, in an area with the highest concentration of quality campsites in Europe.

  The campsite is situated on its own 700-m-long sandy beach, directly on the shopping street of Lido di Jesolo, the second largest holiday resort in Italy, with direct public connections to Venice every 30 min. The campsite also has its own marina and is only 1500 m from the Venice lagoon.

- *Concept*

  The concept of the campsite aims to develop a unique product in a unique location for an upscale family audience, which distinguishes itself from the competition by as many special features as possible. The central position, the widest possible range of services, the maximization of security, the creation of a homogeneous audience, affordability, and very importantly, high sustainability are the main focuses.

- *Affordable luxury*

  New, extremely well-kept infrastructure, a relaxed ambience, a pleasant audience, a unique, comprehensive all-inclusive offer, which includes not only the services of the site (W-LAN, sailing, pedal boats, SUPs, canoes, banana boats, tennis, paintball, pony riding, baby train, language school, sports schools) but also offers free access to Lido di Jesolo's attractions, like the nearby best water park in Italy, 18-hole golf, mini-golf, horse riding, clay pigeon shooting, archery, go-karting, diving, wellness, pirate ship for children. For this all-inclusive package, the campsite received the ADAC Camping Innovation Award in 2006 and has been honored year after year for its high standards. No wonder, because twice a week in the evening an award-winning pianist plays classical music for the guests at the bar.

- *Glamping*

  Campers have 359 sites for caravans and motorhomes at their disposal, including oversized sites with private bathrooms, high privacy, and excellent service. Camping sites, however, are not offered. There are also 131 luxurious "Holiday Homes," for which solutions not specific to camping have been selected and which therefore represent a cross-product. They offer hotel-like comfort and the special feeling of open-air living. These accommodations are attractive for both campers and non-campers (90% of the population). The latter get to know the relaxed and carefree camping atmosphere, which is an essential part of the glamping feeling. The openness, helpfulness, and positive attitude of the guests is exactly what impresses first-time campers. The main difference to the hotel industry is not the quality of the infrastructure, but the special attitude of the customers and employees. It is therefore the task of the host to create ideal

Fig. 9 Photovoltaic system at the Jesolo International Club Camping

conditions for a positive network of relationships that motivates the clientele and binds them to the company.

- *Sustainability*

Jesolo International Club Camping considers sustainability to be an indispensable part of its quality offer and has adapted its infrastructure and processes to this end over time.

Since 2008, the campsite has been operating according to the Ecocamping environmental management system, which regulates the use of water, energy, chemical products and waste, among other things. As a result, water consumption was reduced by 55% and gas consumption by 70%. Since 2007, the printing of advertising material has been dispensed with. Five photovoltaic systems have been installed (for an example see Fig. 9) as well as solar thermal systems and high-performance heat pumps for hot water preparation, 11 electric vehicles are used and only green electricity is purchased. It even has a "Tesla Destination Charger," the free charging of electric vehicles is now part of the all-inclusive package. In order to support the shift toward electric mobility, the campsite is providing its customers with a Tesla Model X car free of charge to test drive. In addition, a Segway school was introduced as a further activity for young people, Segway excursions in the lagoon and lagoon trips with electric boats. To date, 2.5 million euros have been invested in environmental investments, which are offset by savings and additional income of over 200,000 euros annually. Jesolo International has received the "climate-friendly company" award and has become the first CO_2-neutral campsite in the world. As a result, the ADAC awarded the site the Camping Innovation Award 2010.

- *Sustainability-related guest feedback*

The feedback from customers is clear: 97% rate the environmental policy of the site as "very good" or "good" in a recent survey. More importantly, 65% of holidaymakers say that the campsite's environmental policy was "very important" or "important" for their holiday decision. In 2014 it was only 55%.

• *Distribution system and auction of reservation sequence*

Jesolo International Club Camping has developed a special distribution system. Sales are made exclusively to private end customers on their own initiative, i.e., without the mediation of third parties: There are no tour operators, no OTAs, no agencies, no groups. It is sold without exception at the list price (rack rate), there are no discount levels, no long-term rates, and no seasonal guests. In principle, there is excess demand for places to stay overnight. This is one of the reasons why the order of reservations is auctioned on their homepage. At the end of October, about 2000 families will offer more than 1000 Euro in addition to the full camping price in order to secure their desired place for the following summer.

More Examples of Successful Glamping Approaches

Another ecologically oriented glamping provider is Hofgut Hopfenburg in Münsingen in the Swabian Alb biosphere region in Germany. The campsite was designed together with ECOCAMPING and offers a wide range of circus and shepherd wagons, which are loosely distributed in the meadow landscape. One focus is on the conservation of endangered breeds of domestic animals.

Very British and very individual is the stay in "Walcot Hall," a glamping refuge in Shropshire, central England. The spectrum of glamping offers on the grounds of a proud eighteenth-century manor house ranges from the Mongolian yurt to the converted fire truck. "Five o'clock tea" which can only truly be enjoyed with a stiff upper lip is included.

In Tyrol, near Innsbruck, the award-winning Glamping-Platz Ferienparadies Natterer See awaits guests who enjoy the morning view of the picturesque Nordkette (mountain range) from the terrace of their safari lodge after a refreshing swim in the nearby lake or still sitting in their private whirlpool. The glamping site has been awarded the Glamping Star.

In the Rocky Mountains (USA), the glamping guest can experience a very natural glamping adventure at the Paws Up Ranch in Montana since the 1990s. Horse riding, climbing, rafting, fly fishing, and hiking are the preferred activities of the Montana glamper. The Glamping accommodations are reminiscent of the tents of the early trappers and settlers and are equipped inside with shower, toilet, and other civilizing amenities (see Fig. 10, interior view of a glamping accommodation). The operators attach particular importance to a balanced ecological balance of their ranch and work with various organizations on their idea of an environmentally sustainable and environmentally friendly tourist offer.

If you like it a bit more exotic, a glamping offer in Vietnam is certainly a good choice. For example, at Lak Lake in Lien Son, northeast of Ho Chi Minh City. Here

Fig. 10 Interior of the Cliffside Camp Honeymoon Suite, The Resort at Paws Up, Montana (Credit: Paws Up)

the glamping guest enjoys the luxurious comfort of his "tent cabin" with a terrace and a wonderful view over the lake and the adjoining forest. The restaurant belonging to the complex spoils the guest with Vietnamese specialties.

In the heart of the Chianti region lies the Glamping Village Orlando in Chianti, also awarded the Glamping Star. On 14 hectares, the glamping guest is offered a unique selection of different glamping objects in the middle of an oak forest. Safari tent, Airlodge, Lodgesuite, Cottage Next, Cottage Clever, Air-Dreamer, Coco Suite, Bohemian Lodge, and CubeSuite are the glamping accommodations, all equipped with shower and WC. The restaurant focuses on regional products, the nearby winery invites you to an exclusive dinner with wine tasting. Excursions to Siena, Florence, and Arezzo complete the cultural offer.

In the Franconian town of Pleinfeld, Waldcamping Brombach offers its visitors a large selection of glamping accommodations: from sleeping barrels to circus wagons, from penthouse pods to safari tents, there is something for every glamping taste. Also, the culinary needs do not miss out. The local restaurant offers international and regional Franconian cuisine.

Back to the glamping roots, the Serengeti Park Hodenhagen is located between Hamburg and Hanover. Here you can observe the wild animals of the African savannah in the luxurious safari tent at the park's Victoria Lake and go on a photo safari in a jeep.

This is of course also possible in Africa itself, more precisely in Marakele National Park. At the foot of the Waterberg Mountains in Marataba Safari Lodge, the glamping guest can expect 15 safari tent suites including a swimming pool, open-air restaurants, and of course lots of wildlife and safari adventures (see Figs. 11 and 12).

A very special glamping experience is offered by the newly opened Germanenland at the Alfsee. Here the glamping guest sleeps in a very authentic looking Germanic House and enjoys the luxury and comfort of an exclusive mobile

Fig. 11 Interior view of the luxury tent suite at Marataba Safari Lodge, South Africa (Credit: MORE)

Fig. 12 Exterior view of the luxury tent suite at Marataba Safari Lodge, South Africa (Credit: MORE)

home. In addition, the Germanenland offers a distinctive and exclusive sauna and bathing landscape, which is housed in Germanic modeled longhouses. In addition, events (Germanic Games) and a variety of restaurants round off the historic Germanic experience at the Alfsee.

Glamping: "Manufacturer"

The development of the "safari tent manufacturer" Luxetenten from the Netherlands shows how dynamic the glamping trend really is. While the Dutch had a modest annual production of about 30 more or less luxurious tents in 2009 and mainly supplied the European market (Netherlands, Italy, Germany, France), Luxetenten now serves glamping customers all over the world, exporting to New Zealand and China. The annual production is reaching almost four-digit ranges, the tents are available in various sizes—up to 300 m^2 of floor space—and with all imaginable equipment, including whirlpool and fireplace. But other manufacturers of camping rental properties are also registering a rapidly increasing demand for glamping options. The glamping specialist NATURWAGEN & LODGES, based in Ellerhoop in Schleswig-Holstein, has sparked a real glamping boom in the German-speaking camping market in recent years and now supplies more than 80 camping sites and holiday parks in Germany, Austria, and Switzerland with rustic glamping pods, romantic circus, and shepherd wagons as well as original troll and BUGG huts. Tendency: strongly increasing. In close cooperation with Vacanceselect founder and glamping pioneer Loek van de Loo, the Italian manufacturer Crippa-Concept developed numerous and highly diverse glamping objects ranging from the luxurious lodge suite with free-standing bathtub to the two-storey air lodge with a sky roof offering an unhindered view of the stars. Several thousand glamping objects originate from this collaboration. The glamping trend has also arrived in the manufacturing of mobile homes. Instead of plastic and elastomers, mobile home producers are increasingly focusing on sustainable production and recyclable, resource-saving materials. When designing the interior and exterior spaces, mobile home manufacturers such as the Dutch ARCABO and the French Louisiane Group (TAOS model) are inspired by glamping.

6 Megayacht Charter

Markus Krall

Imagine a villa. Living space? Say 900 m^2. Partitioning? One suite for you, a few more for your guests, salons, terraces, and a gym plus pool. And now imagine you are not bound to one place. The villa swims and takes you to the most beautiful spots in the world. Welcome to the world of big yachts!

Superyachts are the last great freedom in this day and age. They combine luxury, safety, and mobility. And they signal true exclusivity at a time when this term is being used in an inflationary manner. Superyachts are unique, uniqueness is their principle. They measure 40, 60, or even 170 m, have tailor-made interiors, and are at a technical level far removed from the mainstream. Exceptional personalities drive these superyachts and they are built by highly specialized shipyards.

Around 7000 yachts with a length of more than 30 m are currently floating around the globe. If you take the official length from which a yacht becomes a mega yacht— i.e., 24 m—it is much more; but serious statistics have not yet been collected; you can assume there are at least 10,000 units. In addition, just under 800 further yachts are under construction at the approximately 100 shipyards focusing on superyachts. Europe, in particular, plays a major role here: 12,800 yacht meters (350 projects) are currently being built in Italy, around 4000 yacht meters (65 projects) in the Nether-lands and 1700 yacht meters (55 projects) in England. Among the top 5 nations are Turkey (3600 m) and Taiwan (1800 m). With Lürssen, Abeking & Rasmussen and Nobiskrug, Germany also has quite renowned addresses. In Bremen and Rendsburg, however, pure custom-built yachts from 70 m in length are manufactured, with which of course no large throughput can be achieved as with semi-custom formats of 25 m in length. However, if one were to rank by volume and not by yacht length, Germany would rank second worldwide (Figures according to Global Order Book/ Boat International Media and own research).

Here the average volume of a delivered yacht is about 7000 gross tons, in Italy it is about 400 gross tons. Also, the price per gross ton is much higher. The largest yachts built in Germany, such as the 155-m-long DILBAR, required a roughly estimated investment of 600 million euros from the client. However, only those directly involved know about exact budgets of this magnitude.

By the way, yacht ownership entails even greater obligations than, for example, a vacation property or a second home. On almost every yacht there is a permanent crew—on 25-m yachts, there are between two and three, on 50-m yachts 10–14 and on 100-m yachts even 70–100 employees. So, a yacht can be compared to a small business that has to be managed continuously. In addition, there are, for example, costs for berths, repairs, and maintenance. The annual cost of maintaining a yacht is between 5 and 10% of the purchase price, depending on the condition of the yacht and the owner's demands.

This is another reason why another branch of the mega yacht industry is devel-oping steadily and well—the charter business. The B2B booking platform of the Mediterranean Yacht Brokers Association (MYBA), yachtfolio.com, currently lists 1629 manned yachts between 20 and 99 m; weekly prices range between 15,000 and 1 million euros. For many customers, a charter trip is the more economical and uncomplicated way to enjoy a yacht—especially since many yacht owners also do not use their own yacht for more than 2 or 3 weeks per year. Exceptions also confirm the rule here, of course: After all, a number of owners are constantly circumnavigating the world with their (sailing) yachts.

Every yacht that is available for charter has a Central Agent. It is the first port of call for all customers and brokers (with customers) when it comes to chartering this yacht. The Central Agent markets the yacht to B2C (charterer) and B2B customers (broker), manages the booking calendar, and handles the administrative tasks for the owner. Some Central Agents manage a fleet of over 20 ships, the world's largest fleet is listed by the Swiss broker Ocean Independence with more than 100 yachts and 15 branches between Hong Kong and Fort Lauderdale. The Central Agent can, of course, book customers for the yachts he manages himself—as an in-house deal, so to speak—but these cases occur rather seldom, usually the inquiries come from external brokers, the so-called retail brokers. In the event of a contract being concluded, the commission is divided as follows: 25% for the Central Agent, 75% for the retail broker. At first glance, the retail business is far more lucrative, but also far more unpredictable. Moreover, there is a tendency among the large brokerage houses to divide the commission equally, as the workload for the Central Agents is steadily increasing.

In addition to the weekly price, which—roughly speaking—includes the yacht plus crew plus meals, the charterer also pays an Advanced Provisioning Allowance (APA), an advance on the incidental costs incurred during the stay. These include fuel for yachts, dinghies and water toys, mooring fees, customs formalities, and fees for communications and shipping agents. Food and drinks are also paid for by the APA. Thirty percent of the charter price is common in the industry; the exact accounting is the responsibility of the captain of the yacht. A delivery fee may also be applicable if guests wish to embark at a port other than the home port of the yacht or the port where the previous charter ends. In any case, the charterer in the EU member states must pay VAT, which can be reduced to 6.6% depending on the itinerary (source: MYBA).

The first charter trips were organized at the beginning of the 1970s. In Germany, for example, Moncada Yachts, an Italian yacht dealer who had previously represented several brands as a sales broker and wanted to develop this new business segment, gained a foothold. A yacht of 20 m was already considered to be large and was valued at around 2000 D-Mark per day; the 53-m-long "Gaul," which was watered in the mid-1970s, was considered to be a giant. The worldwide charter fleet consisted of 100–200 yachts, there was no serious documentation. We communicated with the customers via letter and telex, with the yachts via Norddeich radio or other stations. "It was a wild time," recalls Adelheid Chirco, who then as now works as a charter broker and has witnessed the complete development of the market. "Except for the period following the terrorist attacks in New York, the industry has actually always grown or at least remained stable." Only the needs of the customers have changed. While voluminous Jongert sailing yachts were in demand in the 1980s, especially among Chirco's German-speaking clientele, this relationship reversed over the years. "From the year 2000 at the latest," says Chirco, "the market was dominated by motor yachts. The desire for comfort and volume continues to rank first today. How many cabins there are on board and what the decor of the yacht looks like, especially inside, are perhaps the most important decision criteria for or against a yacht." However, the customer structure had hardly changed. Around 80% of all charter trips arranged by Chirco are undertaken by families. There are often three generations on board for whom such a journey is the ideal way to reunite families. "Those who book with us often don't have a nine-to-five job," says Chirco (sources: Ocean Independence AG and personal interview).

Sixty percent of the time the ships travel in the Mediterranean, followed by Croatia with a 30% share and then Greece/Turkey. The most popular yacht size for years has been a 30–40-m-long motor yacht, which can cost between 50,000 and 100,000 euros per week. Most guests (from Ocean Independence) go on board in Cannes, followed by Palma de Mallorca, Naples, Monaco, Ibiza, St. Tropez, Tivat, Olbia, and Palermo (source: Ocean Independence AG).

While the market as a whole is stable, booking conditions have changed drastically. Customers or their brokers increasingly book at short notice, one could almost speak of a last-minute business. "In the past," says expert Chirco, "charter trips were planned on a long-term basis. Today bookings sometimes come only three days before the start of the journey; and we are talking here about high six-figure sums." The fact that the crews need at least 48 hours between two charter tours for cleaning and equipping the yacht is increasingly met with incomprehension. In addition, unreliability is increasing: even with signed contracts, there are sometimes no shows or attempts to reduce travel time shortly before departure.

So, social trends are reflected even in absolute high-end tourism such as super yachting. Brokers and owners have to come to terms with this as well as with the fact that more and more booking platforms are emerging which, however, do not fulfill any advisory function. Nearly all reputable brokers organized in the Mediterranean Yacht Brokers Association advise to contact a reputable agency for charter bookings, because they are informed about the latest legal regulations and can compile the best itineraries from many years of experience. In addition, brokers—unlike booking portals—personally visit the yachts and can professionally classify crews and equipment.

Whether a booking behavior analog to portals such as booking.com or airbnb.com will ever establish itself can still be doubted at present. However, brokers, shipyards, and suppliers to the yacht industry now also have to take new marketing paths in order to reach Millennials and Generation Y, respectively. Facebook, Instagram, Landing Pages, Influencer-Marketing, and the organization of hip Customer-Only-Events can be found wherever a CEO is in charge who is not closed to the future, looks after customers himself, and does not operate a top-down management.

7   The Art of Hospitality: About the Luxury of Touching People in Their Entirety[2]

Cucumber Instead of Caviar

Andreas Caminada

According to Maslow's pyramid of needs, food is one of the fundamental basic needs, but it can also be an art form that goes far beyond the intake of calories and becomes an experience for all the senses. It is not that people talk about cooking for nothing. International top gastronomy has become incredibly competitive. Even though there are no limits to creativity, there are still trends, developments, politically and ecologically correct and incorrect (think of the discussions about frog legs, turtle soup, shark fins, or foie gras), which must be taken into account when designing offers. Top gastronomy has become an indispensable part of the luxury world. Just good cooking is not enough anymore. It is more about taking the guest on a culinary journey and touching all the senses. It is about creating a holistic culinary experience.

Hospitality

The creation of a culinary experience that goes far beyond cooking requires a team effort: the well-rehearsed interaction of the kitchen brigade, service team, back of house, and front of house. So, people are at the center of the operation.

The word hospitality contains the words "guest" and "friendship," and both have to do with relationships between people. Consequently, the philosophy of a (luxury) restaurant should focus on the human being. An organization must develop a spirit in which people feel well taken care of, regardless of whether they are an employee, a guest, a business partner, or a hiker. When he crosses the threshold to the restaurant, he must feel welcome.

The People

Nothing works without a team. Therefore, the focus must be on the people who are active in the organization, who make up the organization. Motivation and the promotion of development opportunities for each individual are important. This investment in people is the basis for the team spirit. Then everyone gets down to it, even beyond the limits of their job description. It goes without saying that a dishwasher will help a guest with a suitcase or a receptionist will serve a coffee in the lounge. So, everyone becomes a host and everyone in the house feels the team spirit.

---

[2]With the editorial collaboration of Marc Aeberhard.

Today more than ever, luxury means creating this feeling of togetherness. The personal and individually tailored service is becoming more and more important.

But this attitude does not come by itself, it must be encouraged, and people must be trained accordingly. And this has a highly motivating effect; even internally, new goals have to be set again and again and opportunities for further development have to be exploited. This attitude and energy are the most important capital. Being a leader also means assuming responsibility and giving something to the industry, to people.

The driving force behind this commitment is the creation of an atmosphere of well-being, security, and love. A one-sided focus on profit would make this mindset impossible.

Products and Cooking

Although we (the three Michelin stars restaurant Schauenstein) can look back on a huge development in the kitchen, our foundations are the good old "Pauli" and the classic French cuisine. They are and remain the basis (also) of our training. But standstill means regression and therefore we tirelessly strive for further development, improvement, and perfection. The exchange with other chefs is very important to experience and learn what is going on in the world. What are new trends? What are current discussions, both in the kitchen and in society? Just a few years ago it was the ultimate "luxury" to process the craziest ingredients from the most remote corners of the world. They have gotten away from that by now. Today, the trend—and I think it will continue for quite some time—is regionalism. Real, honest stuff from the neighborhood.

It is important to tell authentic stories. And stories are always connected to a place. That is why we attach great importance to such stories when selecting our products, as the example of the Tsar's apple shows: Here in Domleschg one of the best apple varieties has been growing for centuries. This apple is known for its outstanding taste and was exported exclusively to the Zarenhof in St. Petersburg. Although nobody in the high Russian aristocratic circles knew that this apple came from Switzerland, everyone knew that it came from the Domleschg fruit valley.

The castle in which our restaurant is located is also marked by history and stories. Daniel Bonifaci, who lived here in the house in the late sixteenth and early seventeenth centuries, created a kind of etiquette for young people. And later it was my great uncle, the Chur bishop Rest Giusep Caminada, who went in and out of the castle here. Now it is up to us to add stories, episodes, and experiences, sometimes smaller, sometimes larger, to keep the house and its soul alive.

Of course, langoustines and halibut have their place on the menus, but more importantly, they are rooted in their immediate surroundings. We work very closely with the local farmers here. We now bake bread with special grains, which only grow here, use ancient plums and vegetables, and have even experimented with artichokes. These are now thriving so well in the local microclimate that the yield has already multiplied within a few years. Recently, for example, we have been trying to grow our own figs. The aim is to illustrate the seasonal diversity of agricultural products, thus sending a clear signal of honest sustainability and to obtain the best quality at all times in close cooperation with producers, fruit growers, Chrüüter-Fraueli, but also cheese-makers, butchers, etc. For us, authenticity is not the buzz-word that has been somewhat worn out in the meantime, but an honest reality that is lived every day.

The worst thing is mainstream. It is therefore very important that every chef does his best with the typical specialties of his region. Let everyone do what they do best.

Among my favorites are very simple products like cucumber, onion, and tomato. The difficulty lies in turning these simple ingredients into a top dish. It does not always need luxury products, but the dramaturgy of the menu must be right. That is the real luxury! It is about working with different textures, conjuring up warm and cold on the plate at the same time, crispy and soft, bringing excitement to the plate. With a staged sequence of the menu, we tell a story. The goal is to create one's own identity, one must create one's own handwriting, find one's own way, and remain faithful to it. It is not about following trends but finding your own kitchen language and then following it without stopping. It is particularly important to constantly develop yourself further.

Often the question arises as to how extensive a menu must be, and the answer is not easy. There are no clear rules. But of course, it is not about 20–30-course sequences. Our menus are usually six to eight course, but the intermezzo, the upgrading through small attentions here and there, the little surprises, the excitement and curiosity, joy and delight trigger, decorate the overall experience. And depending on the situation, a small course can be added. Flexibility is a key competence, both in terms of seasonality and the availability of products in the right quality, but especially in terms of the expectations, wishes, and dreams of our guests.

Creating Experiences: The Great Luxury of Having Time

The guest not only comes to eat but also spends time with us. And we want to make this time span—may it be shorter or longer—a highlight, a very special experience. The stay here in the house should give the guest the opportunity to feel the combination of service, region, the other people, the food, and the location as a whole.

Flexibility

Such an attitude demands a lot from the team; it requires maximum flexibility, a high expenditure of time, a lot of energy, full concentration, and maximum presence. This makes our everyday life very, very intense. But we are rewarded with the biggest gift a restaurant can have: with happy guests who do not want to leave anymore. Lunch guests from Hawaii were still there at 2 am in the morning. Situations like this happen again and again, and this shows that it is not just about a meal, but rather about the time they spend in Schauenstein. The guest is visiting us and should arrange the time with us in such a way that he takes a maximum experience home with him. For us, this means true, dedicated, lived hospitality. It is our understanding of the true art of hospitality. And perhaps also the secret behind our team spirit.

At Schloss Schauenstein we have created a small island where guests can escape from everyday life.

On the one hand, flexibility means making every experience so special and unique that the guest gets the feeling that it has been tailor-made for him or her. Flexibility is not only important for our guests but also for us with regard to our orientation toward the future.

The New, Old Marketplace Idea

The town of Fürstenau is a challenge, but also a great luck. The secluded location in the canton of Graubünden, away from the major urban centers such as Zurich, Basel or Bern, means that Schloss Schauenstein and our restaurant have to make a daily effort to ensure that the guest's journey to us has been worthwhile. The history of the place helps us. According to a decree issued by King Charles IV in 1354, Fürstenau was granted town charter, even though the town has only a few houses and a small castle. And so we became the smallest city in the world.

It is this diversity—partly given by the situation and history, partly created by us—that enables us to exchange on different levels. It is the exchange of products, thoughts, experiences, values, services, feelings and experiences, the appreciation, devotion and respect for history, nature and people that count. It is only in this dynamic symbiosis of hard and soft facts that individual parts become a large whole that has to be created, maintained, preserved, and carefully developed anew every day. This is our true understanding of luxury gastronomy in this day and age.

Literature

Benedictine monastery Disentis (2014) 1400 years, Disentis
Buck M, Ruetz D (ed) (2018) ITB world travel trends report 2017/2018. https://www.itb-berlin.de/Presse/Downloads/Publikationen/. Accessed 20 Jan 2019
Ministry of Tourism, Sultanate of Oman (2016a) Statistics. Zugegriffen. http://www.omantourism.gov.om/wps/portal/mot/tourism/oman/home/media/statistics/!ut/p/a1/hc9ND4IwDAbgX8OVFlGc3iagIuJHMAF3MWDmJAFmcMrrfdxovxq_e2jxv0wKDFFidXQuRqULWWXnvmbNzFuP5yvcsXBILMYhJ5MSxa-Oyp8FWA_xSFP_lE2AP4k7otNuf68wodDEI_XW4sSMLqfUGuqSDgTeaev1BpLc5T_DjhhkwUcr88c-W1rlNBLCGH3jDG_PS6PFRqdN5aKCBbduaQkpRcnMvKwM_RY7yrCB91RDzGk5VisWqSogi9AYHnizg/d15/d5/L2dJQSEvUUt3QS80SmlFL1o2XzlLN0VTMjA2ME9MQjcwSVNVSDFTR0IzOE0w/. Accessed 12 Sept 2018

Ministry of Tourism, Sultanate of Oman (2016b) Our mission—our vision. Zugegriffen. http://
www.omantourism.gov.om/wps/portal/mot/tourism/oman/home/ministry/about/vision/!ut/p/
a1/jZFbT4MwFIB_Da_0dLDS-VYuDgREU4msL4aZjmEYXRiOv28lS4xusp23c_
J9OTckUIFEWx7rquxr1ZbNdy7I2yJ2Aj4DAlniOhDxPMR86Vo0BQ2sNAD_BIPRJ4_
3yVPgY8goBu3TlHDuWZDNT763ZKHtJNpwYw-iOHiOX6wUA8O39f_
xbTqDyHdD31mkuhW5zZ8Y8Ir_isSITG3wBzgfcQSmbjh5Rc%2D%
2DAugtHpCoGrUeP7pi7dqiFRKd3MhOduZnp8vbvt8f7gwwYBgGs1KqaqT5rnYGXFK26tC
j4jeJuGzRfpfneQF19DFvjsmGfQHOhjfZ/d15/d5/
L2dJQSEvUUt3QS80SmlFL1o2X09SUVNGTkU0MDAxSzQwwSUlMOTlHRkozMEUz/.
Accessed 14 Nov 2018
Schönbächler D (1999) The benedictine abbey of disentis. Swiss Cultural Guide GSK, Bern
Statista (2018) Luxury-oriented consumers in Germany by information interest in holidays and
travel in comparison with the population in 2018. https://de.statista.com/statistik/daten/studie/
939325/umfrage/umfrage-unter-luxusorientierten-konsumenten-zum-interesse-an-reisen/.
Accessed 20 Jan 2019
Wenzel E (2012) Trend researcher: Glamping is the tourism trend 2020. https://glampinginfo.
wordpress.com/2012/01/08/trendforscher-glamping-ist-der-tourismus-trend-2020/.   Accessed
26 Nov 2018

Marc Aeberhard founded Luxury Hotel & Spa Management Ltd. in Zurich in 2004 and has been
acting as Managing Director ever since. The company has access to a global network of travel trade
partners, lifestyle and travel media and (U)HNWI and works closely with public relations and sales
and marketing agencies in Frankfurt, Munich, Paris, Dubai, Milan, New York, Hong Kong, and
London. Furthermore, the native Swiss is a member of the consulting networks of the Gerson
Lehmann Group, USA and Hotellerie Suisse, Bern. He also takes on an active role in the "Luxury"
task force of the management of ITB Berlin, Germany. As author and co-author of various specialist
publications, his name can be found regularly. He also holds guest lectures in Berlin, Istanbul,
Lausanne, Lucerne, Munich, Singapore, Stuttgart, Thun, Vienna, Worms, Zurich, etc. The graduate
hotelier graduated with distinction from the Ecole Hôtelière de Lausanne (EHL) and previously
completed his studies in business administration as lic.rer.pol. (MBA) at the University of Bern. The
luxury hotelier has more than 20 years of experience in the fields of hotel opening, management,
and renovation/refurbishment in Abu Dhabi, Germany, France, Maldives, Morocco, Seychelles, Sri
Lanka, Switzerland, Thailand, Ukraine, and Cyprus of small hotels in high and top end. Many of the
hotels have been awarded international prizes. All projects are based on the definition of New
Luxury and work according to the principles of Triple Bottomline.

Andreas Caminada grew up in Sagogn in Graubünden and completed a cooking apprenticeship
at Hotel Signina in neighboring Laax. After completing his apprenticeship in 1998, he worked in
several companies and top restaurants in Switzerland and neighboring countries. Since 2003,
Caminada has been leaseholder and chef de cuisine in Fürstenau, Grisons. His boutique hotel
Schloss Schauenstein has nine rooms and suites and the restaurant of the same name is awarded
three Michelin stars and 19 Gault Millau points. In 2015, he launched a second restaurant brand
called "IGNIV by Andreas Caminada," which since then can be found in Bad Ragaz and St. Moritz.
Andreas Caminada founded the foundation "Fundaziun Uccelin" in 2015 to promote cooking and
service talents. Since 2018, the top chef has also been the host at "Casa Caminada" with its
restaurant, shop, bakery, and ten guest rooms.

Sergio Comino has been the Director of the Jesolo International Club Camping for many years.
He attaches particular importance to combining the highest quality with maximum sustainability.
For years, the campsite has been voted the best place in Europe by various media and organizations.
At the same time, operation is climate-neutral due to the massive use of renewable energies.

Dirk Gowin is Managing Director of *Select Luxury* in Berlin. The long-time product manager at the luxury travel specialist Windrose Finest Travel went into business for himself in 2018 and, together with the consolidator Aerticket, bought the Berlin tour operator *Select—exklusives Reisen*. Under the new brand *Select Luxury Travel* he now wants to take off and has gathered a lot of experience in his team.

Marcus Krall is responsible at Ocean Independence AG for PR & New Business in the D-A-CH (German-speaking) region. Before that, he was editor-in-chief of the superyacht magazine Boote Exclusiv for 10 years.

Brett McDonald (Owner, Flame of Africa)

Thomas Reimann is a travel journalist from Hamburg, who has occupied himself with the glamping trend for 10 years. He is regarded as a recognized rental camping and glamping expert. Thomas Reimann runs the glamping site glamping-inspektor.de beside his activity as ECOCAMPING press spokesman.

Marco Walter is co-founder and managing director of ECOCAMPING, the initiative for sustainable camping tourism in Europe. ECOCAMPING has been in existence for 20 years and manages a network of more than 220 campsites that are committed to sustainable management, including numerous facilities with glamping facilities.

Summary and Outlook

Roland Conrady, David Ruetz, and Marc Aeberhard

This book closes a gap in the landscape of German-language publications. Despite the unmistakable market presence of luxury travel and luxury demand, there is a blatant lack of research and publications on luxury tourism, especially in German-speaking countries.

The definition of luxury has changed over time. The last 20 years in particular have shown, along with the social changes from a material to a post-hedonistic social imprint, that there is a great deal of uncertainty in the market about what is meant by luxury. An unmanageable variety of terms and incomplete attempts at definition contribute significantly to disorder and disorientation in the markets.

The lack of orientation leads to different coping strategies in the market. The spectrum ranges from progressive innovation to fear-driven, persistent status quo politics. In the meantime, however, a rethinking in product design and communication has begun. Yet there are huge gulfs between words and deeds. There is a lack of consistent implementation of the findings of trendsetters and everyday life in the mass market.

Cultural, social, and societal differences in the various markets of the world (Europe versus Arabia versus China versus India versus Russia, etc.) lead to different stages of development in the tourism (luxury) understanding and consequently to a cacophony in the market (supply/demand) and in marketing (communication and sales mix).

R. Conrady (✉)
University of Applied Sciences Worms, Worms, Germany
e-mail: conrady@hs-worms.de

D. Ruetz
Head of ITB Berlin, Messe Berlin GmbH, Berlin, Germany
e-mail: ruetz@messe-berlin.de

M. Aeberhard
Luxury Hotel & Spa Management Ltd., Zürich, Switzerland

© Springer Nature Switzerland AG 2020
R. Conrady et al. (eds.), *Luxury Tourism*, Tourism, Hospitality & Event Management, https://doi.org/10.1007/978-3-030-59893-8_10

This book is opening the door and exemplifies the various facets of the luxury phenomenon, allowing the most diverse protagonists to have their say.

The introductory chapter contains Roland Conrady's explanation of the ubiquitous term "luxury." Core elements are abundance/waste and satisfaction of needs/desires. This is followed by an examination of the connections between luxury and business. It becomes apparent that the implications of a modern understanding of luxury—in which immaterial aspects play a major role—for the economy are still little researched. This is particularly true for tourism markets. Research gaps in luxury tourism are named.

In the second chapter, Roland Conrady analyzes the macro-environment of the (luxury) tourism market. This consists of five environments: the political-legal, the economic, the socio-cultural, the technological, and the ecological-natural environment. It can be seen that the future development of (luxury) tourism markets can be plausibly derived from the analysis of the environmental factors of the five environments. It can be assumed that future developments in the five environments will greatly favor the luxury tourism segment.

The third chapter analyzes the luxury phenomenon in detail. First, David Bosshart shows that the phenomenon of luxury is also subject to a life cycle. In highly developed economies, luxury has reached the seniority phase, in which less is often more, in which time, space, and leisure and the ability to experience, decode, and enjoy the essential become significant. This chapter also includes a detailed description of the size and functioning of various luxury markets by Antonella Mei-Pochtler and Hannes Gurzki. Tourism and the cars/yachts segment are by far the largest luxury market segments, with sales of around €400 billion each. In the future, however, the emphasis will shift toward tourism: travel is dominating the wish list of luxury, as Dorothea Hohn points out.

The fourth chapter provides behavioral explanations of luxury consumption. Hasso Spode first takes the reader on a journey through time to define luxury. He describes how this term has changed over the centuries and how the understanding of luxury has become more democratic, especially in the post-war period—but even today luxury is understood as an abundance of a scarce good, although the definition of what is scarce has fundamentally changed. Spode says that people often do things simply because they are fun, and that being in the game is a special quality of just being oneself. In their contributions, Hannes Gurzki and David Woisetschläger talk about luxury as the extraordinary, as a contrast to the ordinary, and explore the question of human motifs and behavior—from esthetics to self-realization. In another article, they explain the motivations and factors that trigger a consumer decision for luxury goods. It is about creating the need for the unattainable, the path to its fulfillment and the further development of the need for the ever new. Marc Aeberhard writes that fear is an important driving force for human creation. It is about overcoming existential fears, such as the feeling of social isolation or failure in the economic process. Pastor Stephan Hagenow writes of luxury as a foretaste of paradise and critically warns that luxury should also be understood as a journey to oneself, and formulates the ten commandments of luxury tourism.

The fifth chapter examines the marketing management of luxury providers. Keiko Kirihara and Marc Aeberhard discuss the role of globalization versus regionalization on the one hand, but also the importance of luxury brands and their influence on purchasing decisions compared to the consumer goods industry and tourism on the other—taking into account socio-cultural factors and demographic parameters. The journalist Juliet Kinsman focuses on the importance of modern media, which not only serve as a means to the end of "been there, done that," but can also be used specifically for self-reflection and the creation of meaning, thus making a lasting contribution to the responsible handling of being. Marc Aeberhard also elaborates in his chapter on management instruments in the luxury hotel industry. He makes a strict distinction between high-end and top-end hotels and underlines the outstanding importance and the necessary intensive engagement with people, who are always the focus of attention in the provision of services, but also in demand. Magda Antonioli Corgliano and Sara Bricchi highlight the pioneering role of luxury tourism. They show how status, brand recognition, but also authenticity or uniqueness characterize luxury. The change in definition from tangible to intangible criteria becomes significant. In short, it is about anything that money cannot buy.

The sixth chapter highlights important segments of the supply side of the tourism market: accommodation (Marc Aeberhard), air transport (Stephan Grandy and Roland Conrady), cruises (Thomas Illes), road transport (Jens Wohltorf/Adam Parken), rail transport (Ralf Vogler/Maria Wenske), and tour operators/travel agents (Norbert Pokorny). It is shown how the offers of the respective luxury segments are designed and how they differ from offers of the premium and mass market. The comments provide indications for the further development of offers below the luxury segment.

The seventh chapter analyzes the luxury relevance of selected megatrends in tourism. The most important megatrends in tourism include sustainability, digitalization, demographic change, and health. The contributions focus on their relevance for luxury travel. Four authors, based on their many years of practical experience, give a profound account of the specific characteristics of these megatrends and why they are relevant. Stefan Gössling concisely describes from a human ecological perspective whether luxury tourism is compatible with the sustainable use of resources and ecosystems. Marc Aeberhard discusses in his contribution whether the immaterial part of the phenomenon of luxury is compatible with the demands for more and more digitalization. Jörg Meurer vividly demonstrates how, under demographic change, new target groups for luxury and premium brands are gaining a completely new and exciting relevance. Finally, Mario Krause places the aspect of health as an immaterial factor in relation to luxury (and travel), setting an important accent as a counterpoint to the ubiquitous phenomenon of wellness.

The eighth chapter provides insights into the analysis, design, and future perspectives of luxury features. The concept of luxury must increasingly be understood intellectually. The abbot of the almost 1500-year-old monastery in Disentis, Switzerland, describes how luxury must be defined far beyond the material concept, how he grasps the level of spirituality, and how the goal becomes the journey to one's own self, to one's own center. He sums up how the goal must not be adventure, but experience. Hannes Gurzki emphasizes the outstanding importance of status in society, which goes hand in hand with consumption and possession of luxury, while Verena Zaugg-Faszl addresses the demands of space and architecture as an expression of luxury. Terms such as stealth design, understatement, and what makes a space a good space are explained here. Philipp Schmidt describes the effect of nature on humans. He investigates the question of which substances the body releases during an intensive confrontation with nature and how pure nature creates true feelings of happiness.

The ninth chapter consists of case studies and best practice examples of luxury tourism. Successfully implemented strategies in the field of luxury tourism show that the demanding, experienced guest demands authenticity, sustainability, individual and personal service, discretion, no paparazzi, as well as time and space. Proven experts in the field describe in detail and thematically structured in individual case studies which parameters need to be taken into account. Oman is regarded as an example of a destination that has been combining sustainability and luxury travel for many years. Dirk Gowin explains this phenomenon from the perspective of a tour operator specializing in luxury travel. Marc Aeberhard illuminates two very different hospitality concepts from the point of view of the internationally experienced hotelier—an island paradise in the Indian Ocean with a unique selling point and a monastery hotel in the Swiss Alps as a true retreat, perhaps with "no frills," but with all the more room for the luxury of freedom of thought. Brett McDonald (Botswana) and a consortium of authors led by Sergio Comino (Venice) write about accommodation concepts in nature on the phenomenon of "glamping." Markus Krall, a long-time expert for superyachts, impressively illuminates the phenomenon of large private boats. This chapter is rounded off by a contribution by top Swiss chef Andreas Caminada, whose restaurant was awarded 3 Michelin stars and 19 Gault Millau points, on the role of Food and Beverage in luxury tourism.

This book offers views of the future of luxury tourism at various points. The next stage in luxury development (intellectuality and spirituality) is already beginning to emerge. Research gaps that need to be closed are also named.

One of the most painful gaps today is that the label "luxury" is used almost ubiquitously and the concept of luxury is thus degenerating into an empty phrase. Therefore, the authors will develop the so-called "New Luxury Score Model," with which a classification and systematization of luxury offers will succeed.

In the follow-up book, the authors will build on the collected findings of this book and attempt to go beyond the New Luxury Score model and create generally binding guidelines, definitions, and instructions for luxury tourism in the middle of the twenty-first century.

Roland Conrady   Since 2002, Prof. Dr. Roland Conrady has been a professor at the tourism/ transport department of the Worms University of Applied Sciences. His research and teaching focuses on aviation, tourism and digitalization. Since 2004, Roland Conrady has also been the Scientific Director of the world's largest tourism convention, the ITB Berlin Convention. He was president of the German Society for Tourism Science (DGT) e.V. and is a book author (among others Conrady, R./Fichert, F./Sterzenbach, R., Luftverkehr, Munich 2019). Roland Conrady is a member of various advisory boards of companies and politics. Previously, he was head of the study programme "Electronic Business" and Professor of General Business Administration at the Heilbronn University of Applied Sciences. After graduating as Dr. rer. pol. from the University of Cologne in 1990, he held various management positions at Deutsche Lufthansa AG until 1998.

David Ruetz   (Head of ITB) has headed the ITB Berlin, the World's Leading Travel Trade Show®. Under his direction, ITB Berlin has seen many achievements, including the Innovation Award of the Federal Association of the German Tourism Industry (BTW). As part of the management of ITB, he was significantly responsible for the establishment of the ITB Asia in Singapore (since 2008) and also helped to introduce the ITB China in Shanghai (premiere in May 2017). After his studies in Zurich and Berlin, David Ruetz entered tourism and the event industry and joined Messe Berlin in 2001. He is editor, author and co-author of numerous studies and books on trends and developments in the exhibition, travel and tourism industry, among others the annual World Travel Trends Reports. His interests include digitalization in the trade fair and event sector. As a guest lecturer, he regularly works at universities in Germany and Austria and is also a member of the advisory board of R.I.F.E.L. e. V. (Research Institute for Exhibition and Live Communication) and the IUBH—International University (until 2018) and is a member of the Executive Board of Research Association for Holidays and Travel (FUR).

Marc Aeberhard   founded Luxury Hotel & Spa Management Ltd. in Zurich in 2004 and has been acting as Managing Director ever since. The company has access to a global network of travel trade partners, lifestyle and travel media and (U)HNWI and works closely with public relations and sales and marketing agencies in Frankfurt, Munich, Paris, Dubai, Milan, New York, Hong Kong and London. Furthermore, the native Swiss is a member of the consulting networks of the Gerson Lehmann Group, USA, and Hotellerie Suisse, Bern. He also takes on an active role in the "luxury" task force of the management of ITB Berlin, Germany. As author and co-author of various specialist publications, his name can be found regularly. He also holds guest lectures in Berlin, Istanbul, Lausanne, Lucerne, Munich, Singapore, Stuttgart, Thun, Vienna, Worms, Zurich, etc. The graduate hotelier graduated with distinction from the Ecole Hôtelière de Lausanne (EHL) and previously completed his studies in business administration as lic. rer.pol. (MBA) at the University of Bern. The luxury hotelier has more than 20 years of experience in the fields of hotel opening, management and renovation/refurbishment in Abu Dhabi, Germany, France, Maldives, Morocco, Seychelles, Sri Lanka, Switzerland, Thailand, Ukraine and Cyprus of small hotels in high and top end. Many of the hotels have been awarded international prizes. All projects are based on the definition of new luxury and work according to the principles of the triple bottom line.